DESIGN OF SOLAR THERMAL POWER PLANTS

DESIGN OF SOLAR THERMAL POWER PLANTS

ZHIFENG WANG

ELSEVIER
CIP
ACADEMIC PRESS
An imprint of Elsevier

Academic Press is an imprint of Elsevier
125 London Wall, London EC2Y 5AS, United Kingdom
525 B Street, Suite 1650, San Diego, CA 92101, United States
50 Hampshire Street, 5th Floor, Cambridge, MA 02139, United States
The Boulevard, Langford Lane, Kidlington, Oxford OX5 1GB, United Kingdom

Library of Congress Cataloging-in-Publication Data
A catalog record for this book is available from the Library of Congress

British Library Cataloguing-in-Publication Data
A catalogue record for this book is available from the British Library

ISBN: 978-0-12-815613-1

For information on all Academic Press publications visit our website at https://www.elsevier.com/books-and-journals

Publisher: Joe Hayton
Acquisition Editor: Lisa Reading
Editorial Project Manager: John Leonard
Production Project Manager: Sruthi Satheesh
Cover Designer: Miles Hitchen

Typeset by TNQ Technologies

This book is applicable to the construction design of new solar thermal power plants and existing facility expansions that use water and steam working medium. Steam pressure parameters include secondary MP (medium pressure), MP, and secondary HP (high pressure), the nominal evaporation capacity corresponding to the output power of the receiver is 8—800 t/h, and the capacity of the condensing steam turbine is 1—100 MW.
Heat transfer medium for solar collector fields in the book can be water/steam, synthetic oil, molten salt, air/ceramic, etc. Concentration modes include parabolic trough, power tower, and Fresnel reflectors for concentrated solar thermal power plants.

Contents

Preface xi

1. Introduction
   1.1 General Principles of Solar Thermal Power Plant Design 3
   1.2 Brief Introduction to Solar Thermal Power Generation 7

2. The Solar Resource and Meteorological Parameters
   2.1 The Nature of the Solar Resource 47
   2.2 The Solar Constant and Radiation Spectrum 48
   2.3 Atmospheric Influences on Solar Irradiation 50
   2.4 Calculating Methods for Solar Position 52
   2.5 Distribution of the Solar Resource in Several Typical Areas of China 60
   2.6 Solar Irradiance Prediction Methods 83
   2.7 Distribution of Solar Direct Normal Radiation Resources in China 89
   2.8 Various Special Climate Conditions in the Plant Area 100
   2.9 Measuring Instrument 107
   2.10 Global Direct Normal Irradiance Distributions 112

3. General Design of a Solar Thermal Power Plant
   3.1 Power Plant Design Point 117
   3.2 Heliostat Field Efficiency Analysis for Power Plants 119
   3.3 Thermal Performance of Parabolic Trough Collector 150
   3.4 Basic Data Required by Power Plant Design 192
   3.5 Major Parameters and Principles of Design 193
   3.6 Description of General Parameters of the Power Plant 194
   3.7 Calculation of Annual Power Generation 194
   3.8 Determination of Thermal Storage Reserve 214
   3.9 Main Points for Power Plant Site Plan 216
   3.10 Notices for Concentration Field Layout 223

4. Design of the Concentration System
   4.1 General System Description 225
   4.2 Principles for Concentration Field Layout 229
   4.3 Design of the Solar Tower Power Plant Concentrating Field 243
   4.4 Control Design of the Heliostat Field of a Solar Tower Power Plant 256
   4.5 Solar Field Design of Parabolic Trough Power Plant 264
   4.6 Description of solar Concentrator 273
   4.7 Instantaneous Efficiency 276

4.8 Design of the Parabolic Trough Collector Field 279
4.9 Concentrator Field Control Design of the Parabolic Trough Power Plant 281
4.10 Wind Load Characteristics of the Concentrator 283

5. Design of the Receiver System
5.1 General Receiver System Description 319
5.2 Selection of Materials for the Receiver System 327
5.3 Selection of Pipes and Pumps for Receiver System 330
5.4 Receiver System Control 331
5.5 Design of the Operation Modes of the Receiver System 334
5.6 The Discharge System and Equipment of the Receiver 336
5.7 Vacuum Performance of the Parabolic Trough Receiver Tube 339

6. Thermal Storage Systems
6.1 General System Description 388
6.2 Technical Requirements of Thermal Storage Systems 392
6.3 Thermal Storage Materials and Modes 393
6.4 Categories and Constitutions of Thermal Storage Systems 401
6.5 Selection of Thermal Storage Materials and Tanks 412
6.6 Charging and Discharge Equipment of the Thermal Storage Tank and Respective Process Design 413
6.7 Thermal Storage System Control 414
6.8 Facilities for Thermal Storage System Inspection 415

7. Site Selection, Power Load, and Power Generation Procedures
7.1 Site Selection 417
7.2 Power Load and Power Generation Procedures 423

8. Plant Layout Planning
8.1 Basic Rules 425
8.2 Layout of the Main Buildings and Concentration Field 428
8.3 Communication and Transportation 429
8.4 Vertical Layout 430
8.5 Pipeline Layout 432

9. Main Powerhouse Layout
9.1 Direction of Main Powerhouse 435
9.2 Main Powerhouse and Thermal Storage 435
9.3 Solar Thermal Storage System Layout 435

10. Water Treatment Equipment and System
10.1 Receiver and Evaporator Makeup Water Treatment 437
10.2 Calculation of Water Treatment Equipment 437
10.3 Feed Water and Boiler Water Modification and Thermal System Steam Sampling 438
10.4 Anticorrosion 438

11. Power System
11.1 Power Grid Connection of Power Plant 439
11.2 System Protection 439
11.3 System Communication 439

12. Electrical Equipment and System
12.1 High-Voltage Electrical Installations 441
12.2 Main Electrical Control Room 441
12.3 DC System 441
12.4 Electrical Measuring Instrument 442
12.5 Relay Protection and Automatic Safety Device 442
12.6 Lighting System 442
12.7 Cable Selection and Layout 443
12.8 Overvoltage Protection and Grounding 443
12.9 Electrical Installations in a Dangerous Environment with Potential Explosions and Fire Hazards 443

13. Thermal Automation
13.1 Basic Rules 445
13.2 Control Mode 446
13.3 Thermal Inspection 446
13.4 Automatic Adjustment 446
13.5 Thermal Protection 447
13.6 Interlocking 447
13.7 Power Supply and Steam Supply 447
13.8 Control Room 448
13.9 Cables, Conduits, and Local Equipment Layout 448
13.10 Basic Rules for Building Space Heating 449
13.11 Solar Tower 449
13.12 Heating Network and Heating Station in the Plant Area 450

14. Architecture and Structure
14.1 Basic Rules 451
14.2 Fire Protection 452
14.3 Interior Environment 453

15. Auxiliary and Affiliated Facilities 455

16. Environmental Protection of the Concentrating Solar Power Plant
16.1 Basic Rules 457
16.2 Requirements for Environmental Protection Design 457
16.3 Pollution Prevention and Treatment 458
16.4 Environmental Protection Facilities 458

References **459**
Index **463**

Preface

SOLAR THERMAL POWER PLANT DESIGN

Since the beginning of China's research on solar thermal power (also known as concentrating solar power, CSP) generation in 1996, CSP generation technologies have gone through the entire development process from inception to realization. Ever since "the 11th Five-Year Plan," CSP generation technologies have enjoyed rapid development throughout the country, with the appearance of a large variety of experimental solar thermal collection and storage systems and experimental power plants. Breakthroughs have also been made in the manufacturing techniques of core components and materials; professional CSP generation equipment manufacturers have also appeared. The first China state standard for CSP technology was released in September 2011. Along with a further deepening of technologies during "the 12th Five-Year Plan" as well as commercial development, this reference book for CSP plant design has become a requisite work for understanding solar power plant commercialization. Currently, there is no other reference book in the world that systematically describes the design methods of CSP plants.

The book mainly focuses on CSP technologies, and those mainly consist of power tower and parabolic trough collector technologies and thermal storage. The book not only describes the design of CSP systems, but also explains the design methods and operation modes of key facilities in detail, such as the heliostat, heliostat field, parabolic trough concentrator, parabolic trough receiver tube, and whole-plant distributed control system, or DCS. It also discusses the fundamental basis of designing CSP plants and system as well as the key design points that should be considered.

CSP generation design mainly includes resource evaluation, site selection, design of the optical efficiency of the concentration field, thermal control of the receiver and electrical design, thermal storage capacity, thermal storage charging and discharging design, heat exchanger and evaporation design, whole-plant electrics, whole-plant thermal control instruments, power plant construction, and whole-plant security design. By focusing on the contents just mentioned, the book offers calculation and design methods separately in different chapters and provides examples by integrating with practices of the author, so as to facilitate the understanding of readers.

Solar irradiation serves as the basis for solar power utilization. The evaluation of solar resources used by CSP generation is the most fundamental process for solar power plant design and site selection. Although it has been stated in numerous articles, the author further discusses solar resource evaluation by integrating his own research, especially stressing its relationship with the site selection of thermal power plants.

Solar concentrators, solar receivers, and thermal storage are three major core components for CSP systems. The section of the book that discusses these topics mainly explains equipment application, evaluation methods for equipment performance, and the thinking and methods for equipment design.

In equipment performance evaluation, great difficulties have been encountered in the facula error analysis of the heliostat. The book offers mathematical approaches and test methods corresponding to facula error analyses of various types of heliostats, which basically have been derived from the research achievements of the author and his graduate students.

Heat loss of receivers is the subject at the core of receiver design, with a variety of analysis methods. The book offers a relatively simple calculation method and corresponding examples.

The book was preliminarily defined as a reference book similar to a design manual by specifically referring to China's state standard of "Code for Design of Small-size Thermal Power Plant" (GB 50049), which may be recognized by readers from its general arrangement. However, as writing progressed, the author discovered that books about and references for CSP generation technologies had rarely been seen in China, while many methods were still under development and evolution. Thus, it was determined that the major concepts and methods should be explained more explicitly to benefit readers. By introducing these descriptions and analyses, the book is easier to be understand and greater reference value as well. These gains were made by sacrificing its original style as a "design manual."

The book was mainly composed by Wang Zhifeng with coauthors Guo Minghuan (2.4, 3.2.1, 3.2.2, 3.2.3, 4.2.2, 4.2.3), Li Xin (5.5), Yu Qiang (2.6), Gong Bo (4.7), and others.

The book was summarized and made by the author, his colleagues and students, and the coauthors based on multiple years of research and engineering practice in CSP technology. It is written by following the principles of sharing his own "fruits planted by himself" with readers and thus striving for fewer examples extracted from the achievements of others. The writing work started at the beginning of 2010 with a hope that the book could be finished within a year and be capable of catching up to "Chinese National 11th Five-Year Plan" 863 Program project acceptance at the end of 2011. However, additional progress in research work, especially steam production by the Beijing Badaling power tower solar power plant in July 2011, its power generation in August 2012, and its gradual commissioning, the author's knowledge of CSP generation technologies deepened. The author discovered that content that was as useful as possible would not be sufficient without the author's original theories and experiments. Particularly, CSP generation technologies still remain at the development stage, and many basic concepts and terms have been expressed in diversified ways in the articles and writings of different world-renowned scholars, such as the most important concept in power plant design, namely the "design point." As for similar major content, the author expresses his own opinions in the book by conducting in-depth theoretical and experimental research. Therefore, after 5 years, this book can finally be dedicated to its readers. Henceforth, along with the deepening of research work, the book may be modified regularly so that upgraded "achievements"—for example, improvements in the design and operation mode of the thermal storage unit—can be dedicated to readers and the industry as well.

Special thanks should be given by the author to Mr. Xu Jianzhong, Academician of the Chinese Academy of Sciences, and Mr. Huang Ming, Chairman of Himin Solar Energy Group Co., Ltd., who have made every endeavor possible, for more than a decade, to support the author in conducting research on solar thermal power generation and its practices; meanwhile, the author is also grateful for care and support from his family. During research on solar thermal power generation and its practices, the author has been deeply grateful for tremendous support from the national "863 Program" (2006AA050100), (2006AA050100), "973 Program" (2010CB227100), "the National Natural Science Fund of China," "Beijing Municipal Science and Technology Project," "Knowledge Innovative Project of CAS," "International Cooperation Program of the Ministry of Science and Technology," and the "6th and 7th" Framework Programmes for Research of the European Union.

The book assimilates the development experiences and essence of CSP technologies both at home and abroad and provides them to readers. Yet due to the author's limited knowledge, as well as insufficient practices in CSP plant R&D and construction, many imperfections may exist. It would be greatly appreciated if readers could provide the author with critiques and corrections of these imperfections for use in future editions.

Zhifeng Wang
March, 2018

CHAPTER

1

Introduction

Since the beginning of the 21st century, energy and environmental problems have become increasingly more noticeable. Due to limited nonrenewable fossil energy resources and the severe influences of excessive use of these resources on the environment, excessive greenhouse gas emissions, global warming, and severe deterioration of regional climates and ecological environments appeared and have greatly endangered the living space of humans. The prominent advantages and development potential of concentrating solar power (CSP)—also known as solar thermal power (STP) or concentrated solar power—generation technology have aroused widespread concern. The main challenge it faces right now is to reduce its power generation costs so that it can compete with fossil fuels. In the next two decades, it is estimated that stable and economic CSP generation technology will gradually mature and become strongly competitive commercially. CSP generation technology features stable and constant power output, low costs, and outstanding technical and economic advantages; the development strategy of this technology is of great significance.

The basic process of CSP generation involves concentrators, receivers, thermal storage, thermal power conversion, etc. Thermodynamics, heat transfer, optics, mechanics, materials science, information science, and many other disciplines and interdisciplinary studies serve as the theoretical foundation of CSP generation technologies. Only by mastering these key technologies can we greatly improve the efficiency of the system and further reduce power generation costs so that we may further push forward its large-scale commercialization and development, and realize the effective utilization of solar energy.

For power plant design and operational targets, the following two questions are key points to be solved in terms of CSP research and engineering; they are also the main contents of this book:

1. Optical efficiency and cost of the concentrator. High-density concentration of solar irradiation acts as the basic process of CSP generation. Concentrator costs in solar tower and parabolic trough

Design of Solar Thermal Power Plants
https://doi.org/10.1016/B978-0-12-815613-1.00001-8

systems account for 45%–70% of the primary investment; the annual mean efficiency of a concentration field is normally 58%–72%, so research on the concentration process greatly influences the efficiency and cost of the system.

Energy losses in the concentration process mainly include cosine, reflection, air transmission, and receiver interception losses caused by concentrator errors. In addition, the limits of working environmental conditions and concentrator shelf life, while still ensuring concentrator precision, mean that concentrator cost reductions now face great restrictions. Considering both of these factors, it is necessary to carry out in-depth research on the collection of optical energy and high-precision concentration using aspects of optics, mechanics, and materials science and overcome the influences of concentrator mirror shape aberration and tracking errors on energy flow transmission efficiency as well as the problem of low CSP system conversion efficiency caused by spatial and temporal distribution of the energy flux failing to satisfy the requirement of receiver; an integrated design method of solar beam concentration and thermal absorption based on the highly efficient energy flow transmission must be established.

2. CSP system conversion efficiency and reliability of devices. When the efficiency of an CSP system is increased by 1%, the levelized cost of electricity from CSP generation will decrease by 8%, and the corresponding total capital investment will be reduced by 5%–6%. System efficiency has significant impacts on CSP system costs. Future technical developments shall be mainly based on stable operation of the system, improvements in system efficiency, and development of major technical equipment techniques, system integration techniques, equipment performance evaluation methods and their respective testing platforms, technical standards, and regulations in the large-scale CSP generation system. Conventional thermal power conversion efficiency was improved along with increases in the parameters of the working medium, and the basic approach to improving the efficiency of the cycle was to increase the temperature and pressure of the working fluid. During CSP generation, however, the efficiency of solar receiver system conversion decreases with increases in the temperature of the heat transfer medium, which is also accompanied by intensive unsteadiness in time, nonuniformness in space, and transient strong energy flow impact. Therefore, improvements in thermal power conversion efficiency shall not be accomplished by completely relying on the regular thermal cycle, and the laws of fluid flow and heat transfer processes are also distinguished from regular ones. To greatly improve the efficiency,

TABLE 1.1 Comparison of Solar Thermal Power Generation and Solar Photovoltaic Power Generation

Items to Be Compared	Solar Thermal Power	Photovoltaic Power Generation
Power Generation Mechanism	sunlight energy- thermal energy-power, with working process	Photoelectric effect of materials, sunlight energy-power, without working process
Definition of Efficiency	Annual mean efficiency, Generation energy/annual solar irradiation, kWh/kWh	Peak value efficiency, generation power/input power, kW/kW
Capacity	No scale limits; dish system is suitable for distributed power generation; tower-type and parabolic trough systems are suitable for large-scale projects	No scale limits; can be either separately applied, or applied in large-scale project
Solar Spectrum	300–3000 nm	300–600 nm
Power Quality	Small load fluctuation, high quality	Large fluctuation without power storage impacts on power grid, poor quality
Cost	high	lower

it is not possible to simply apply conventional materials systems. These technical considerations pose challenges to the conventional techniques being applied right now. The development of large-scale, highly efficient CSP cycle technologies requires new and further research on highly efficient concentrator fields, unsteady high-temperature heat transfer and thermal storage mechanisms, materials design, reliability of the CSP-generation system and its recurring effects on the overall system, etc.

3. Differences between CSP generation and solar photovoltaic power generation. The two solar energy power generation modes are compared in Table 1.1.

1.1 GENERAL PRINCIPLES OF SOLAR THERMAL POWER PLANT DESIGN

Design of the CSP plant shall follow the general principles of (1) tallying to national conditions, advances in technology, economic feasibility, and operating in a safe and reliable manner; (2) striving for economic and social benefits, saving energy, engineering investment, and raw materials, and shortening the construction period; and (3) being in line with existing Chinese state standards and regulations for saving land, water conservancy, and environmental protection, as well as exercising requirements for labor safety and industrial hygiene.

1.1.1 Constitution of the Solar Thermal Power Plant

An CSP plant consists of three major units: solar energy collection, thermal energy storage, and a thermal power generation unit. The first two mainly include the irradiation concentrator, the receiver, thermal storage, and the evaporator, whereas the last mainly includes the turbine, the power generator, control of the power cycle, the electricity system, water treatment, and the supply system.

Capacity of an CSP plant shall be determined according to the capacity of the generator unit, which is irrelevant to solar irradiation resources, environmental and meteorological conditions and concentrator power. Power plants of equivalent capacity may correspond to concentration fields (mirror fields) of different sizes.

An CSP plant can be constructed economically by using combined heating and electricity based on solar direct normal irradiation (DNI) resources, the current status of the local power load, and thermal load.

CSP can be complemented by coal, petroleum, or natural gas in a mixed-fuel power plant constructed according to circumstances in areas with an abundant solar resource and coal or petroleum resources.

According to the needs of thermal and power load development in corporate planning, construction of a self-contained heating-type CSP plant with an appropriate scale is suggested.

1.1.2 Selection of Pressure Parameters for Power Generation Units

It is suggested that the water steam pressure parameters of generator units be selected according to unified short-term and long-term construction plans while being in-line with the following rules:

1. For a generator unit with a stand-alone capacity of 1.5 MW and below, a medium-pressure (MP) or lower MP steam turbine is suggested. For one with a stand-alone capacity of 3 MW, a MP steam turbine is suggested. For one with a stand-alone capacity of 6 MW or above, an MP or secondary high-pressure (HP) steam turbine is suggested.
2. For a condensing-type generator unit with a stand-alone capacity of 3 MW, lower MP parameters are suggested; for one with a stand-alone capacity of 6 MW or above, MP or lower HP parameters are suggested.

3. For solar collectors within the same power plant, the same type of collector with the same output parameters should be used; generator units within the same power plant should also use the same parameters. For a parabolic-trough-and-tower mixed-CSP plant, the parabolic trough system should be used as the preheater with the tower used for the superheated part.
4. When designing the concentration field, the influences of sun beam shading and blocking between the reflectors on concentration efficiency shall be considered; attention should also be paid to the land use rate and future expansion needs of the concentration field and thermal storage system. Normally the land coverage of a parabolic trough concentration field is about 2.5 times that of the total aperture area of the concentrators, whereas the land coverage of a solar tower concentration field is 4–6 times that of the total aperture area of the heliostats and also related to the height of tower. In some countries, land quotas are quite complex.

1.1.3 Heat Transfer Fluid of the Receiver

Water/steam, synthetic oil, air, or molten salt can be selected as the heat transfer fluid. The working medium of a steam turbine is water/steam. For a CSP plant that uses steam as the corresponding working medium, water pretreatment equipment must use desalinated water; otherwise, permanent damage may be caused to the reverse osmosis water system.

1.1.4 Schedule Capacity of the Power Plant and Number of Installed Units

New power plants can be designed and constructed all at once or in sections according to incremental load speed based on scheduled capacity. Due to the comparatively large investment, the concentration field corresponding to a power plant can be designed all at once but constructed in sections. The major loop for synthetic oil, the design and construction of the parabolic trough collector field, and the height of the receiver tower in the tower power plant shall be configured to match the intended ultimate capacity of the power plant. A large-scale collector field can be divided into different thermal collection modules, the thermal fluid output of which will flow into the thermal storage unit. In the thermal collection system that directly produces steam, the steam will be discharged to the main pipe of the power plant.

The number of condensing power turbines shall not exceed four in one plant. For a power tower plant with an installed capacity of less than 100 MW, no more than one receiver corresponding to the concentration field shall be installed [1]. For a large-capacity tower power plant, the multitower system shall be considered when designing the concentration field. A single-tower system is recommended for a tower power plant that uses molten salt as endothermic fluid because of great difficulties in the high-temperature molten-salt transmission process, poor reliability, and high pipeline cost.

The turbine and boiler configuration, model selection of main auxiliary facilities, major production process system, and main powerhouse layout in the power plant shall be determined through technical and economic analyses. While satisfying the safe, economical, and reliable operation of the power plant, the system and/or layout can be simplified in an appropriate manner.

1.1.5 Control of Power Plant Influences on the Environment

In designing the power plant, it is necessary to indicate the disposal plan for the concentrator as well as the thermal storage and heat transfer materials, especially thermal storage medium to be used in large amounts. The working medium of water/steam for heat transfer and thermal storage is very environmental friendship.

If landscaping is damaged during concentration field construction, a land restoration program shall be provided.

Wastewater, sewage, light pollution, noise, and all kinds of other pollutants shall be prevented, controlled, and discharged by implementing and executing national laws, regulations, and standards for environmental protection, and the relevant rules for labor and industrial hygiene must be tallied. These items can only be discharged by satisfying the respective standards.

Engineering facilities for pollutant prevention and control as well as labor and industrial hygiene facilities must be designed, constructed, and placed into operation with the core work on a simultaneous basis.

1.1.6 Power Plant Seismic Resistance and Windproof Design

The solar collection system consists of concentrators and receivers. Concentrators use optical equipment with high precision requirements. Any deformation of the foundation or supporting structure of the concentrator will greatly influence the precision of the concentrator and have major impacts on the overall working conditions of the power plant and could even result in scrapping of the concentration field. The seismic design of the concentrator shall be based on the hundred-year earthquake. The design of the receiver tower must be conducted by executing the current China state standard. Seismic resistance shall also be considered during design of the power plant's concentrator.

Wind-resistant design for the concentrator and receiver tower of a CSP power plant shall be conducted according to the hundred-year wind scale of the plant's applicable locality.

1.1.7 Principles of Concentration Field Design

The determination of the concentration field area serves as the key to CSP plant design and is normally calculated by applying the design point method.

Design point is a very important concept for CSP generation design and can be used to determine the parameters of various segments of solar concentration field, the receiver, thermal storage, and power generation. Factors of a design point include time, solar DNI, ambient air temperature, and wind speed, etc. The time selection is normally midday during the spring or autumn equinox; annual mean temperature can be used as the ambient air temperature and annual mean wind speed can be used as the respective wind speed.

In determining unit capacity, two methods can be used for selecting the solar direct normal irradiance that corresponds to the design point:

1. Apply solar direct normal irradiance = 1 kW/m^2 when the designed area of the concentration field is small; if the calculated concentration field area has an irradiance of less than 1 kW/m^2, it is impossible for the needs of the power generation and thermal storage systems to be directly satisfied by field output.
2. Apply the annual mean solar direct normal irradiance of the locality when the designed area of the concentration field is large. The output of the concentration field is normally sufficient to satisfy the energy needs for thermal storage and the steam turbine. In cases where solar irradiance exceeds the annual mean level, a portion of the concentration field shall be closed.

To exert maximum functionality of the concentrator relative to a large one-time investment, the first method is normally adopted for concentration field design.

The annual capacity factor of the CSP plant is determined by the design point and operational mode of the power plant.

Thermal storage capacity is determined by generator unit capacity, the annual capacity factor, and the operational mode of the power plant.

1.2 BRIEF INTRODUCTION TO SOLAR THERMAL POWER GENERATION

1.2.1 Basic Concepts of Solar Thermal Power Plants

With the gradual exploration and consumption of conventional energy resources such as coal, petroleum, and natural gas, the fossil energy sources used by humans and maintained for thousands of years are now facing exhaustion. In addition, with severe pollution in the global environment, more and more governments around the world are planning and energetically exploiting various new energies to ensure a large-scale energy supply and maintain and satisfy rapid economic growth and the interests of their citizens. CSP generation technology is viewed as a low-cost sustainable power supply clean-energy technology.

1.2.1.1 Basic Concepts of Solar Thermal Power Generation Technology

CSP generation is a system that converts solar energy into thermal energy and generates power through thermal–work conversion. The thermal–work conversion system is similar to conventional thermal power generation except that CSP generation also contains a solar-to-thermal conversion process; it uses a solar radiation to thermal–work coupled system. A CSP plant normally consists of thermal collection, thermal storage, and thermal–work power conversion systems. Based on different concentration modes for the CSP generation system, the CSP generation [2] normally can be divided into solar tower (also known as central receiver), parabolic trough, parabolic dish, linear Fresnel reflector, etc. Those that have already reached the commercial application level are mainly concentrating solar tower and parabolic trough types. CSP enjoys the advantages of comparatively mature techniques, low power generation costs, and minor impacts on the power grid; thus it has been deemed the most promising among various renewable energy power-generation modes. Meanwhile, CSP thermal–work conversion is partially similar to that of a conventional thermal generator unit. Existing mature techniques can be utilized, and thus CSP is especially suitable for large-scale applications. In 2018, a total capacity of 5206 MW was connected to the power grid all over the world, with 1048 MW under construction, and 3691 MW under development.

As mentioned in "Technology Roadmaps Concentrating Solar Power," released by the International Energy Agency in September 2014 under a proper policy support, it was estimated that by 2050 the cumulative installed capacity of global CSP generation facilities would reach 1089 GW with a mean capacity factor of 50% (4380 h/year), an annual power generation of 4770 TWh that would account for 11.3% of global power production (9.6% of which derives from pure solar energy), and China's CSP generation would account for 4% of the global amount with an annual power generation of about 190 TWh. In areas with excellent solar resources, CSP generation is expected to become a competitive large-capacity power supply that will undertake peak modulation and medium power load by the year 2020 and basic load power generation by 2025–30.

Based on geographic information system analysis, the potential capacity for the installation of CSP generation in China that meets the basic conditions for CSP generation—direct normal irradiation $\geq$5 kWh/(m^2 day) and surface slope ≤3%—approximates 16,000 GW, which is similar to that of the United States; the potential capacity to be installed with a direct normal irradiation of not less than 7 kWh/(m^2 day) approximates 1400 GW. In terms of annual power generation capacity, China's potential annual CSP generation capacity is 42,000 TWh/year, which means that even if all fossil fuel energy resources become exhausted, China will still enjoy abundant CSP generation resources far beyond those required for self-sufficiency.

China has abundant solar energy resources. Its annual solar irradiation falls approximately in the range of 1050–2450 kWh/m^2; on average, the solar energy that irradiates the 9.6 million square kilometers of land in China every year is equivalent to 1700 billion tons of standard coal. About 300,000 square kilometers of the Gobi Desert in China, which accounts for about 23% of China's total desert area, can be used to develop solar power generation. Based on existing CSP generation technologies and annual conversion efficiency, constructing power plants on China's 70,000 square kilometers of sand would result in annual power generation that would satisfy power demands of China equal to those for all of 2018. China's extremely abundant solar and sand resources are especially prevalent western China, where CSP technologies will play a significant role in economic development, environmental protection, and resource protection. As constantly supported by the 8th, 10th, 11th, and 12th Five-Year Plans of China on science and technology issued by the Ministry of Science and Technology of the People's Republic of China, numerous achievements have been made on parabolic trough, solar tower, and parabolic dish CSP generation systems. In July 2011, the Beijing Badaling megawatt-level solar tower power plant was completed and started to produce steam [3]; it started generating power in August 2012. The power plant can generate power not only by pure solar energy, but also by being connected with fossil fuels in a parallel manner. The Chinese first parabolic trough solar power plant was completed in the Yanqing district of Beijing in 2017.

CSP generation is capable of applying two thermal cycles, namely direct and indirect (double-loop). The former directly drives the steam turbine unit for power generation (Fig. 1.1) by using the steam produced by the receiver. The latter produces steam thermally using the working media-water or fluid with a low boiling point in the auxiliary system through thermal exchange during the thermal cycle of the main system and thus driving the steam turbine unit for power generation (Fig. 1.2). Compared with a conventional thermal power plant, the most intuitive difference between the two is that the conventional boiler is replaced with thermal collection and storage facilities in CSP generation, whereas the thermal cycle mode and respective equipment applied for thermal–work-power conversion are basically the same as those used in conventional power plants. In comparing an CSP power plant's acquisition mode with

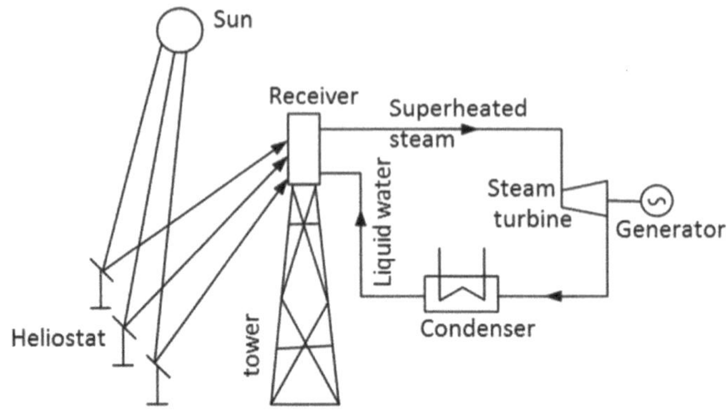

FIGURE 1.1 Schematic diagram of water/steam tower power generation system.

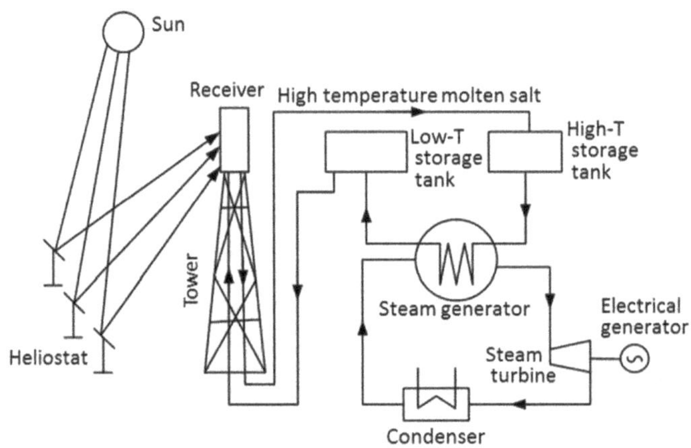

FIGURE 1.2 Schematic diagram of molten-salt tower power generation system.

that of a conventional thermal power plant, the biggest difference lies in the unstable source of thermal energy. As solar irradiation itself features time discontinuity, it may be greatly influenced by weather conditions. Thus the thermal process demonstrates an unstable state, frequent variations, and complexity that lead to nonlinearity, time variation, and uncertainty of multivariable coupling and result in a variety and complexity of operational modes and control means. Along with the development of large-scale thermal storage technology, it is possible to realize large-scale stable operation. For a tower power plant, a thermal collection system that consists of concentrators and receivers does not exist in conventional power plants; in contrast to a boiler, the collection system is quite complex with multiple variables, loops, and operational modes.

A major characteristic of CSP generation technology is thermal collection. The concentration ratio is one of the most important parameters for

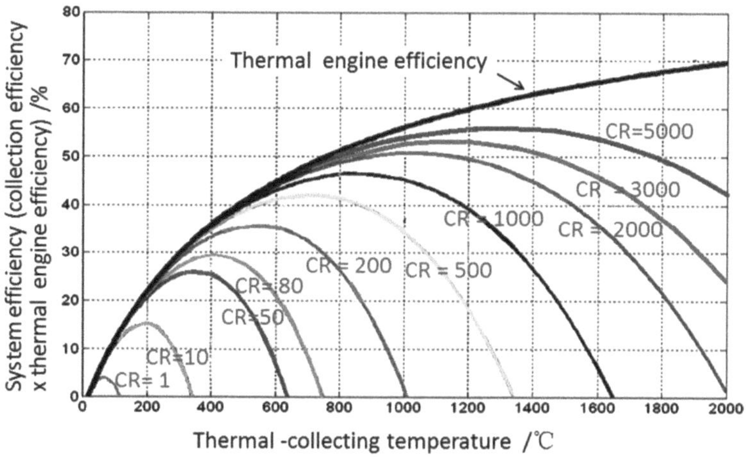

FIGURE 1.3 Relationship of solar thermal power generation system efficiency, thermal-collecting Temperature, and concentration ratio.

CSP generation system design. The greater of the concentration ratio, the more possible it is to achieve a higher maximum temperature (Fig. 1.3). The concentration ratio is the ratio of mean radiation flux density that gathers on the surface of the receiver's aperture to the solar direct normal irradiance that enters the aperture of the concentration field. Annual power generation is a key factor that determines the benefits of an CSP plant. The annual power generation of an CSP plant is the product of the CSP plant's annual efficiency and that solar direct normal irradiance that has been cast on the aperture area of the concentration field. Thus the CSP plant's annual efficiency and solar direct normal irradiance at the construction site of the CSP plant are two extremely critical factors. The CSP plant's annual efficiency (which can also be deemed the system efficiency) is determined by the thermal collection efficiency and the efficiency of the thermal engine. As shown in Fig. 1.3, based on a certain concentration ratio along with increases in the thermal collection temperature, the system efficiency curve will demonstrate a "saddle point," which is mainly caused by the increased efficiency of the thermal engine along with the increment of the thermal collection temperature. However, due to increased heat losses by the receiver, thermal collection efficiency decreases after reaching a certain level. Therefore, in the CSP generation system, simply increasing the working temperature of the system is not advised; instead, the concentration ratio and thermal collection temperature should be comprehensively considered by applying the high-ratio daylight concentration and high-performance absorber techniques.

Based on the concentration mode, CSP generation technologies can be divided into two systems, point focusing and line focusing, with the point focusing system mainly including solar tower (also known as central receiver system) and parabolic dish/Stirling solar power generation, and the line focusing system mainly including parabolic trough and linear Fresnel reflector solar power generation. In these four forms of CSP generation technology, parabolic dish/Stirling engine power generation technology enjoys the highest concentration ratio (1000–3000), followed by solar tower (300–1000), whereas the line focusing system's parabolic trough (70–80) and linear Fresnel reflector (25–100) concentration ratios are comparatively low.

1.2.1.2 Characteristics of Solar Thermal Power Generation Technology

CSP generation is by its nature a way to utilize solar thermal energy. Its generation principle is a clean and green energy utilization method. The development of CSP generation technology is of great significance for the sustainable development of human economies and societies. Compared with other energy utilization methods, CSP generation enjoys certain unique development advantages:

1. Resource request: always available for use. Compared with other renewable energies, the solar resource is inexhaustible and always available for use. China is a nation with extremely abundant solar resources. Solar irradiation received by its land areas approximates the equivalent of 1700 billion tons of standard coal, resulting from an annual sunshine duration that exceeds 2200 h. Total irradiance exceeds 5000 MJ/m^2 that is abundant or comparatively abundant over a vast area including: Alxa League in western Inner Mongolia and Ordos, Hexi Corridor in western Gansu, Qinghai, Tibet, and Hami and Turpan of Xinjiang. Accounting for over two-thirds of the total area of China, these areas have excellent conditions for solar energy utilization. In particular, Gansu, Hexi Corridor, Qinghai, and Tibet possess certain water resources and are sparsely populated, thus enjoying the potential for development of large-scale CSP plants. In addition, the Gobi Desert, deserted land, abandoned saline land, and desert land in western China are vast in area. For example, the portion of the Hanggin Banner of Inner Mongolia along the south coast of the Yellow River that is suitable for the development of CSP generation is as large as 10,000 ha; the area has rich surface water resources, and construction of a two million-kW CSP plant with annual power generation of up to 10 billion kWh is possible. Dunhuang of Gansu has a flat Gobi Desert of over 5000 square kilometers. After implementation of the "Transferring Water from Dahaerteng River to Danghe River" project (a water conservancy project), a one million-kW CSP plant can be constructed. Thus in

terms of developable and exploitable resources, solar resources are superior to wind, biomass, geothermal, hydro, and other renewable energies (Table 1.2); in terms of exploitable areas, solar availability is broader than that of geothermal, oceanic, and like energies.

2. Environmental influences: extremely low. The entire CSP generation process doesn't produce pollutants or greenhouse gases; compared with conventional fossil fuel power generation, it is a clean energy utilization form. Meanwhile, during utilization and exploitation of resources, the ecological environment is not damaged or influenced; compared with wind, hydro, geothermal, oceanic, and similar energies, it enjoys the advantage of being environmentally friendly. Furthermore, by viewing the overall life cycle, energy consumption level, and environmental influences of the entire process of CSP generation from equipment manufacturing, to power generation, to scrapping, it can be seen that they are equivalent to those of other renewable energy utilization forms. Compared with the manufacturing and scrapping of solar panels for solar photovoltaic power generation, CSP generation's energy consumption and pollution levels are greatly reduced. The carbon dioxide emissions of an CSP generation system during its life cycle are extremely low. Based on 2009 technology, the carbon dioxide emissions of an CSP plant during its life cycle were about 17 g/kWh, which is far below those of coal (776 g/kWh) and natural gas combined-cycle (396 g/kWh) power plants. Thus CSP generation uses renewable energy power generation and utilization with minimal environmental impacts.

TABLE 1.2 Exploitation Potentials of Renewable Energy Resources in China [4]

Type		Exploitable Resources
Solar energy		1700 billion tons of standard coal
Wind energy		1 billion kW, including 300 million kW on land
Hydroenergy		Economically exploitable = 400 million kW
		Technically exploitable = 540 million kW
Biomass energy	Biomass power generation	300 million tons of straw + 300 million tons of forest waste
	Liquid fuel	50 million tons
	Methane	80 billion m^3
	Total	–
Geothermal, excluding medium and low temperature		6 million kW

3. Output characteristics of power generation: smooth. Due to different power generation principles, the output characteristics of CSP generation are superior to those of solar photovoltaic and wind power generation. This is especially true for the thermal generator unit through thermal storage units, as it is capable of generating power on a significant, smooth basis while reducing output fluctuations. Based on different thermal storage modes, the utilization hours and power generation of the power plant can be improved by varying degrees, which can improve the adjustment performance of the plant. Furthermore, the output characteristics of CSP generation are normally ameliorated through afterburning or in combination with conventional thermal power generation so that it can be used during the night for constant power generation, including for stable output and undertaking basic load operation.
4. Characteristics for power grid connection: flexible and steady. An CSP plant with thermal storage and afterburning facilities is distinguished from other energy sources, such as wind power and solar photovoltaics, that experience fluctuating power supply. Thermal storage facilities can be used to generate power in a smooth manner, improve the flexibility of the power grid, make up for the fluctuation characteristics of wind power and photovoltaic power generation, and improve the capability of the power grid by eliminating fluctuating power supply. Meanwhile, an CSP generation system that has been equipped with thermal storage facilities converts partial solar energy during daytime into thermal energy and stores it in a thermal storage system; at night or when peak regulation of the power grid is required, it can be used to generate power and satisfy power grid requests while ensuring more stable and reliable power output. Photovoltaic power generation directly converts optical energy into power, and the rest of the energy can only be stored using batteries, for which the costs are far greater than they are for solar-concentrating power generation (which uses thermal storage). Thus it is easier to store the excessive energy to realize constant and stable power generation and peak regulation. It is the most important and obvious advantage of CSP generation against wind power, solar photovoltaics, and other renewable energy power generation modes; it is conducive to stable operation of the power system and is easier to connect to the power grid. Furthermore, because it drives a steam turbine for power generation through the production of superheated steam, CSP generation has the same power-generation mode as that of conventional thermal power, so it will not have a negative impact on the power grid. It provides reactive power using a generation mode that is friendly to the existing power system.

1.2.1.3 Comparison of Solar Thermal Power and Solar Photovoltaic Power Generation

1. Energy storage. Energy storage in CSP power generation has clear advantages, in that thermal storage techniques are more mature and much less expensive than other power storage techniques. Currently, solar photovoltaic power plants are still unable to realize the expectation of all-weather power generation, whereas CSP has already accomplished this target. The remarkable feature of CSP generation technology is the application of a thermal storage system, which is also a major advantage of CSP generation versus photovoltaic power generation. The thermal storage system (Fig. 1.4) accumulates excess thermal output by the receiver under intensive solar irradiation for use in case of clouds, overcast conditions, and peak times while realizing: (1) generation capacity buffering; (2) controllable power output; (3) stable power output; (4) improvement in annual availability and the increment of full-capacity generation hours; and (5) improvement in the effectiveness of the CSP plant as well as a reduction in power generation cost. Research has indicated that an CSP plant with a thermal storage system can improve its annual availability from 25% (in the case of no thermal storage) to 65%. Thus thermal storage technology serves as a key factor in the competition between CSP generation and any other renewable energy power generation. By applying a long-duration thermal storage system, CSP generation is capable of satisfying the requests of the basic load power market in the future. Currently, the longest thermal storage time by a power plant has exceeded 15 h.

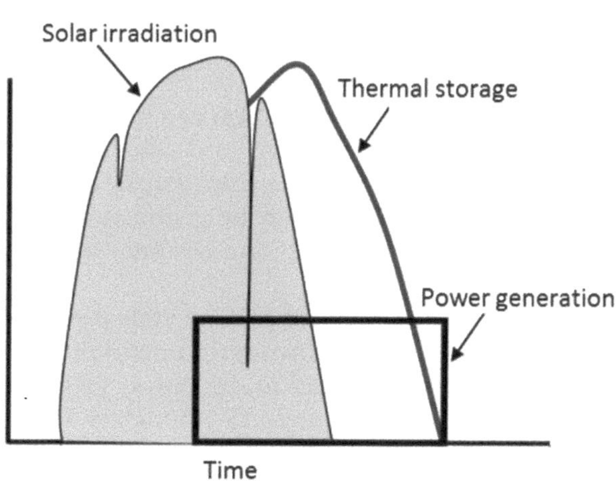

FIGURE 1.4 Schematic diagram of thermal storage techniques in solar thermal power generation.

FIGURE 1.5 Kuraymat parabolic trough/natural gas combined-cycle testing facilities. *Picture: Iberdrola, 2011.*

In addition to utilizing thermal storage technology, a CSP generation system can conduct a combined cycle with coal, fuel, natural gas, biomass, and similar forms of power generation (Fig. 1.5) to overcome its disadvantages of discontinuity and instability, realize all-weather uninterrupted power generation, and achieve optimal technical economy. The Beijing Badaling CSP testing power plant can perform complementary operational experiments between gas/fuel power plants and parabolic trough and solar tower power plants (Fig. 1.6) [5].
Also, CSP generation can be combined with the thermochemical process to realize highly efficient solar thermochemical power generation. Waste heat of CSP generation can be used for seawater desalinization (Fig. 1.7), space heating, and other uses, resulting in integrated utilization. Recently, some scientists have also proposed using CSP generation technology for the gasification and liquidation of coal and for forming long-distance gas or liquid fuels:

2. Power grid access. The instability of photovoltaic power generation has created major challenges for power grid operation, whereas CSP power generation can be connected to the power grid just as in conventional thermal power generation without any adverse effects. CSP power generation is a green power supply for the basic load.

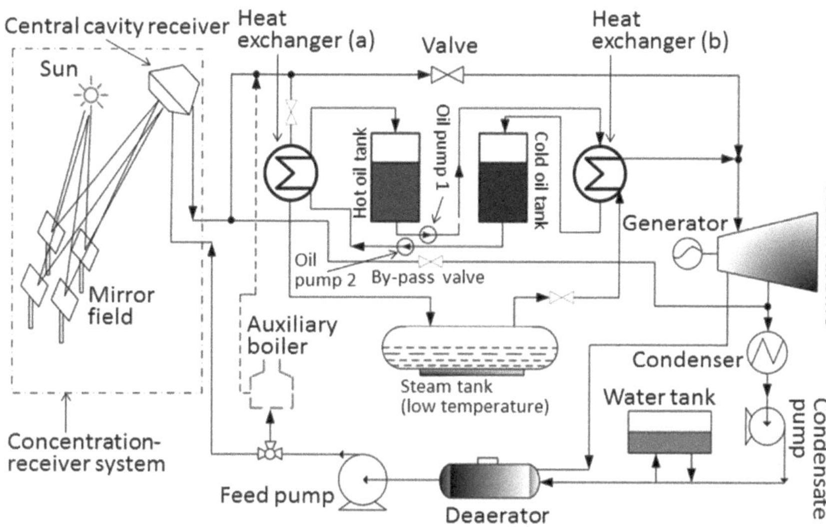

FIGURE 1.6 Complementary scheme of Beijing Badaling solar thermal power testing power plant. *Provided by the Institute of Electrical Engineering, Chinese Academy of Sciences, 2012.*

FIGURE 1.7 Solar thermal power generation and seawater desalinization combined-cycle system. *Picture provided by the Institute of Electrical Engineering, Chinese Academy of Sciences, 2013.*

3. Power dispatching. CSP generation can also serve as a power-dispatching source for the power grid like a pumped-storage power plant; under a considerable peak–valley price mechanism, it will generate greater economic benefits. This also accelerates the cost payback period of the power plant.
4. During the energy conversion process, photovoltaic power generation only requires one optical–power conversion, whereas CSP power generation requires the secondary conversion of optical to thermal to power. Although this has increased the difficulty of system integration, thermal production as the intermediate link of CSP plant operation has also expanded the application scope of CSP power-generation techniques. For example, the superheated steam produced in CSP generation can be used for combined power generation with conventional coal, gas, and biomass power plants. Furthermore, the thermal energy produced by CSP generation can be deemed a by-product for use in seawater desalinization, industrial thermal and air-conditioning, etc.
5. Raw material supply. Photovoltaic power generation mainly consists of photovoltaic solar panels. Currently, those that have been widely applied are crystalline silicon cells and CdTe thin-film cells, the raw material supplies of which could be tight. Market prices could rise, especially for those elements of CdTe cells that use rare metals that are subject to triggering large price fluctuations. Raw materials of the CSP power plant, on the other hand, are mainly commodity items such as glass, steel, concrete, and other common materials in sufficient supply; major price fluctuations in these raw material supplies are unlikely.

1.2.2 Main Technical Forms of Solar Thermal Power Generation

1.2.2.1 Solar Tower Power Generation

Solar tower power generation (Fig. 1.8) is a system that transmits solar irradiation to the receiver mounted on the tower and acquires the high-temperature heat transfer medium through multiple heliostats by tracking movement of the sun, generating power directly or indirectly through the thermal cycle using a high-temperature heat transfer liquid [6]. Solar tower power plants mainly include a heliostat, a receiver tower, a receiver, thermal storage, and a generator unit. Under the working state of the solar tower thermal power plant, all heliostats in the concentration field reflect daylight through double-axis tracking of the azimuthal angle and altitude angle to the receiver mounted on the solar tower in order to thermal the heat transfer fluid inside the receiver of

FIGURE 1.8 Solar tower power generation of Gemasolar plant owned by Torresol Energy, Spain. Picture provided by SENER, Spain, 2018.

receiver. The concentration ratio of solar tower power generation falls into the range of 300–1000, and thus it is easy to realize a comparatively higher system operation temperature. Furthermore, solar tower power generation systems feature a short heat transfer path, small heat losses, and high collection efficiency. Therefore, the solar tower power generation system features comparatively higher comprehensive optical–power conversion efficiency.

Based on different thermal transfer media in the receiver, the system operational mode and performance characteristics of the power plant may be distinguished from each other. Thermal transfer media that are currently available mainly include water/steam, molten salt, and air. In a water/steam power plant system, high-temperature high-pressure steam generated by the receiver can be directly used to drive the steam turbine to generate power; it enjoys the advantage of a thermal-absorbing medium that is the same as the working medium, and annual mean efficiency can exceed 15%. A molten-salt power plant system uses an indirect thermal cycle power generation system, which requires the use of a molten-salt/steam generator to indirectly produce high-temperature high-pressure steam to drive the steam turbine to generate power. Compared with the water/steam power plant system, a molten-salt system can realize supercritical, ultrasupercritical, and other high-parameter operational modes due to low pipeline pressure during high-temperature operation and thus further improve the efficiency of the solar tower thermal power generation system. It is also convenient for storing energy and thus is a technology that enjoys an extremely efficient standardization prospect. An air receiver power plant normally applies the Brayton cycle thermal power generation mode in which air passes through the receiver and becomes 700°C above-high-temperature hot air before entering the gas turbine; the hot air drives the compressor to work and realize power output, which greatly reduces gas consumption, and its operation efficiency can exceed 30%. In addition, it can realize water-free operation and serves as a major research direction for the development of a highly efficient solar tower thermal power plant in the future.

1.2.2.2 *Parabolic Trough Solar Power Generation*

Parabolic trough solar power generation (Fig. 1.9) is a technology that concentrates solar irradiation in the receiver tube mounted at the focal line of the paraboloid through linear parabolic mirrors that track the movement of the sun and thermal the heat transfer liquid for power generation. Key equipment of a parabolic trough power plant mainly includes a concentrator, a receiver tube, and thermal storage. The parabolic trough power plant is the first (1980s) thermal power generation technology to realize commercial operation, with a maximum power plant capacity of up to 80 MW while still ensuring stable operation. Certain problems with the parabolic trough power generation technology are the low concentration ratio of the paraboloid mirror (70–80), difficulty raising the working temperature of the heat transfer liquid, and restrained system efficiency.

FIGURE 1.9 Parabolic trough solar power generation. *Picture provided by the Institute of Electrical Engineering, Chinese Academy of Sciences, 2017.*

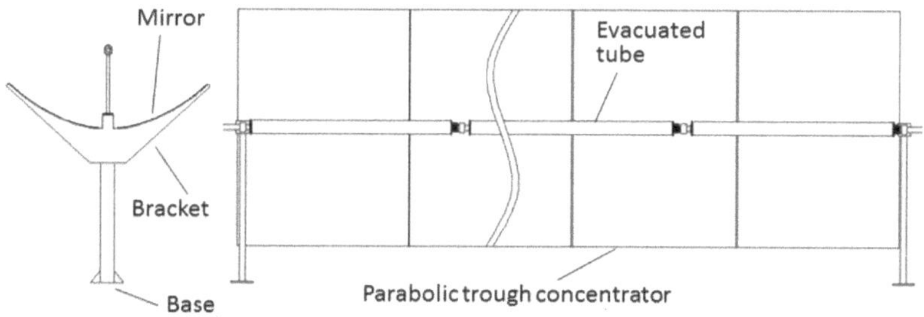

FIGURE 1.10 Structural diagram of parabolic trough solar collector.

As shown in Fig. 1.10, a parabolic trough solar power collector consists of a parabolic trough concentrator that tracks the movement of the sun and a receiver tube mounted at the focal point of the paraboloid. The parabolic trough concentrator uses a single-axis tracking concentrator, namely a concentrator with a mirror element revolving in a one-dimensional manner by surrounding a single axis to track the movement of the sun. The surface of the parabolic mirror is the trajectory formed by a line moving along a certain parabolic curve while parallel to the fixed line. Thus with a parabolic trough concentrator that tracks the movement of the sun, DNI is constantly concentrated on the surface of the receiver tube and creates a focal line so that the heat transfer liquid inside the receiver tube can be heated. High-temperature and high-pressure steam is then generated directly or through an oil—water heat exchange system in order to participate the thermal cycle power generation system and drive the steam turbine to function and generate power or provide the requested steam for industrial processes. The heat transfer liquid of the system transfers thermal energy and is normally water/steam, synthetic oil, or molten salt.

The parabolic trough concentrator is a key component that receives and reflects solar radiation and consists of the base, bracket, mirror, power machine, transmission system, and control system. A typical parabolic trough concentrator is made up of multiple units connected in series along the axis and equipped with a power, transmission, and control system. Normally for a parabolic trough concentrator with small radiation areas, hydraulic or mechanical transmission can be applied; for one with large radiation areas, only hydraulic transmission can be applied.

A bracket is connected to the mirror through fixtures to support and ensure the stability of the parabolic mirror surface; its structure can be categorized as torque tube, torque box, and space truss types; the materials are normally metals, such as steel or aluminum products, and the processing pattern is mainly welding and punching.

Structures of the mirror can be categorized as single-layer or composite. The single-layer structure is an ultraclear glass hot-bending parabolic trough surface that is coated with silver, whereas the composite structure

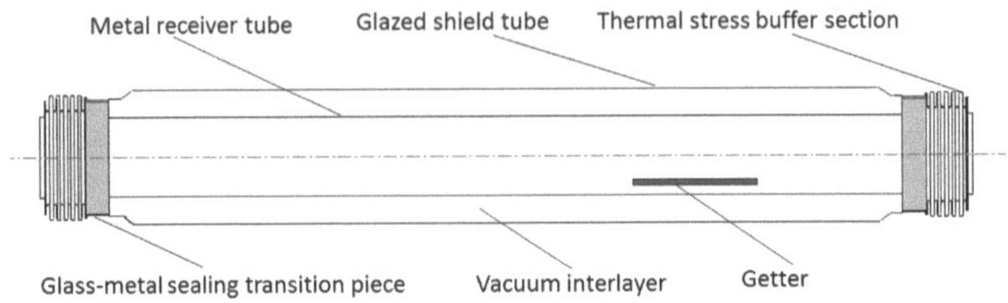

FIGURE 1.11 Structural diagram of receiver tube of parabolic trough solar collector.

consists of a backboard and adhesive and reflective materials. The backboard functions to create a parabolic surface, and it can be made from steel plate, aluminum plate, float glass, and fiberglass. Reflection materials can be thin glass mirror, metal film, or firm composite materials. Adhesive materials can be PVB, neutral organic silicone, etc., in which the aluminum reflector has a high reflection rate created with an aluminum plate through the use of surface finishing and oxidation protection. A silver-coated polymer mirror is a reflection surface with a high reflection rate that is created by coating with silver on one side of the high-transmittance, strong weather-resistance polymer film; it is equipped with multiple layers of protective film that are attached to the bottom of the curved surface to create a curved mirror.

As shown in Fig. 1.11, the receiver tube of the parabolic trough solar collector is a core component of the parabolic trough collector and is typically about 4 m long. The interior tube is a commercial-type metal receiver tube with an external diameter of 70 mm, whereas the exterior tube is a glazed shield tube with an external diameter that falls in the range of 115–125 mm. Due to the metal receiver tube and glazed shield tube having different coefficients of expansion and thermal intensities during operation, high-temperature-resistant glass and metal sealing pieces are required as transition pieces to ensure an airtight connection. In addition, a metal corrugated pipe is used as the thermal stress buffer section to relieve the longitudinal thermal expansion difference between the metal receiver tube and the glazed shield tube. To ensure degree of vacuum degree in the vacuum interlayers of the receiver tube, a getter must be mounted between the metal receiver tube and the glazed shield tube. Furthermore, with any focusing solar irradiation, the seal undertakes great thermal stress that may easily invalidate the sealing of glass and metal. Therefore, thin-walled materials with good reflection performance are required as a solar shade to block radiation while reflecting it to the metal receiver tube.

Both the thermal properties and life of the parabolic trough receiver tube are determined by the vacuum degree of the vacuum interlayer. If the vacuum environment is damaged, not only will the respective heat losses rapidly increase, but also the selective receiver film of the metal receiver tube surface will deteriorate due to oxidation, which may result in severe reduction of the receiver tube's optical efficiency. Under the special working conditions of high temperature and strong radiation, CSP performance and vacuum life can only be ensured when the materials and properties of these components satisfy certain requirements:

1. Glazed shield tube. Due to day–night alternation and temporary cloud occlusions, alternating stress may be generated at the seal, which thus requires high hardness and thermal stability as well as corrosion resistance. Materials that are widely applied at the present include borosilicate glasses such as Pyrex glass, the expansion coefficient of which is 3.3×10^{-6}/K while featuring high hardness, good optical properties, and acid and alkali corrosion resistance. It also has the disadvantages of having no corresponding sealing metal, and its softening temperature approximates 820°C, and thus the temperature is extremely high during sealing operations.
2. Metal receiver tube. The temperature of the metal receiver tube under concentration effect will be much higher than 400°C. Thus it is necessary that it is equipped with high-temperature and corrosion resistance. To eliminate the influences of axial expansion on the collector bracket, the expansion coefficient shall be as small as possible. Due to thermal and gravitational influences, downward deflection may occur, so there must be a sufficient distance between the exterior wall and the interior wall of the glass tube. Currently, high-temperature-resistant 316L stainless steel is normally used with an external diameter of 70 mm, a wall thickness of 3–5.5 mm, a standard length of 4060 mm, and a mean roughness of less than 0.5 μm.
3. Glass-metal sealing transition piece. A certain sealing alloy is applied to solve the inconsistency of the expansion coefficients of the interior metal tube and exterior glass tube. Therefore, both expansion coefficients shall be as close to each other in value as possible in order to satisfy matched sealing and easier welding to the corrugated pipe.
4. Thermal stress buffer section. This buffer is required in order to compensate the expansions of the metal receiver tube and the glazed shield tube. Thus it is necessary that is has good flexibility, excellent tension fatigue strength and life, high-temperature resistance, and acid and alkali corrosion resistance. The respective length shall be as short as possible to increase the effective concentration length of the receiver tube.
5. Getter. A getter is used to absorb the residual gases in the vacuum interlayer after sealing and the released gases of components under high-temperature working status to ensure a satisfactory

vacuum state. A getter accomplishes the target of absorbing residual gases by mainly by relying on physical and chemical absorption.
6. Selective absorption film. According to its working mechanism, it can be categorized as optical interference, intrinsic absorption, metal ceramic, or multilayered gradient film. As a general requirement, for temperatures below 400°C, its absorptivity shall be not less than 95%, and its reflectivity shall be less than 14%. The most widely applied selective absorption film is composite material absorption film, including multilayered gradient metal ceramic film and double-layered absorption film. The multilayered gradient metal ceramic film has a metal substrate, and the absorption layer is made of metal and dielectric gradient film, whereas double-layered absorption film creates two absorption layers and one or two dielectric antireflection layer(s) on the high-reflectance metal substrate to achieve low reflectance without reducing the absorption rate.

As shown in Fig. 1.12, in 2017, a 9000-m^2 parabolic trough solar collector was completed at the Beijing Badaling CSP experimental base. The collector was arranged horizontally along the 3000-m^2 north–south axis and 6000-m^2 west–east axis with a tracking length of 300 m. The bracket was mounted by applying a torque tube structure and selecting an independently developed sandwich-structure glass mirror, the technical parameters of which are shown in Table 1.3.

As shown in Fig. 1.13, the collector is of torque tube-type, the support arm is made from rectangular steel pipe by welding, and the parabolic mirror is made by gluing together the hot-bending glass paraboloid and the ultrathin glass mirror. A parabolic trough solar power collector contains 24 pieces of vacuum receiver tubes; the metal absorber pipe inside

FIGURE 1.12 Beijing parabolic trough solar power collector. *Provided by the Institute of Electrical Engineering, Chinese Academy of Sciences, 2017.*

TABLE 1.3 Parameters for Parabolic Trough Solar Power Collector

Item	Parameters	Item	Parameters
Total area/m^2	9000	Glazed shield tube wall thickness/mm	3
Aperture area/m^2	9000	Single-piece receiver tube length/mm	4060
Aperture width/m	5.76	Total length of receiver tube/mm	97,440
Focal length/mm	1.71	Total length of collector/m	1500
Glass thickness of mirror/mm	4	Tracking precision/(°)	±0.1
Thickness of glass mirror/mm	3.2	Maximum operating temperature/°C	400
Exterior diameter of metal receiver tube/mm	70	Maximum operating pressure/MPa	1.6
Exterior diameter of glazed shield tube/mm	120	Tracking axis direction	3000 m^2 north–south 6000 m^2 west–east

Provided by the Institute of Electrical Engineering, Chinese Academy of Sciences, 2017.

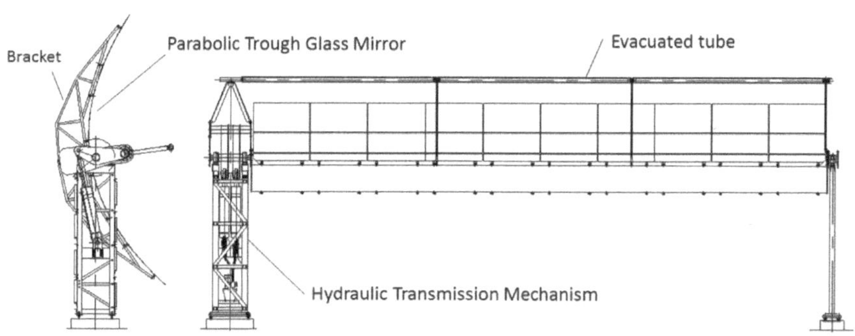

FIGURE 1.13 Structural diagram of parabolic trough collector. *Provided by the Institute of Electrical Engineering, Chinese Academy of Sciences, 2010.*

the cover glass tube is made from 316L stainless steel with a high-temperature-resistant metal ceramic selective absorption coating on the exterior surface. The thermal stress buffer section is made from stainless steel corrugated pipes, the displacement of which is calculated based on the thermal expansion differences generated by the metal receiver tube at 450°C and the glazed shield tube at 0°C when the length is 4 m. Heat transfer oil is used as the heat transfer fluid inside the receiver tube, the type of which is selected according to minimum ambient air temperatures in different seasons: Dowtherm A by Dow Chemical is used, the main ingredients of which are diphenyl and diphenyl ether.

1.2.2.3 Dish-Stirling Solar Power Generation

Dish-Stirling solar power generation (Fig. 1.14) is a system that concentrates solar beam radiation on the generator mounted at the focal point by utilizing a parabolic dish concentrator to generate power through the Stirling cycle. Key components of the dish-Stirling solar power generation system include the parabolic dish concentrator, Stirling generator, and transmission system. Both dish-Stirling power generation and solar tower power use technology that incorporates point focusing concentration and

FIGURE 1.14 Dish-stirling solar power generation. Picture provided by the (Oriental Great Ocean New Energy Technology Development Co., Ltd., CHINA, 2017).

a thermal collection method. For dish-Stirling, the concentration ratio is 600–3000, operational temperature is up to 750°C, and solar dish-Stirling net efficiency of converting peak solar energy into power is up to 30%. The dish-Stirling system features less power, normally 5–50 kW. Thus it can be used independently as the distributed power generation system in remote areas as well as being incorporated into an MW-level power plant for grid-connected power generation.

1.2.2.4 Liner Fresnel Reflector Solar Power Generation

Linear Fresnel reflector (LFR) solar power generation (Fig. 1.15) is a system that concentrates solar beam radiation into a receiver tube mounted at the focal point of the Fresnel mirror through the FLR mirror tracking of the movement of the sun and generates high-temperature working media for thermal cycle power generation. Major components of LFR power generation include the liner reflective mirror, receiver tube, and transmission system. The LFR power generation system is a simplified parabolic trough power generation system. The parabolic trough concentrator is replaced by a surface mirror; the mirror features a small distance to ground, low wind load, a simple structure, an intensive layout, and higher land-use efficiency; furthermore, vacuum treatment for the receiver tube is not necessary, thus reducing technical difficulties and costs. The total cost of the system is comparatively low. However, due to the system's low concentration ratio, the operational temperature stays low, resulting in low system efficiency as well.

Multiple CSP generation modes are compared in Table 1.4.

FIGURE 1.15 Linear Fresnel reflector solar power generation. Picture provided by Himin Solar Energy Co., Ltd., CHINA, 2010.

TABLE 1.4 Comparative Performance of Three Concentrating Solar Thermal Power Generation Systems

Performance Parameters	Parabolic Trough	Solar Tower	Dish-Stirling
Installed capacity	50–600 MW	10–600 MW	5–25 kW
Working temperature of receiver/°C	400	565–1200	750
Maximum efficiency/%	20	23	30
Annual mean efficiency/%	12–16	16–20	12–25
Commercialization experiences	Rich	Yes	No
Thermal storage conditions	Yes	Yes	No
Power of fossil fuel hybrid thermal source	Yes	Yes	Yes
Potential of Brayton–Rankine combined cycle	No	Yes	Yes
Future mean power generation cost (levelized electricity cost), US¢/(kWh)	6	6	10

1.2.3 Basic Terms

This section describes the basic concepts of CSP generation that are frequently used, which will be helpful in providing designers with clear knowledge of these concepts.

Firstly, "design point." "Design point" is an important yet hard-to-understand concept associated with CSP plant design. There is no design point in the conventional thermal power and photovoltaic power systems.

The design point is used in a solar power generation system to determine the parameters of the solar thermal collection and power generation systems, including year, day, hour as well as the corresponding weather conditions and solar direct normal irradiance.

The design point is associated with a specific hour and corresponding solar irradiance and ambient air temperature. It can be used to clarify the area of concentration field, capacity of the steam turbine generator, capacity of thermal storage on a quantitative basis, and relationship among these crucial factors. Normally, a design point is not defined based on peak value and extreme solar angle under local weather conditions, and wind speed is not considered. For a large-scale power plant with a thermal storage system, the design point is typically defined by considering the output capacity of the collector field, which is equivalent to the thermal power of the steam turbine generator under full load operation. Examples are as follows:

A. Design point of some CSP plant
Time: midday of spring equinox
Solar irradiance and environmental conditions: solar direct normal irradiance = the mean solar direct normal irradiance for multiple years at the locality on the spring equinox; ambient air temperature = mean ambient air temperature for 30 years at the locality

B. Design point of some CSP plant
Time: midday of spring equinox
Solar irradiance and environmental conditions: solar irradiance = 1000 W/m^2; ambient air temperature = mean ambient air temperature for 30 years at the locality on the spring equinox

Differences between A and B are as follows:

1. When solar irradiance exceeds the set value of the design point, the thermal output of the collector field in plan A can be transmitted to thermal storage, which means that in cases of partial thermal output of the collector field being transmitted to thermal storage, there will be no impacts on full-load operation of the steam turbine. For plan B, solar irradiance has already been defined as the maximum value possible on the earth's surface, which does not exist.
2. When solar irradiance is less than the set value of the design point, the steam turbine cannot operate under full load. Thanks to the design of plan B, collector field output can never directly drive the steam turbine to operate under full load; instead, full-load functioning of the steam turbine relies solely on the operation of the thermal storage system.

Based on the above, plan A is more optimized than plan "B." However, with plan "B," the collector field's energy output will never be "excessive," yet such an outcome is possible with plan A. For areas with an extremely nonuniform seasonal distribution of solar irradiance, the annual mean solar irradiance is low but the transient solar irradiance is high, and this might result in collector field output that exceeds the requested level of the steam turbine and thermal storage under certain weather conditions. At that moment, part of the concentration field will be closed, resulting in wasted investment; for example, in Hainan Province in China. For arid and semiarid areas in northwestern China, the daily mean solar direct normal irradiance is comparatively even, for which plan "A" is suitable; for Hainan, plan "B" is more appropriate.

1.2.3.1 Optic Terms

1. Absorber.
   Element of the receiver absorbing radiant solar energy and transferring it to a fluid in the form of heat.
2. Concentrator aperture area.
   The maximum projected area of solar irradiation intercepted by the concentrator, which is actually the sum of all mirror areas in a concentrator.
   This is different from the contour area. The concentrator contour area contains the clearance between reflective glasses and is normally larger than the aperture area.
3. Solar collector net aperture area.
   The area of the perpendicular projection over the aperture plane of the solar collector reflecting/refracting components. In a line focusing system it is this surface plus the part of the perpendicular projection of the steel receiver tube onto the aperture plane that does not overlap, provided that the sun-oriented side of the receiver is absorbing radiation. For LFR and heliostat: the net aperture area of a Fresnel collector or heliostat is defined as the sum of the net collecting areas of its mirror segments. The net aperture area of a mirror segment is the perpendicular projection of the reflective mirror area over its aperture plane when they are in horizontal position.
4. Solar collector gross aperture area.
   The area of the flat surface defined by the outer perimeter of the collector, including the gaps between adjacent reflectors. This definition may be used for modules, heliostats, heliostat fields, parabolic dishes, LFRs, etc., as well as complete concentrating collectors.
5. Optical concentration ratio.
   Ratio of the average irradiance integrated over the receiver area to irradiance incident on the solar collector aperture, also called as flux concentration ratio.
6. Geometric concentration ratio.
   The ratio of the collector aperture area to the receiver aperture area.
7. Receiver net collection area.
   The maximum receiver flat area that accepts concentrated solar radiation. It is given by the sum of the products of the active length and diameter of the receiver elements that compose the receiver.
8. Solar field
   The part of the CSP plant that collects and concentrates beam solar radiation. In CSP plants with a parabolic trough collector or Fresnel linear collectors, the solar field is composed of a set of solar collectors and their piping interconnections and headers. In a central receiver

plant, the solar field is composed of the heliostats. In CSP plants with parabolic dishes the solar field is composed of the parabolic dishes.

9. Parabolic trough collector
   A line-focus solar collector that concentrates solar radiation by means of a reflector with a parabolic cross section. It is composed of a set of elements that altogether can track the sun as a single unit.
10. Linear Fresnel collector
    A line-focus solar collector that uses reflectors composed of at least two longitudinal segments with parallel axes to concentrate solar radiation onto a fixed receiver.
11. CSP plant
    Synonymous with solar thermal power and concentrating solar power plants. A facility that applies solar concentration and thermodynamic processes to convert direct solar radiation into electricity suitable for distribution and consumption. The facility may include further sources of thermal energy, such as fossil fuel or biomass, in parallel with solar radiation.
12. Concentrator
    The part of a solar collector composed of reflecting or refracting elements that concentrate and redirect beam solar radiation onto the receiver.
13. Useful solar irradiation
    The integral of the useful radiant solar power over the time interval considered, measured in kWh (1 kWh = 3.6 MJ).
14. Heat transfer fluid
    Fluid used to carry heat from one system component to another in a CSP plant.
15. Heliostat
    A system that reflects beam solar radiation toward a predetermined fixed target by means of a single reflecting element or a set of reflecting elements (facets) controlled by a two-axis solar tracking system.
16. Solar collector aperture area.
    The maximum projected area that accepts solar radiation. The geometrical dimensions of heliostat and parabolic trough concentrators are separately shown in Fig. 1.16 (in which the side length of a single heliostat is c) and Fig. 1.17. For a heliostat, the concentrator aperture area $= 64 \times c^2$, and the concentrator's contour aperture area $= a \times b$. As shown in Fig. 1.17, a parabolic trough concentrator consists of 7×12-meter-long units with a unit length of c and a total length of a with concentrator aperture area $= 7 \times b \times c$ and concentrator contour aperture area $= a \times b$.

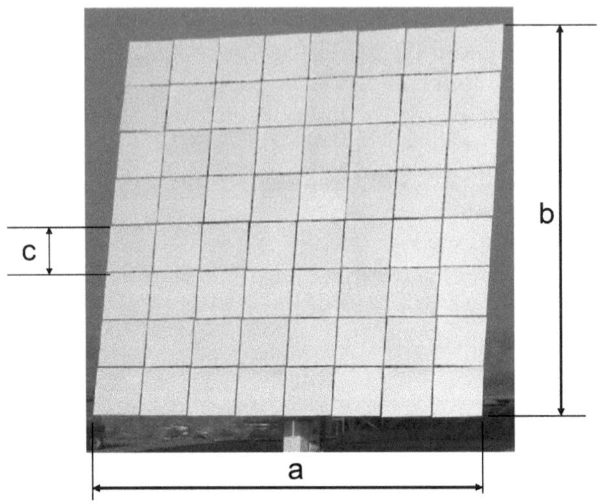

FIGURE 1.16 Geometrical dimensions of heliostat.

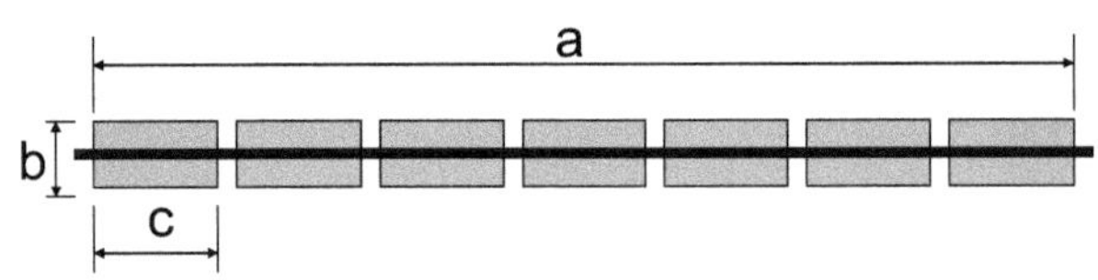

FIGURE 1.17 Geometrical dimensions of parabolic trough concentrator.

17. Receiver aperture area. The maximum receiver flat area that accepts concentrated solar radiation. This is the area of the flat surface defined by the outer perimeter of the receiver, including nonactive zones (if any) between adjacent receiver elements composing the receiver. For receivers without a secondary concentrator and composed of several parallel tubes, it is given by the product of the total length of each tube and the total width of the receiver. For receivers without a secondary concentrator and composed of a single tube, it is given by the product of the total length and the diameter of the receiver tube (excluding the glass cover, if any). For receivers with a secondary concentrator, it is given by the product of the total length of the receiver and the width of the aperture area of the secondary concentrator. For cavity receivers, it is the flat surface associated with the aperture of the cavity.
18. Concentrator performance requirements.
While the concentrator is receiving and reflecting solar energy, there exist specular reflectance losses, including specular loss, Cosine loss, shading and blocking loss, atmospheric attenuation loss,

and spillage loss. Based on this, the concentration field layout design must consider the causes for these losses and mitigate them through a reasonable layout of concentrators to collect more solar irradiation through the receivers.

a. Specular loss. Based on the need for concentrating efficiency, specular reflectance on the reflective surface of the concentrator is normally high, about 0.93–0.94. However, as a heliostat is exposed to atmospheric conditions while functioning, environmental factors such as dust and humidity will contribute to decreases in specular reflectance.
Fig. 1.18 shows the measured result of one heliostat in the Beijing Badaling CSP plant, based on which we can see that influences of dust accumulation contribute to a decrease in specular reflectance from 94.6% on Aug. 23, 2011, to 45.5% on Oct. 10, 2011. On Oct. 13, 2011, due to rain, the specular reflectance climbed back to 82.1%.

b. Cosine loss. In order to reflect sun beam onto a fixed target, the surface of the heliostat is not always perpendicular to the incident light and may create certain angles. Cosine loss is generated because of the reduction of the heliostat surface area against the sun beam projected area caused by such an inclination. The value of cosine efficiency is in proportion to the cosine value of the angle between the heliostat surface's normal direction and incident solar radiation. When arranging heliostats in a concentration field, heliostats must be arranged in areas with the greatest cosine efficiencies possible.

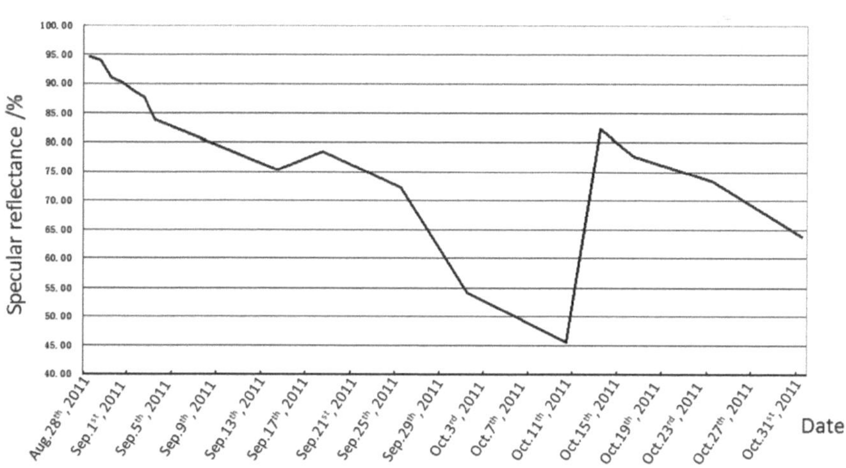

FIGURE 1.18 Influence of dust on the reflectance of the heliostat. *Provided by the Institute of Electrical Engineering, Chinese Academy of Sciences, 2012.*

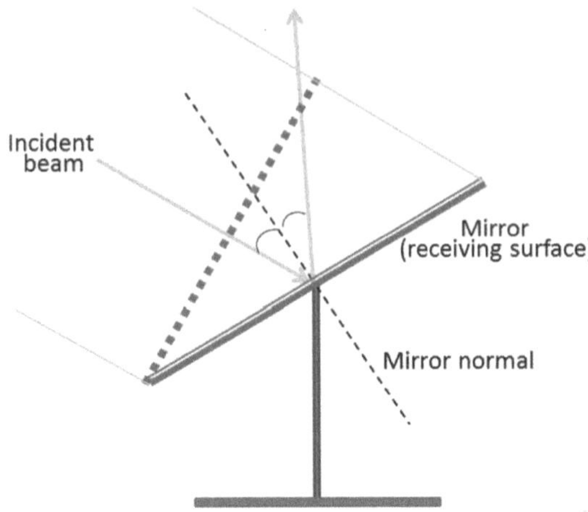

FIGURE 1.19 Cosine loss of heliostat.

Fig. 1.19 shows the ratio of solar irradiation received on certain areas of the surface to the maximum received solar irradiation, which is equivalent to the cosine value of the angle between the incident beam and the normal direction of the receiving surface.

c. Shading and blocking losses. As shown in Fig. 1.20, shading loss occurs when the reflective surface of a heliostat is under the shadow of one heliostat or several neighboring heliostats. Due to the shading of the frontal mirror, the rear heliostat might not receive any solar radiation. Such circumstances are especially bad during winter when the height of the sun is comparatively low.

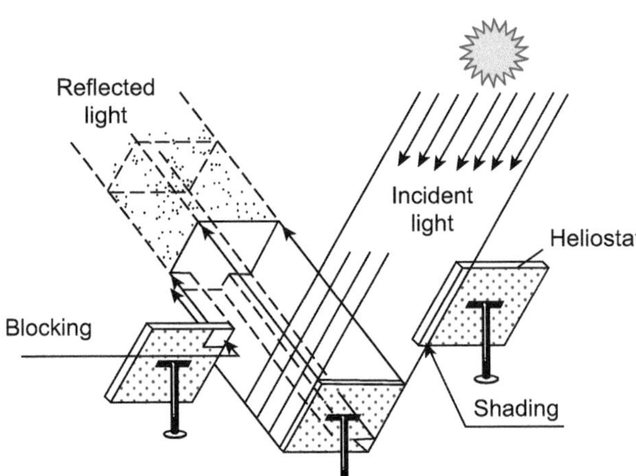

FIGURE 1.20 Shading and blocking losses.

FIGURE 1.21 Influence of blocking from frontal heliostat on rear heliostat's reflected light. *Provided by the Institute of Electrical Engineering, Chinese Academy of Sciences, 2013.*

Shadow from the receiver tower or other objects will cause certain shading loss for the heliostat concentration field as well. Fig. 1.21 displays the influence of blocking from the frontal heliostat on the rear heliostat's reflective light. Fig. 1.22 shows the influence of receiver tower shadow on the concentration field. Although the heliostat is not under any shadow, the shade of the rear side of neighboring heliostats may cause a situation where the reflective solar irradiation is not received by the receiver, the corresponding loss of which is referred to as blocking loss. In Fig. 1.21, the bright band at the upper section of the heliostat is caused by the rear heliostat's reflective light being blocked by the frontal heliostat. The frontal heliostat blocks the path of solar radiation between the rear heliostat and the receiver. Values of shading and blocking losses are relevant to the time when solar energy is received as well as the position of the heliostat itself, which are calculated mainly based on the projected area of neighboring heliostats on the calculated heliostat along the solar incident light direction or along the reflected solar beam direction of the receiver mounted on the tower. Normally, it is necessary to consider shading and blocking on the calculated heliostat caused by several neighboring heliostats. For partial heliostats, it might be possible for the overlapping of shading and blocking losses, which should be taken into consideration during calculation. When designing a heliostat concentration field free of blocking, the distance between heliostats necessarily increases, and the

FIGURE 1.22 Influence of receiver tower shadow on concentration field. *Provided by the Institute of Electrical Engineering, Chinese Academy of Sciences, 2013.*

mean distance between heliostats in the concentration field and tower increases as well. Thus the heliostat's optical efficiency decreases, as does the annual efficiency of the entire concentration field. Thus blocking-free design does not equate to optimized design.

Considering the causes of shading and blocking losses, heliostats should not be mounted too closely to each other. Based on this point, it is possible to properly reduce mutual blocking by restraining the distance between neighboring heliostats.

To realize comprehensive utilization of land, plants can be grown at the bottom of the heliostat. In this case, it is necessary to analyze the influence of the heliostat on the solar irradiation receiving of the ground surface. As shown in Fig. 1.23, this is the mean shading rate of the ground surface in the heliostat concentration field of the Beijing Badaling solar tower power plant from 8:00–16:00 on March 21. The shadow at the outer margin of the figure indicates no shading.

d. Atmospheric attenuation loss. When solar radiation is reflected from the heliostat to the receiver, the energy loss of solar irradiation caused by attenuation during atmospheric propagation is referred to as attenuation loss. The degree of attenuation is normally relevant to the height of the sun (which

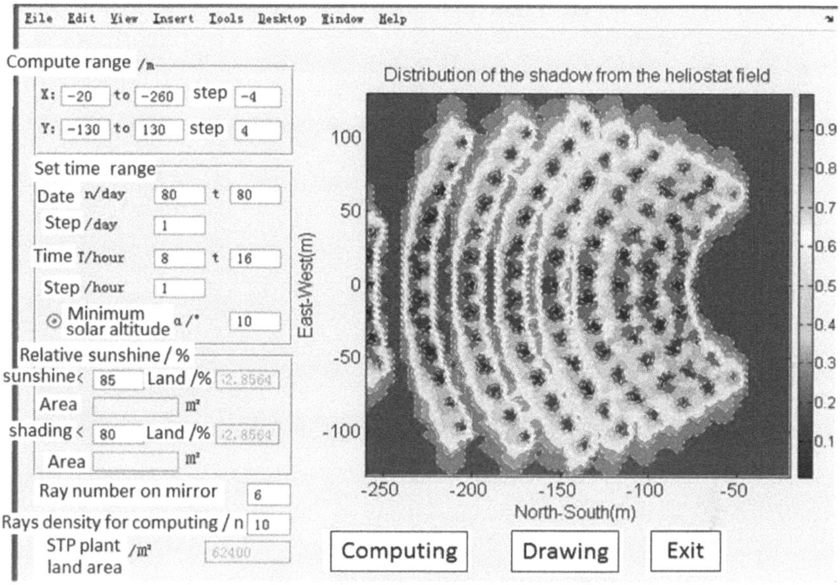

FIGURE 1.23 Analysis of heliostat concentration field shadow.

changes over time), local elevation, factors caused by atmospheric conditions (such as dust, moisture, and carbon dioxide content), and distance. The further the heliostat is from the receiver, the greater the attenuation loss. Thus the heliostat concentration field layout should be restrained to a certain range that is not too distant from the receiver. Fig. 1.24 indicates the solar radiation loss (atmospheric attenuation loss) when aerosol concentration in the air is high at the Beijing Badaling solar tower power plant, from which the solar beam caused by solar radiation scattering can be clearly identified.

e. Spillage loss. This refers to the loss of solar radiation energy reflected from the heliostat that overflows to the atmosphere without reaching the surface of the receiver.
The size of a facula generated by a heliostat on the surface of the receiver aperture is mainly relevant to the heliostat's mirror shape error, tracking control error, and solar cone angle. Furthermore, it is related to the relative position of the heliostat against the receiver and also changes with the position variation of the sun. All of the above factors influence the concentration effect of the heliostat, which is likely to result in the generation of larger faculae by the reflective solar beam of the heliostat on the surface of the receiver aperture and overflow from the receiver aperture to the atmosphere (refer to Figs. 1.25 and 1.26).

FIGURE 1.24 Atmospheric attenuation loss. *Provided by the Institute of Electrical Engineering, Chinese Academy of Sciences, 2011.*

FIGURE 1.25 Spillage loss. *Provided by the Institute of Electrical Engineering, Chinese Academy of Sciences, 2013.*

Therefore, a range of heliostat concentration field layouts should be defined while considering the concentration performance of the heliostat, the dimensions of the receiver aperture and other factors to ensure that the heliostat on the ground is capable of concentrating reflective solar beam radiation within the receiver aperture to the largest extent possible.

FIGURE 1.26 Spillage loss of Gemasolar solar tower power plant owned by Torresol Energy, Spain. Picture provided by SENER, Spain, 2018.

Various losses are calculated in Table 1.5.
Solar radiation obtained on the receiver in the end shall be the sum of all the energy projected by the heliostat on the receiver in the entire concentration field:

$$E = \sum IA\eta_{\cos}\eta_{\text{ref}}\eta_{S\&B}\eta_{\text{att}}\eta_{\text{int}} \tag{1.1}$$

The optical efficiency of the heliostat concentration field is:

$$\eta_{\text{field}} = \eta_{\cos}\eta_{\text{ref}}\eta_{S\&B}\eta_{\text{att}}\eta_{\text{int}} \tag{1.2}$$

Features of software for calculating the annual mean optical performance of the heliostat concentration field are compared in Table 1.6.

19. Mirror surface errors of concentrator. This is the error caused by the inconsistency of the actual reflective surface and the theoretical reflective surface of the concentrator, including canting position

TABLE 1.5 Calculation of Various Losses

Factor	Calculation	Factor	Calculation
Solar direct normal irradiance/(W/m^2)	I	Atmospheric attenuation loss (η_{att})	$I\eta_{\cos}\eta_{\text{ref}}\eta_{\text{S\&B}}\eta_{\text{att}}$
Cosine loss ($\eta_{\cos}$)	$I\eta_{\cos}$	Spillage loss (η_{int})	$I\eta_{\cos}\eta_{\text{ref}}\eta_{\text{S\&B}}\eta_{\text{att}}\eta_{\text{int}}$
Shadow and shade losses ($\eta_{\text{ref}}, \eta_{\text{S\&B}}$)	$I\eta_{\cos}\eta_{\text{ref}}\eta_{\text{S\&B}}$	Concentration field aperture area (A)	$IA\eta_{\cos}\eta_{\text{ref}}\eta_{\text{S\&B}}\eta_{\text{att}}\eta_{\text{int}}$

TABLE 1.6 Feature Comparison of Softwares for Calculating Annual Mean Optical Performance of Heliostat Concentration Fields

Model	RCELL	WINDELSOL	ASPOC
Arrangement pattern	Radial staggered/ north–south staggered/ Radial cornfield/ north–south cornfield	Radial staggered	Radial staggered
Performance calculation	Annual calculation time points: 200 Minimum solar altitude: 0°	Annual calculation time points: 29 Minimum solar altitude: 15 degrees	4-Degree polynomial fit for 5-day energy data 4-Degree polynomial fit for a daily 9 h energy value
Definition of unit and quantity	North–south square of 11 × 11 or 21 × 21 units Dimensions of the unit are related to the height of the tower Performance calculation for the center of each unit No quantity restriction for unit	12 × 12 Units with specified space Radial size of each unit is related to the height of the tower Performance calculation for each unit	8 × 10 Units with specified space Energy calculation for the heliostat at the center of each unit
Annual energy	University of Houston's energy flux model Truncation factor is not related to time Shadow and shade losses in annual mean values (not related to the height of the tower)	Improved Houston energy flow model Use annual mean truncation factor Calculate annual mean shadow and shade losses (not related to the height of the tower)	Gauss energy flow distribution model Use the hour truncation factor Calculate the hourly shadow and shade losses related to the height of the tower
Density	Different radial space and circumferential space for corresponding units Circumferential spaces of various units are equivalent to each other during optimization and are determined through minimum unit energy cost optimization Radial space is determined according to terrain and shading design	To be provided with space parameter from the external source	Radial space and circumferential space are determined by three individual variables
Memo	Calculate shadow and shade losses of 48 neighboring heliostats	Consider shadow and shade losses of 12 neighboring heliostats	Consider shadow and shade losses of neighboring heliostats within a radius of 50 m

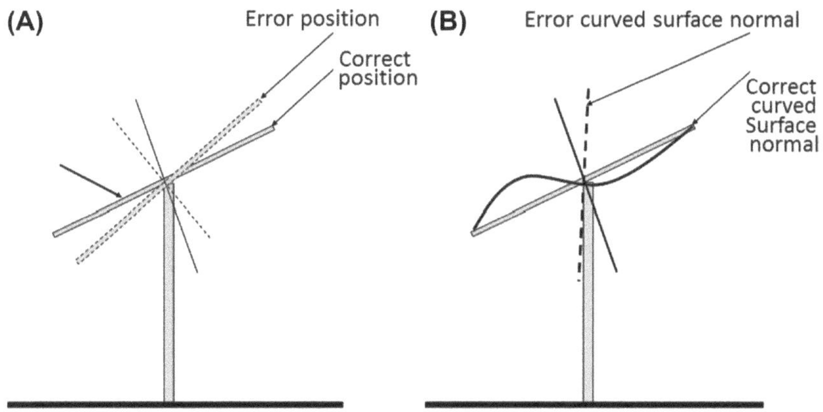

FIGURE 1.27 Concentrator mirror surface errors. (A) Canting position error. (B) Mirror slope error.

error and slope error, in which the incidence point position of solar radiation is inconsistent with the expected position and is subject to position error; this position error is mainly caused by installation, namely support structure location error. Fig. 1.27A indicates an incident solar direct radiation position error caused by altitude angle error due to inappropriate installation of the support structure of the reflective surface.
A surface slope at the incidence point inconsistent with the theoretical value is subject to slope error, namely the normal error on the reflective surface, which is related to the mirror manufacturing process, field assembly, temperature, gravity deformation, wind force, and other error factors (refer to Fig. 1.27B).

20. Efficiency of concentration field. This is the ratio of solar radiation energy (kWh) reflected or transmitted by the concentration field into the aperture of the receiver to the total direct normal solar radiation energy (kWh) on the mirror surface of the concentration field within unit time. According to Eqs. (1.1) and (1.2), due to the position variation of the sun for the parabolic trough concentrator and heliostat, the efficiency of the concentration field changes according to different solar angles, whereas the efficiency of the dish concentrator does not change with the solar angle.
21. Annual efficiency of the concentration field. This is the ratio of the solar radiation energy (kWh) reflected or transmitted by the concentration field into the aperture of the receiver within a year to the total incident solar direct normal radiation energy (kWh) on the mirror surface of the concentration field.

1.2.3.2 Thermodynamic Terms

1. Parabolic trough surface. This is the trajectory of the line parallel to the fixed line and moving along a certain parabolic curve, as shown in Fig. 1.28.
2. Receiver efficiency. This is the ratio of the total energy in the receiver obtained through heat transfer medium to the total energy that enters into the aperture of the receiver within unit time.
3. Receiver peak flux density (unit: W/m^2). This is the maximum radiation energy flux density received on the receiver surface. This value is crucial in receiver design, which determines the material (allowable energy flux density), thermal transfer structure, and mechanical structure of the solar receiver.
   Figs. 1.29 and 1.30 indicate the flux density distribution of the solar concentrator at different places against the focal point, in which the peak flux density at a place 20 cm from the focal point is 80 kW/m^2 and peak flux density at the focal point is 350 kW/m^2.
   The relationship between the position of the aperture of the receiver and the absorber can also be determined by applying the method indicated in Figs. 1.29 and 1.30.
4. Rated thermal power of receiver (unit: W). This is the output thermal power of the receiver at the design point. Rated value refers to the corresponding value at the design point.
   This value depends on one of the most important parameters for calculating the thermal balance of the system. The type of receiver, thermal storage, heat exchanger, and steam turbine are selected by using this data. The thermal storage operational mode of the system is also related to this value.

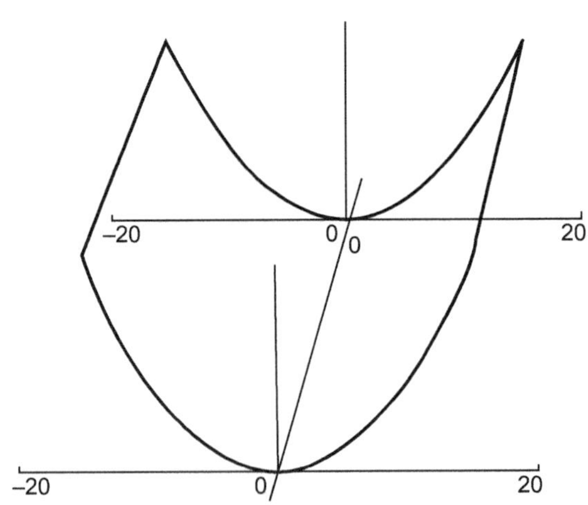

FIGURE 1.28 Parabolic trough surface.

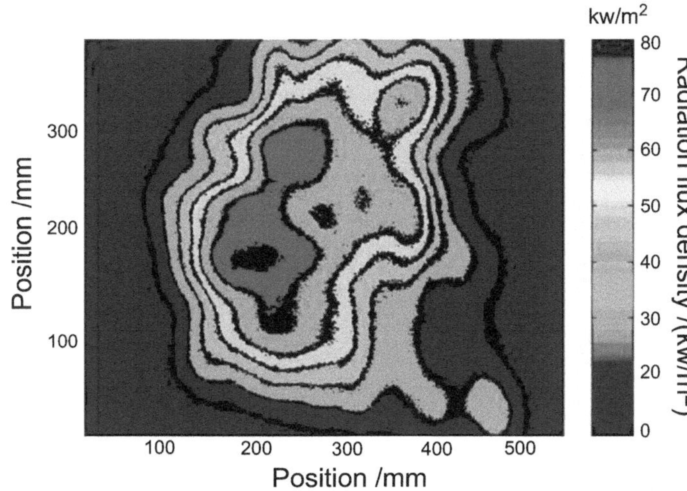

FIGURE 1.29 Energy flux distribution of a solar concentrator on a target plane located 20 cm from focal point.

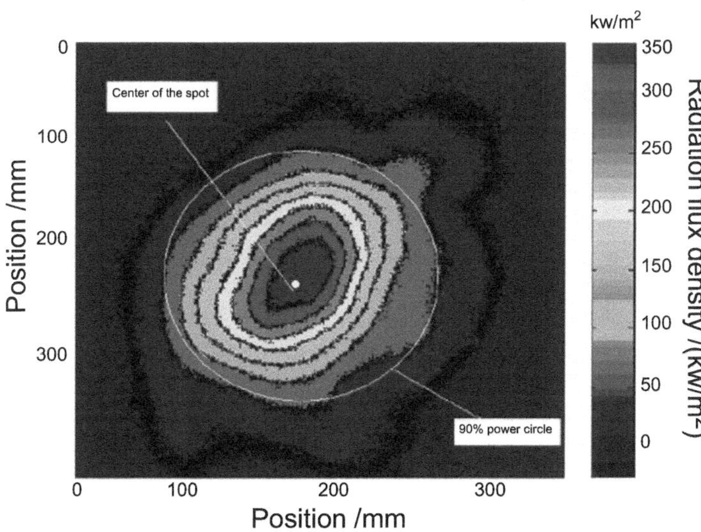

FIGURE 1.30 Energy flux distribution of the solar concentrator on the target plane located at focal point.

Table 1.7 shows data for the PS10 tower power plant located in Spain. The rated optical efficiency of the concentration field is distinguished from the annual mean optical efficiency of the concentration field. When designing thermal storage according to

TABLE 1.7 Relationship Between Rated Thermal Power and Annual Mean Efficiency at PS10 Power Plant

Item	Data	Item	Data
Rated optical efficiency of the concentration field	77%	Annual mean shading and blocking efficiency of the concentration field	>95.5%
Annual mean optical efficiency of the concentration field	64.0%	Rated output solar radiation power of the concentration field	55.0 MW
Mean cosine efficiency of the concentration field	>81%	Annual mean output solar radiation power of the concentration field	45.7 MW

the rated output and annual mean output, different capacities and operational modes can be obtained, which normally are calculated based on the rated thermal power of the receiver.
Besides the properties and thermal parameters of the concentrator and receiver themselves, major external factors that influence this value include position, irradiance, ambient air temperature, and wind speed.

5. Receiver net thermal power (unit: W). This is the energy of the receiver transmitted to the working fluid within unit time.
6. Thermal storage capacity (unit: J). This is the parameter for describing the amount of thermal storage.
7. Emission ratio. This is the ratio of the radiant existence of the radiator to that of the full radiator (black body) under the same temperature, which is applicable for a specific wavelength or a certain wavelength range.
8. Reflection. This is the process of radiation being reflected by the incident surface into the incident medium under the condition of no wavelength or frequency variation.
9. Reflectance. This is the ratio of the reflected and incident radiation flux by the mirror.
10. Absorption. This is the process of radiation energy being converted into other forms of energy by interaction with matter.
11. Absorptance. This is the ratio of the radiation flux absorbed by the panel to the incident radiation flux, which is applicable for a specific wavelength or a certain wavelength range, and is recommended as 300–2500 nm.

1.2.3.3 System Terms

1. Design point power. This is the rated output electrical power of the system at the design point.

2. Collector field efficiency. This is the ratio of the energy obtained by the heat transfer medium from the collector field to the incident solar direct normal radiation energy on the aperture of the collector field. This value represents the ratio of energy, not the ratio of power. A collector field consists of a concentration field and corresponding receivers.
3. Collector field annual efficiency. This is the ratio of the total thermal energy obtained by the heat transfer medium from the collector field within a year to the total incident solar direct normal radiation energy on the aperture of the concentration field.
4. CSP plant annual efficiency. This is the ratio of annual power generation of the CSP plant to the solar direct normal irradiance transmitted to the mirror surface of the concentration field. Comparisons between several efficiency values of the Beijing Badaling tower power plant and the 10 MW Solar Two tower in the United States are shown in Table 1.8.

TABLE 1.8 Comparison of Annual Mean Performances of Solar Thermal Power Generation Systems

Item	Badaling Tower	Solar Two Tower	Item	Badaling Tower	Solar Two
Power plant scale/MW	1	10	Annual mean efficiency of the concentration field/%	68.7	50.3
Heliostat area/m^2	100	40/95	Receiver efficiency/%	85	76
Total aperture area of the concentration field/m^2	10,000	81,400	Steam turbine efficiency/%	22.3	32.6
Thermal storage capacity (full-load generating duration)/h	1	3	House power consumption rate of power plant/%	12	17
Operational temperature of steam turbine/°C	400	565	Annual mean power generation efficiency/%	8.35	7.9
Operational pressure of steam turbine/MPa	2.8	10			

FIGURE 1.31 Ivanpah tower power plant under commissioning (September, 2013).

Decreased efficiency of Solar Two in the United States is caused by the heliostat being located too far from the receiver tower after the area of the concentration field was enlarged, sometimes even to the south of the tower with an extremely low optical efficiency. Thus the efficiency of the concentration field is low. Furthermore, if the large-scale heliostat concentration field is poorly designed, the spillage loss of the receiver might be great, which results in a decrease in the efficiency of the collector field. Special attention should be paid to this when designing a large-scale tower power plant. The Ivanpah tower plant is under commissioning (Fig. 1.31) with a total capacity of 392 MW; it consists of three receiver towers, with each receiver tower corresponding to a capacity above 130 MW.

5. Energy storage utilization factor. This is the percentage of available thermal energy to the total thermal storage capacity in the energy storage system.
6. Solar multiple. Ratio of the net thermal power of the solar field at the design point to the thermal power required from the solar field to run the power block at rated power, which reflects differences between thermal collection system capacity and power generation system capacity. This parameter at a specific design point can be used to define the rated capacity of the steam turbine and thermal storage.

CHAPTER

2

The Solar Resource and Meteorological Parameters

2.1 THE NATURE OF THE SOLAR RESOURCE

Total solar *resource* reserve refers to solar radiation energy in the form of electromagnetic energy that reaches Earth and is directly or indirectly utilized by humans, whereas total solar *energy* reserve refers to all solar radiation energy in the form of electromagnetic energy that reaches Earth, whether utilized by humans or not.

Solar resource technical exploitable capacity refers to that part of the total solar resource reserve that has been and is to be exploited under current technical conditions without considering economic and other conditions.

Solar resource economic exploitable capacity refers to that part of the total solar resource reserve that has been and is to be exploited under local economic conditions at present and that is technically exploitable within a foreseeable period, the exploitation costs of which can compete with those of other energies.

2.1.1 Advantages of Solar Resource Utilization

The solar resource refers to solar energy that can be directly or indirectly (through photothermal, photoelectric, and photochemical conversions, photo bio-electric and the like) utilized by humans, with several advantages:

1. Ubiquity. The sun shines over Earth and solar energy is ubiquitous. It can be utilized on-site without requiring transportation, which creates a huge advantage for solving energy supply problems in remote areas, villages, and islands that are not conveniently located.

Design of Solar Thermal Power Plants
https://doi.org/10.1016/B978-0-12-815613-1.00002-X

2. Harmlessness. Solar energy utilization has no waste residue, waste materials, wastewater, waste gas emissions, noise, or the production of other hazardous substances; it will not pollute or harm the environment.
3. Long-lastingness. As long as the sun exists, there is solar radiation energy. Thus solar energy has an inexhaustible supply and is always available for use.
4. Enormousness. Solar energy is energy from inside the sun that is produced through continuous nuclear fusion reactions. The solar energy that Earth receives each second is equivalent to nearly 5 million tons of standard coal, which equates to 130 trillion tons of standard coal per year and equates to more than 10,000 times the world's annual energy consumption at present.

2.1.2 Disadvantages of Solar Resource Utilization

1. Small energy density, namely small capacity density. During midday on a clear day, at the spot on the ground perpendicular to the direction of solar radiation, the received solar energy density is about 1 kW/m^2. As a source of power, such energy density is deemed quite low. Thus under high-temperature conditions, solar energy utilization normally requires a set of solar energy collection equipment of considerable size and covering a large land area while featuring a significant amount of material, a complex structure, and high costs. All of these have had a negative impact on solar energy promotion and utilization.
2. Instability. Solar direct radiation energy that reaches a specific ground area is extremely unstable due to weather and seasonal factors, creating difficulty for large-scale utilization.
3. Discontinuity. The level of solar direct radiation energy reaching the ground changes throughout the day and night; because of this, most solar energy equipment cannot function at night. To overcome the difficulties caused by solar direct radiation's absence at night, energy storage equipment must be developed and equipped so that it can collect solar energy during clear days and store it to be utilized at night or on rainy days.

2.2 THE SOLAR CONSTANT AND RADIATION SPECTRUM

2.2.1 Solar Irradiation Expression

1. Solar irradiance (W/m^2) is a physical parameter for describing the degree of solar radiation, namely solar radiation energy in W/m^2 perpendicularly projected on a unit of area within a unit of time. It is a parameter most commonly used for solar photovoltaics and thermal utilization.

2. Radiant intensity (W/sr, where *sr* refers to steradian) is the quotient of the radiation power leaving the point radiation source (or radiation source panel) and the respective solid angle element within the solid angle element in a given direction.
3. Radiance ($W/[sr \cdot m^2]$) is the quotient of the panel's radiant intensity on a specific point on the surface in a given direction and the orthographic projection area of the respective panel on the surface perpendicular to the given direction.
4. Spectral irradiance (W/m^3) is the quotient of irradiance within the range of infinitesimal wavelength and the respective wavelength range.
5. Radiant exposure (MJ/m^2; 1 MJ $= 10^3$ kW s $= 0.28$ kWh) is the total irradiation or accumulated irradiation value within a certain period (such as a day or month).
6. Global radiation is the sum of downward direct solar radiation and scattered solar radiation received on a horizontal surface within a 2π solid angle.
7. Solar altitude refers to the altitude angle at the solar disk center, namely the angular distance from the horizon at the observation point along the azimuth circle where the sun is located to the solar disk center.
8. True solar time is time calculated based on the sun's actual position in the sky and is also known as "apparent time." The time when the sun passes through the local meridian line is called the midday of the local true solar time (with an hour angle of zero). The interval between two passes of the sun through the local meridian line is deemed a true solar day. A true solar day is not always the same length.
9. Because true solar time varies by specific day, for practicability one must use the average day of solar days for an entire year, also known as the "mean solar day."
10. The mean solar day is 24 h on average and is referred to as the "mean solar time."

2.2.2 Solar Radiation Spectrum

The sun is the central body of the solar system and can be deemed a full radiator with a surface temperature of 5777K. It is the source of light and thermal energy for Earth; it continuously transfers an enormous amount of thermal energy to Earth in the form of radiation. Solar radiation is resolved into monochromatic elements that are distributed based on wavelengths or frequencies in sequence from short to long wavelengths, including cosmic lights, γ- and X-rays, and ultraviolet, visible, infrared, and radio radiation. As in energy science, the wavelengths of commonly used solar radiation fall into a range of 0.15–4 μm, which can be divided into three major regions: the ultraviolet region with a shorter wavelength, the infrared region with a longer wavelength, and the visible light region with a wavelength falling in between. Regions vary by wavelength: less than 0.4 μm is considered ultraviolet, over 0.76 μm is considered infrared, and from 0.4 to 0.76 μm is considered visible light. Among solar radiation that reaches the ground, the ultraviolet region accounts for about 8.03% of global solar radiation, visible light accounts for 46.435%, and infrared accounts for 45.54%.

2.3 ATMOSPHERIC INFLUENCES ON SOLAR IRRADIATION

Because of the atmosphere, solar radiation energy that finally reaches Earth's surface has been influenced by various factors. Generally speaking, major influencing factors include solar altitude, air mass number, atmospheric transparency, geographic latitude, sunshine duration, and elevation:

1. Solar altitude. Due to the spectral selectivity of atmospheric depth, energy percentages for the various wavelengths of global solar radiation are differentiated by their different solar altitudes. For energy with a solar altitude of 90 degrees within the solar spectrum, infrared accounts for 50%, visible light for 46%, and ultraviolet for 4%; with a solar altitude of 30 degrees, infrared accounts for 53%, visible light for 44%, and ultraviolet for 3%; with a solar altitude of 5 degrees, infrared accounts for 72%, visible light for 28%, and ultraviolet for approximately 0%.
2. Air mass number (AM). Earth's atmosphere has an average depth of 100 km. When solar radiation travels through the atmosphere, it is reflected, scattered, and absorbed, and the respective spectral intensity distribution and corresponding total irradiance undergo certain changes; the degree of these changes is determined by the amounts of atmospheric substances that radiation has passed through. The ratio of the distance traveled by solar radiation through the atmosphere to the distance traveled by solar radiation through the atmosphere while at the zenith is referred to as AM. For example, AM0 refers to the solar radiation that reaches the atmospheric surface of Earth before entering the atmosphere, namely that the air mass that solar radiation travels through is equal to zero. Normally, the distance that solar radiation travels through the atmosphere when the sun is at the zenith, namely shining perpendicular to the equatorial sea level (the spring equinox/autumn equinox) is called one air mass. When the sun is at other positions, AM always exceeds one. For example, at 8:00–9:00 a.m., AM is around two to three. As AM increases, the distance that solar radiation travels through the

atmosphere increases as well, and as a result it undertakes more attenuation and less energy reaches the ground.
Assuming Earth is a perfect sphere with a mean atmospheric depth of 100 km and a mean radius of 6400 km, it can be calculated that at a position with a latitude around 48 degrees, the solar radiation spectrum will be AM1.5, whereas at the pole ($\alpha = 90$ degrees) it will be AM11.4. The distance that solar radiation travels to pass through the atmosphere increases with increments in latitude, as does the atmospheric influence. Thus areas with higher latitudes normally tend to have smaller irradiances.

3. Atmospheric transparency. Atmospheric transparency is a parameter used to describe the transmittance of solar radiation. During clear weather, atmospheric transparency is high, so more solar radiation reaches the ground. During overcast and stormy skies, atmospheric transparency is quite low, so less solar radiation reaches the ground. Currently, atmospheric transparency in China can be categorized into six levels; level 1 means that the area's atmospheric transparency has reached its maximum, namely that solar irradiance is at its highest level, while levels 2 through 6 decrease in sequence.
4. Geographic latitude. When atmospheric transparency is unchanged, the atmospheric distance gradually increases from low latitude to high latitude, and solar radiation energy weakens correspondingly from low latitude to high latitude.
5. Sunshine duration. Sunshine duration is one of the most commonly used physical parameters for describing the solar resource. Presently, all operating meteorological stations can carry out a sunshine duration observation, which observes the sunshine duration of a specific area (the cumulative time for the ground observation site under solar direct irradiance that is equivalent to and above 120 W/m^2). The unit is an hour, which can be as precise as 0.1 h. The longer the sunshine duration, the greater the global radiation received by the ground.
6. Elevation. Generally speaking, the greater of the elevation, the better the atmospheric transparency and the greater the solar direct radiation.

Sun–Earth distance, topography, terrain, and the like also influence solar radiation. For example, the mean temperature when Earth is at perihelion is 40°C higher than it is when Earth is at aphelion. Another example is that for the same latitude, basin area has a higher temperature than surface area, and sunny slope is hotter than shaded slope.

To sum up, many factors can influence ground solar radiation, yet the amount of solar radiation in a specific area is determined by the foregoing factors in a comprehensive manner.

China's topography is high in the west and low in the east, with three ladderlike distributions:

The top-level region in China's topography is the Qinghai–Tibet Plateau. The Qinghai–Tibet Plateau has a mean elevation of over 4000 m and land coverage of approximately 2.3 million km^2; it is the largest plateau in the world. It is located in southwest China; a series of huge mountains with continuous snowy peaks spreads over the plateau from north to south, including the Kunlun, Altyn Tagh, Qilian, and Tanggula Mountains; Karakoram and Kailas Ranges; and Himalayas. This region also has the most abundant solar resources.

Second-level region. Beyond the Kunlun and Qilian Mountains to the north and Minshan, Qionglai, and Hengduan Mountains to the east of the Qinghai–Tibetan Plateau, the elevation rapidly drops to about 1000–2000 m, with some partial areas as low as 500 m. In this second-level region, the Greater Khingan Range and Taihang Mountains through Wushan Mountain, and further to the Wuling and Xuefeng Mountains in the south, serve as the boundary of the eastern margin. Here spreads a series of high mountains, plateaux, and basins with elevations over 1500 m from north to south, including the Altai, Tian Shan, and Qinling Mountains; Inner Mongolian, Loess, and Yunnan–Guizhou Plateaus; and Junggar, Tarim, Qaidam, and Sichuan Basins. Except for the Yunnan–Guizhou Plateau and the Sichuan Basin, this is basically the region with the second-most-abundant solar resources.

Third-level region. Over the Greater Khingan Range and Xuefeng Mountains, this area directly reaches the coast in the east. Hills and plains in the region have elevations below 500 m. In the third-level region, from north to south, spreads the Northeast China, North China, and Middle–Lower Yangtze Plains; extensive areas of low mountains and hills lie to the south of the Yangtze River and are generally referred to as the Southeast China Hilly Regions. In the former area, elevations are all below 200 m; in the latter, most areas have elevations between 200 and 500 m; only a few hills reach or exceed an elevation of 100 m. This is basically the region with the third-most-abundant solar resources.

2.4 CALCULATING METHODS FOR SOLAR POSITION

During solar thermal utilization, the normal requirement is to consider solar radiation as a black body radiator with a temperature of 6000K and a wavelength of 0.3–3 μm. Solar radiation that reaches the ground is mainly influenced by astronomical and geographical factors such as longitude and latitude, elevation, solar declination angle, solar hour angle, air quality, and weather conditions. Solar radiation can be categorized as either direct or scattered. Solar concentration mainly utilizes direct radiation, namely solar radiation that has not been scattered in the atmosphere. Solar direct radiation is described by direct normal irradiance (DNI, measured in W/m^2), which can be measured with a pyrheliometer that automatically tracks and aligns with the sun. To improve the efficiency of solar radiation utilization, most solar concentrators adopt a single-axis or double-axis revolving method of tracking the movement of the sun. Such concentrators are referred to as tracking concentrators.

2.4.1 Solar Angle

1. *Declination angle*, δ, is the included angle of the line connecting Earth's core to that of the sun and the equatorial plane of Earth.
   Its value varies yearly and changes daily. The respective variation range is $\pm 23°27'$ (refer to Figs. 2.1 and 2.2).
   The approximate declination angle for a specific day can be calculated using the following equation:

$$\delta = 23.45° \sin\left(360° \times \frac{284 + n}{365}\right) \tag{2.1}$$

   in which n is the date serial number that refers to the nth day of the year; for example, $n = 1$ refers to January 1.
   The date serial number "n" can be easily obtained through calculation according to Table 2.1.
2. *Solar time* is based on the time of apparent movement of the sun in the sky. Midday (12:00 noon) in solar time (AST) is when the sun passes perfectly through the local meridian line. At that moment, the sun is at its zenith for the day. Solar time can be converted from commonly used local standard time (LST) using the following equation:

$$\text{AST} = \text{LST} + \text{ET} - 4(\text{SL} - \text{LL}) \tag{2.2}$$

   in which LST refers to local standard time (unit: min); ET is the corrected value (unit: min); SL is the longitude of the spot where the

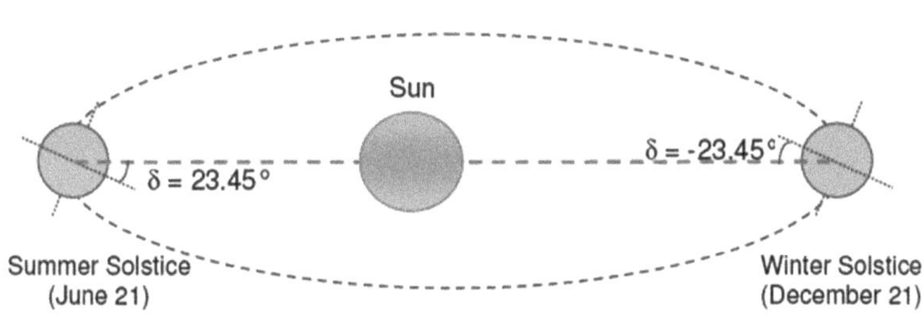

FIGURE 2.1 Variation of declination angle within the Sun's annual operational cycle.

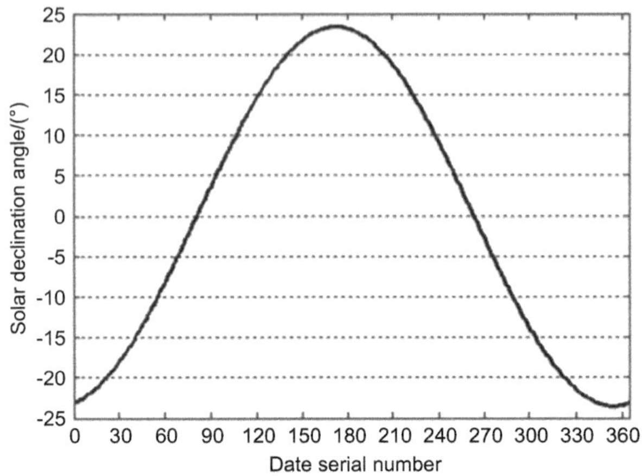

FIGURE 2.2 Variation of solar declination angle during the year.

measuring point has been located in mean time; and LL refers to local longitude, which is positive for east and negative for west.

Two corrections have been made in Eq. (2.2); the first corrects the procession and revolving speed variation during Earth's revolution around the sun, and the second corrects the difference between local longitude and the longitude of the measuring point at mean time:

$$\mathrm{ET} = 9.87 \sin(2\mathrm{B}) - 7.53 \cos(\mathrm{B}) - 1.5 \sin(\mathrm{B})$$

in which $\mathrm{B} = 360°(n - 81)/364$ and n is the date serial number (refer to Table 2.1).

Fig. 2.3 indicates the variation curve of ET over time within a year.

3. *Duration of sunshine*, H_{dl}, represents the difference between the daily sunrise and sunset times and is determined by local latitude and daily solar declination angle. Its calculation is:

$$H_{\mathrm{dl}} = 2\cos^{-1}(-\tan\phi \tan\delta)/15 \tag{2.3}$$

in which ϕ refers to local latitude.

4. *Solar hour angle*, w, is the angular deviation of the sun corresponding to the local meridian line caused by rotation of Earth. At midday solar time, $w = 0°$; w has a negative value in the morning and a positive value in the afternoon. The variation speed of the solar hour angle is 15 degrees per hour. The solar hour angle w can be calculated with the following equation:

$$w = 0.25°(\mathrm{AST} - 720) \tag{2.4}$$

TABLE 2.1 Corresponding Relationship Between Date and Date Serial Number

Month	**1**	**2**	**3**	**4**	**5**	**6**	**7**	**8**	**9**	**10**	**11**	**12**
n (ith day of the month)	i	$i+31$	$i+59$	$i+90$	$i+120$	$i+151$	$i+181$	$i+212$	$i+243$	$i+273$	$i+304$	$i+334$

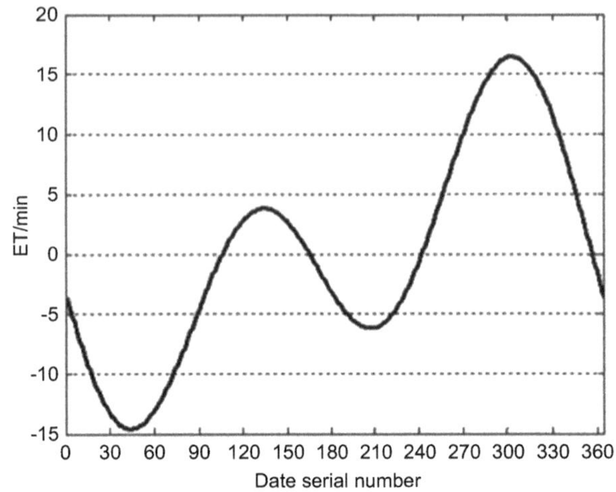

FIGURE 2.3 Variation of corrected value (ET) during the year.

5. *Solar constant*, G_{sc}, refers to unit area solar irradiance out of Earth's atmosphere perpendicular to the direction of radiation propagation with a mean Sun–Earth distance, the value of which is 1353 W/m^2. Along with the minor change in Sun–Earth distance, solar irradiance G_{on} out of the atmosphere on the normal surface of the sun and Earth also undergoes certain changes. According to the measurement, the value of G_{on} changes within the range of ±3%. G_{on} on the *n*th day of the year can be determined through the following equation:

$$G_{on} = G_{sc}\left(1 + 0.033 \cos \frac{360° n}{365}\right) \tag{2.5}$$

6. *Solar zenith angle*, θ_z, is the included angle between line from one specific spot on the ground toward the center of the sun and horizontal ground normals. θ_z can be calculated through the following equation:

$$\cos \theta_z = \cos \delta \cos \phi \cos \omega + \sin \delta \sin \phi \tag{2.6}$$

7. *Solar altitude*, α_s, is the included angle between the line from one specific spot on the ground to the center of the sun and its projection line on the horizontal ground; it is the complement of zenith angle, namely:

$$\alpha_s = 90° - \theta_z \tag{2.7}$$

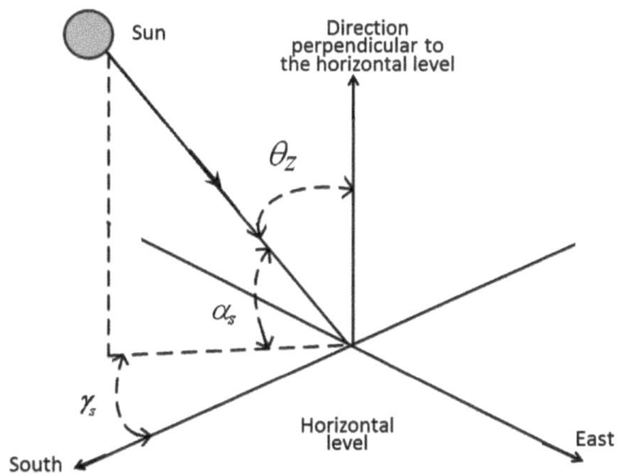

FIGURE 2.4 Schematic diagram of solar zenith, altitude, and azimuth angles.

8. *Solar azimuth angle*, γ_s, is the included angle between the projection line of solar vector on the horizontal ground from one specific spot on the ground to the center of the sun and the south direction, the basic calculation equation of which is as follows:

$$\gamma_s = \arccos\left(\frac{\sin\delta\cos\phi - \cos\delta\cos\omega\sin\phi}{\cos\alpha_s}\right) - 180^\circ \tag{2.8}$$

$$\gamma_s = -\gamma_s, \text{ if } \sin\omega > 0 \tag{2.9}$$

To be intuitive, Fig. 2.4 indicates the geometrical relationship of the solar zenith, altitude, and azimuth angles mentioned above.

9. *Solar flare angle* is the flare angle of the profile of the sun corresponding to one specific spot on the ground, which is also known as the solar radiation divergence angle. As shown in Fig. 2.5, due to the eccentricity of Earth's orbit, the Sun—Earth distance changes within a range of 1.7%.

For a Sun—Earth mean distance of 1.495×10^{11} m, the cone angle of the sun is 32′. The cone angle of the sun is used to indicate that the incident solar direct normal radiation light onto one specific spot on the ground is not a parallel beam. Thus, when designing the specular surface shape of

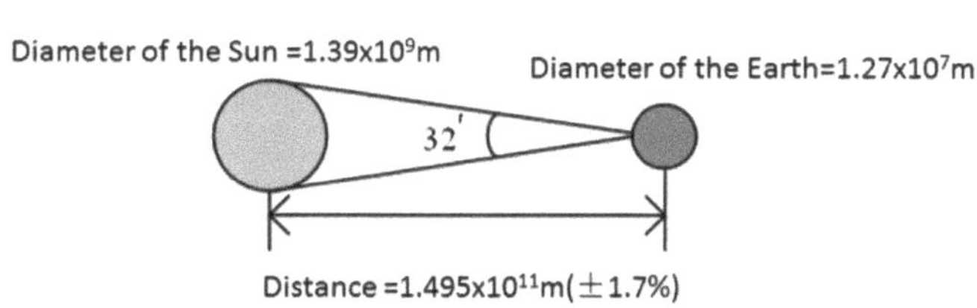

FIGURE 2.5 Basic geometrical relationship of the Sun and Earth.

the solar concentrator and analyzing its concentration performance, influences of the cone angle on the concentration performance of the concentrator must be considered.

Methods for calculating solar position vary; with different algorithms, solar position values of varying precision can be obtained. A highly precise solar position algorithm may consider more factors and tends to be more complex. Eqs. (2.1)–(2.9) are the most basic methods for solar position calculation without considering the influences of multiple practical factors on solar position calculation, such as planetary perturbation on Earth, axial procession of equatorial mean pole surrounding the ecliptic pole, nutation of periodic motion of the equatorial true pole surrounding the mean pole, and atmospheric refraction. The solar position equation applied in astronomy proposed by Meeus can have a precision as high as 0.0003 degrees, yet it requires massive calculation. Some solar utilization facilities having low requirements for solar position precision use the solar position equation with a high calculation speed for precision in the range of 0.008–0.01 degrees. The Beijing Badaling 1-MW solar tower thermal power plant (Badaling), in its heliostat tracking control programs, applied a solar position algorithm proposed by Roberto Grena that considers both precision calculation and time consumption calculation, and solar position precision fell within a range of 0.0027 degrees; solar altitude and azimuth angle at present hours can be obtained through calculation. The input parameters of the algorithm mainly include local longitude, latitude, atmospheric pressure, ambient air temperature, date, and local time.

2.4.2 Calculation of Tracking Angle

1. *Light-receiving surface slope* β, is the included angle of the sloped light-receiving surface against the horizontal surface, $0° \leq \beta \leq 180°$, where $\beta > 90°$ refers to the surface facing downward.
2. *Incidence angle*, θ, is the included angle between the solar incident beam and the normals of a specific surface, the calculation equation of which is as follows:

$$\begin{aligned}\cos\theta &= \sin\delta\sin\phi\cos\beta - \sin\delta\cos\phi\sin\beta\cos\gamma + \cos\delta\cos\phi\cos\beta \\ &\quad \cos\omega + \cos\delta\sin\phi\sin\beta\cos\gamma\cos\omega + \cos\delta\sin\beta\sin\gamma\sin\omega \\ &= \cos\theta_z\cos\beta + \sin\theta_z\sin\beta\cos(\gamma_s - \gamma)\end{aligned} \tag{2.10}$$

In Eq. (2.10), γ refers to the azimuth angle of the light-receiving surface normal against the horizontal surface, and south by west is deemed the forward direction.

For a horizontal surface with $\beta = 0°$, according to Eq. (2.10), $\theta = \theta_z$; namely, zenith angle is the incidence angle of the incident solar beam against the horizontal surface.

For a vertical surface with $\beta = 90°$, Eq. (2.10) turns into:

$$\cos\theta = -\sin\delta\cos\phi\cos\gamma + \cos\delta\sin\phi\cos\gamma\cos\omega + \cos\delta\sin\gamma\sin\omega \tag{2.11}$$

For a surface that revolves by surrounding a horizontal east–west axis, only adjusting the slope during the midday so that the incident solar radiation can be perpendicular to the surface, the corresponding solar incidence angle calculation during daytime is as follows:

$$\cos\theta = \sin^2\delta + \cos^2\delta\cos\omega \tag{2.12a}$$

The corresponding daily fixed surface slope angle is:

$$\beta = |\phi - \delta| \tag{2.12b}$$

The azimuth angle of surface normal is 0 or 180 degrees, which is determined by local latitude and solar declination angle; namely:

$$\begin{cases} \gamma = 0°, & \text{if}(\phi - \delta) \geq 0 \\ \gamma = 180°, & \text{if}(\phi - \delta) < 0 \end{cases} \tag{2.12c}$$

For a surface that continuously revolves by surrounding a horizontal east–west axis and is required to revolve to the point where the solar incidence angle is of minimum value at a specified time, the corresponding incidence angle equation is:

$$\cos\theta = \sqrt{1 - \sin^2\theta_z\sin^2\gamma_s} = \sqrt{1 - \cos^2\delta\sin^2\omega} \tag{2.13a}$$

The corresponding equation for the surface slope angle is:

$$\tan\beta = \tan\theta_z|\cos\gamma_s| \tag{2.13b}$$

The azimuth angle of surface normal is 0 or 180 degrees, namely:

$$\begin{cases} \gamma = 0°, & \text{if } |\gamma_s| \leq 90° \\ \gamma = 180°, & \text{if } |\gamma_s| > 90° \end{cases} \tag{2.13c}$$

For a surface that continuously revolves by surrounding a horizontal north–south axis and is required to revolve to the point where the solar incidence angle is of minimum value at a specified time, the corresponding incidence angle equation is:

$$\cos\theta = \sqrt{\cos^2\theta_z + \cos^2\delta\sin^3\omega} \tag{2.14a}$$

The corresponding equation for the surface slope angle is:

$$\tan\beta = \tan\theta_z|\cos(\gamma - \gamma_s)| \tag{2.14b}$$

The azimuth angle of surface normal can be either −90 or 90 degrees, which is determined by the symbol of solar azimuth angle; namely:

$$\begin{cases} \gamma = 90^\circ, & \text{if } \gamma_s \geq 0^\circ \\ \gamma = -90^\circ, & \text{if } \gamma_s < 0^\circ \end{cases} \tag{2.14c}$$

For a surface that continuously revolves by surrounding a north–south axis parallel to Earth's axis and is required to revolve to the point where the solar incidence angle is of minimum value at a specified time, the corresponding incidence angle equation is:

$$\cos\theta = \cos\delta \tag{2.15a}$$

The continuously changing surface slope angle is:

$$\tan\beta = \frac{\tan\phi}{\cos\gamma} \tag{2.15b}$$

The corresponding azimuth angle of surface normal is:

$$\gamma = \tan^{-1}\frac{\sin\theta_z \sin\gamma_s}{\cos\theta' \sin\phi} + 180^\circ C_1 C_2 \tag{2.15c}$$

In which:

$$\cos\theta' = \cos\theta_z \cos\phi + \sin\theta_z \sin\phi \tag{2.15d}$$

$$C_1 = \begin{cases} 0, & \text{if } \left(\tan^{-1}\dfrac{\sin\theta_z \sin\gamma_s}{\cos\theta' \sin\phi}\right) + \gamma_s = 0^\circ \\ 1, & \text{else} \end{cases} \tag{2.15e}$$

$$C_2 = \begin{cases} 1, & \text{if } \gamma_s \geq 0^\circ \\ -1, & \text{if } \gamma_s < 0^\circ \end{cases} \tag{2.15f}$$

A surface that continuously conducts double-axis tracking is always capable of incident solar radiation perpendicular to the surface, and thus:

$$\begin{cases} \theta = 0^\circ \\ \beta = \theta_z \\ \gamma = \gamma_s \end{cases}$$

2.5 DISTRIBUTION OF THE SOLAR RESOURCE IN SEVERAL TYPICAL AREAS OF CHINA

In order to facilitate the understanding of readers, we would now like to introduce the solar energy resources in Beijing, Lhasa, Golmud, Dunhuang, Turpan, Guizhou, Hainan, and Harbin.

2.5.1 The Solar Resource in Beijing

The coordinates of Beijing are 115°23′17″E, 39°54′27″N, which is based on the geographic coordinates of Tiananmen Square; the elevation of Tiananmen Square is 44.4 m. Declination of the axis line of Beijing is 6°17″ by west. The city starts at 39°28′N in the south and reaches 41°05′N in the north; and starts at 115°25′E in the west and reaches 117°30′E in the east by striding across latitude 1°37′ from north to south and longitude 2°05′ from west to east. As Beijing is located in the middle latitude zone, the city features an obvious warm temperate zone with a semihumid continental monsoon climate, which has had profound impacts on Beijing's other natural factors.

2.5.1.1 Solar Altitude Sunrise Time, and Sunset Time of Beijing

Beijing is located near 40°N and enjoys a solar altitude change of 46°52′ during the year. Around midday, the solar altitude changes from 26°34′ on the winter solstice (December 22) to 73°26′ on the summer solstice (June 21), while sunshine duration changes from 9 h 20 min to 15 h 1 min. Solar radiation varies greatly during the year and serves as the foundation for the temperature alternation and division of the four seasons in Beijing.

Sunrise and sunset times are determined by the position of the sun in the sky. The sun rises latest and sets earliest on the winter solstice; conversely, the sun rises earliest and sets latest on the summer solstice; sunrise and sunset times during spring and autumn fall between these times.

2.5.1.2 Solar Radiation in Beijing

Monthly solar radiation for Beijing is shown in Table 2.2. From January, monthly global radiation starts to increase, enjoys a large increase from March to May, reaches its maximum levels in May and June, and decreases after June. Because July belongs to a rainy season, the monthly global radiation drops quite quickly, followed by decreases from September through November, reaching the bottom in December.

Beijing's global annual radiation is 4702–5707 MJ/m^2. The two high-value areas are in the Badaling Basin and around the northeastern section from Tanghekou to Gubeikou, with global annual radiation as high as 5707 MJ/m^2; a low-value area is located around Xiayunling in the Fangshan District, with global annual radiation of 4702 MJ/m^2. For the year, the variation of solar global radiation shows a unimodal distribution. From January to May, along with an increase in solar altitude and daytime hours, monthly global radiation gradually increases, reaching its maximum by May; from June to December, it decreases along with monthly reductions in solar altitude and daytime hours, reaching its

TABLE 2.2 Monthly Solar Radiation of Beijing (MJ/m^2)

Site	January	February	March	April	May	June	July	August	September	October	November	December	Annual
Observatory	284.7	339.1	510.8	573.6	695.0	674.1	582.0	540.1	494.0	397.7	276.3	238.6	5606
Gubeikou	297.3	351.7	519.2	565.2	678.3	665.7	582.0	552.7	489.9	401.9	284.7	259.6	5648.2
Badaling	301.4	360.1	523.4	561.0	686.6	665.7	577.8	535.9	489.9	401.9	284.7	259.6	5648
Changping	284.7	334.9	502.4	552.7	665.7	653.1	548.5	519.2	481.5	393.6	268.0	251.2	5455.5
Fangshan	276.3	330.8	489.9	548.5	665.7	644.8	548.5	519.2	477.3	381.0	263.8	238.6	5384.4
Chaoyang	272.1	322.4	489.9	535.9	657.3	644.8	535.9	506.6	477.3	376.8	259.6	230.3	5308.9
Xiayunling	234.5	284.7	410.3	481.5	678.3	565.2	477.3	448.0	401.9	318.2	230.3	205.2	4735.4

bottom level in December. Among the four seasons, summer (June–August) enjoys the most solar radiation, whereas winter (December–February) enjoys the least; solar radiation in spring (March–May) is slightly less than that of summer, and solar radiation in autumn falls between the winter and summer levels.

2.5.1.3 Sunshine Duration of Beijing

The annual mean sunshine duration in Beijing falls in the range of 2000–2800 h; in most of the area, the value is around 2600 h (refer to Table 2.3). Annual sunshine distribution is consistent with solar radiation distribution; the maximum value is obtained in Badaling County and Gubeikou at over 2800 h, whereas the minimum value is obtained in Xiayunling with sunshine duration of 2063 h.

TABLE 2.3 Annual Sunshine Duration of Beijing (h)

Area	Hours
Haidian	2620
Chaoyang	2554.8
Shijingshan	2473.3
Tongxian	2722.7
Changping	2641.4
Mentougou	2621.4
Zhaitang	2594.1
Santai	2733.6
Daxing	2769.3
Shunyi	2792.3
Fangshan	2606
Xiayunling	2063.2
Badaling	2813.2
Foye Ding	2491.3
Pinggu	2711.3
Madaoliang	2690.7
Tanghekou	2812.4
Gubeikou	2822.9
Huairou	2731.5
Miyun	2788

Spring has the longest sunshine duration of the year, lasting for 230–290 h; in summer, which is the rainy season, sunshine duration is shortened to around 230 h; in autumn, monthly sunshine duration lasts for 190–245 h; winter has the year's shortest sunshine duration at 190–200 h (refer to Table 2.4).

2.5.1.4 Sunshine Percentage of Beijing

Sunshine percentage refers to the ratio of actual sunshine duration to astronomical value at the same location and within the same period; the larger the percentage, the more days with clear sky. Beijing has sufficient sunshine. During a regular month, the respective sunshine percentage is over 60%; only during July and August is the percentage lower, within a range of 50%–60%. The sunshine percentages of Gubeikou, Tanghekou, and Badaling Basin are the highest in the city, whereas the western area's is the lowest. Among the four seasons, winter enjoys the highest sunshine percentage, whereas summer has the lowest value, and the spring and autumn values fall between those of the other two. Monthly sunshine durations and percentages for Beijing are shown in Table 2.4.

Daily utilization hours of solar direct radiation on a vertical surface in Beijing are greatest during spring and autumn, averaging 6 h per day; summer takes second place with 2–3 h available for daily utilization on average during July and August, the decrease largely a result of the rainy season. Daily utilization hours of global solar radiation on a horizontal surface are greatest during spring; summer takes second place, and winter has the least utilization hours.

At any time, the longer the continuous sunshine duration, the more effective the solar energy received by the solar receiver. When sunshine is interrupted constantly, the solar energy corresponding to this time may be ineffective energy. For example, under the condition of 6 h of continuous sunshine, all types of solar receivers can function effectively. The 6-h continuous sunshine duration in Beijing is as much as 2287 h for the year; the number in spring reaches as much as 661 h, which is 7.2 h per day on average, whereas those of other seasons are all below 550 h, which is below 6 h daily on average. From the perspective of the ratio of continuous to actual sunshine duration during winter, the sunshine duration in Beijing during spring and winter that can be effectively utilized by solar receivers is quite long. If still referring to the standard of a 6-h duration of continuous sunshine, the sunshine duration in these seasons that can be effectively utilized by solar receivers accounts for about 85% of the actual sunshine duration over the same period, whereas in summer it is only 70%.

2.5.1.5 Measured Value of Daily Mean Solar Direct Normal Irradiance of Badaling

Fig. 2.6 shows measured data for Badaling during 347 days in 2009. Based on the statistics, daily mean direct radiation for the year was 324 W/m^2.

TABLE 2.4 Sunshine Duration and Percentage of Beijing

Item	January	February	March	April	May	June	July	August	September	October	November	December	Annual Average
Sunshine duration/h	204	198	237	251	290	276	230	230	245	229	193	192	278
Sunshine percentage/%	68	66	64	63	65	62	51	55	56	67	65	66	63

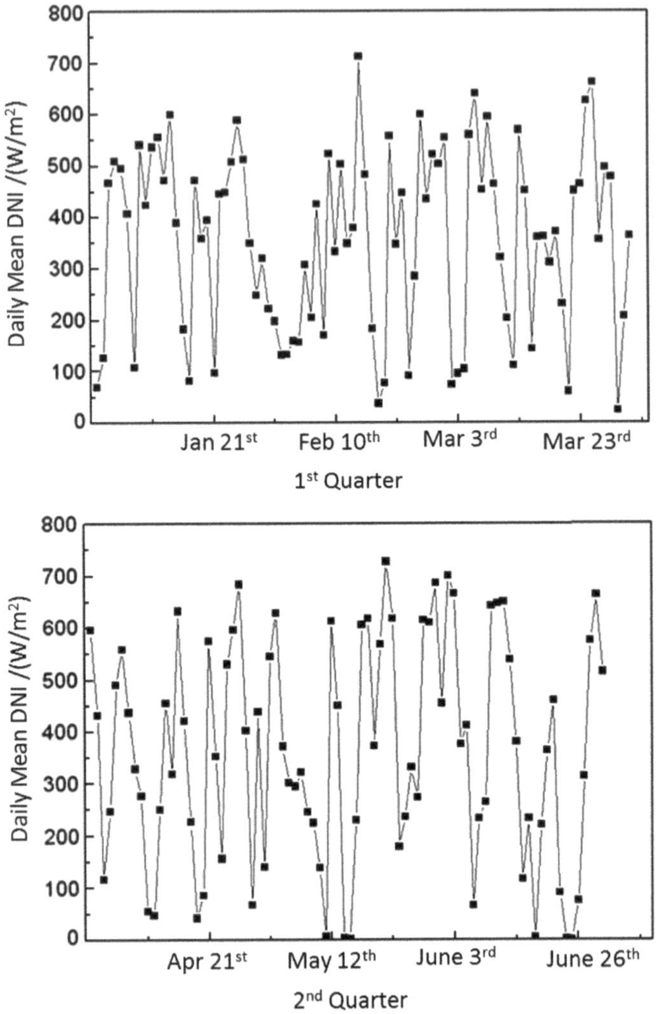

FIGURE 2.6 Daily mean direct normal irradiance for Badaling during the four seasons of 2009.

We can see from Fig. 2.7 that in Badaling, 28 days in 2009 had a daily mean DNI over 600 W/m^2, accounting for 8.1% of the yearly total; 49 days with a daily mean DNI within the range of 500–600 W/m^2 accounted for 14.1% of the yearly total. Days in the year with a daily mean DNI over 300 W/m^2 accounted for 55%.

Fig. 2.8 and Table 2.5 show the number of days during the four seasons in Badaling with satisfactory, typical weather conditions. Fig. 2.8 indicates daily mean solar DNIs for representative days typifying the

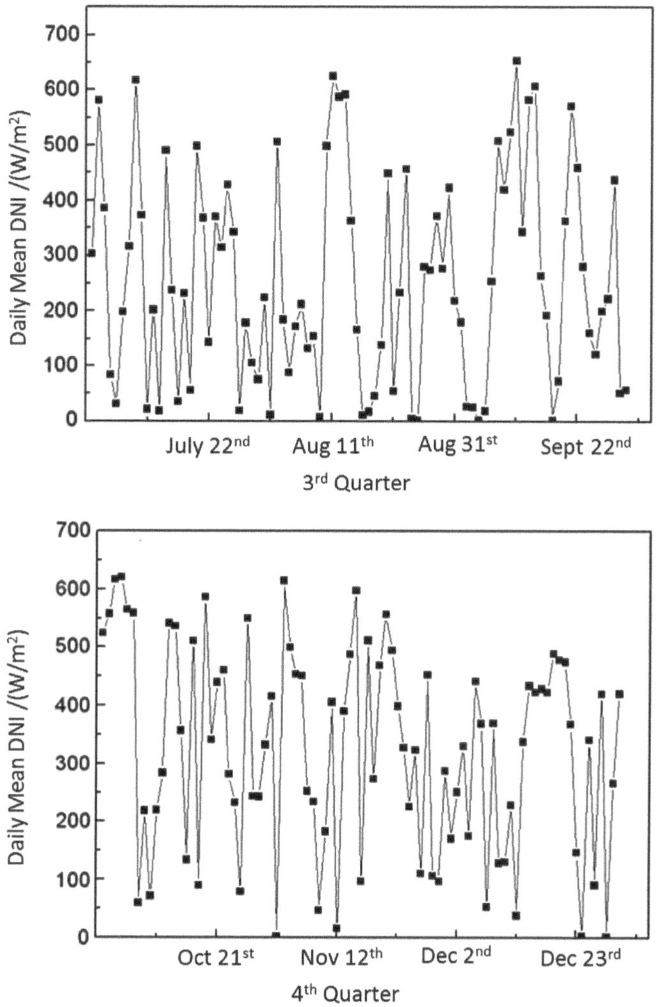

FIGURE 2.6 cont'd.

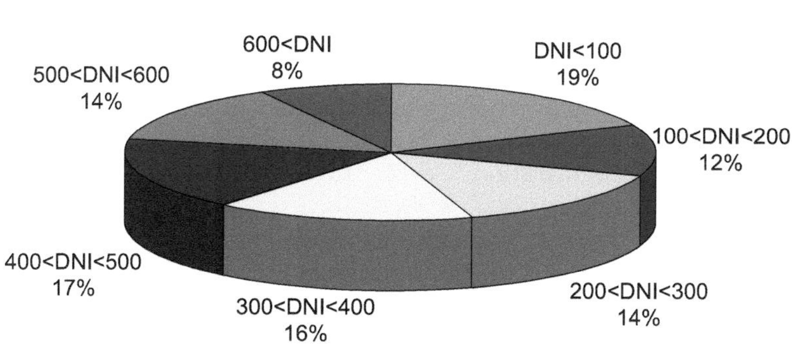

FIGURE 2.7 Daily mean direct normal irradiance for the year.

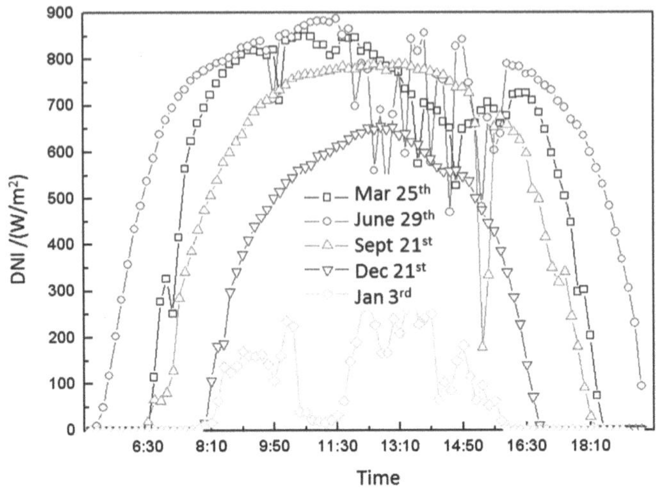

FIGURE 2.8 Typical weather of the four seasons in Badaling, Beijing.

TABLE 2.5 Typical Daily Mean Direct Normal Irradiance and Sunshine Duration of the Seasons

Time	Daily Mean DNI	Effective Daily Mean DNI (>300 W/m^2)/(W/m^2)	Effective Sunshine Duration (>300 W/m^2)/h	Sunshine Duration/h
Spring (March 25)	651	713	11	12
Summer (June 29)	656	710	13	14
Autumn (September 21)	571	662	8.5	12
Winter (December 21)	47	547	7	8

DNI, direct normal irradiance.

four seasons in Badaling, Beijing, which are summarized in Table 2.5. Sunrise and sunset times are shown in Fig. 2.8 as well. When DNI is more than 300 W/m^2, the solar resource is of a certain significance for thermal power generation. Therefore, in Table 2.5, "effective daily mean DNI," "effective sunshine duration," and "sunshine duration" have been specifically summarized.

As indicated in Table 2.5, sunshine duration varies greatly by season. In summer, the sun rises at 5:00 a.m. and does not set until 7:30 p.m., whereas during winter, the sun rises just after 8:00 a.m. and sets around 4:30 p.m.

2.5.2 The Solar Resource in Lhasa

The capital of the Tibet Autonomous Region, Lhasa is located slightly southeast of the middle of Tibet; it stands on the midstream valley plain of the Lhasa River branch of the Yarlung Zangbo River, enjoying a plateau temperate zone with a semiarid monsoon climate. Its coordinates are 91°08′E and 29°40′N with an elevation of 3658 m. It is a plateau arid-climate region featuring strong radiation, comparatively low temperatures, moderate precipitation, and thin air; the sunshine duration here is over 3000 h, giving the city the nickname "Solar City." In summer, the temperature is not high. Arid seasons can be clearly distinguished from wet. The rainy season, which exhibits a tight precipitation period, consists mainly of days with clear daytime skies and rainy nights.

2.5.2.1 Weather Conditions in Lhasa [7]

Lhasa has a high elevation that results in thin air as well as comparatively low temperatures that vary greatly between day and night. The mean temperature in June is 15.9°C, while the mean maximum is 23.2°C; it has the year's highest temperature. In January the mean temperature is −1.6°C, while the mean minimum is −9.1°C; it has the year's lowest temperature. The multiyear extreme maximum temperature is 29.9°C, and the extreme minimum temperature is −16°C, which respectively appeared during June and January (refer to Table 2.6).

2.5.2.2 Solar Radiation and Sunshine Duration of Lhasa

Lhasa is located within the Qinghai–Tibet Plateau temperate zone semiarid monsoon climate region, enjoying annual sunshine duration of 3000 h and featuring arid and windy weather in winter and spring; the annual frost-free period for the area lasts only 100–120 days. The annual precipitation is 200–510 mm and concentrated in June–September with frequent night rains. Relatively speaking, during March–October the weather is warm and humid. From 1961 to 1970, the annual mean sunshine duration was 3005.7 h, and the sunshine percentage was 68%; there are 108.5 clear and 98.8 overcast days on average for the year, and annual global solar radiation is 6680–8400 MJ/m^2 (refer to Table 2.7).

2.5.3 The Solar Resource in Golmud

Golmud is located in the hinterland of the Qinghai–Tibet Plateau south of the boundary of the Haixi Mongol and Tibetan Autonomous Prefecture. The area consists of the Qaidam Basin Midsouth Area and the Tanggula Mountains that have not been mutually connected, with coordinates of 35°10′–37°45′ N and 90°45′–95°46′ E with a total area of about 8.1×10^4 km^2. The urban area is located on the alluvial plain of the

TABLE 2.6 Basic Weather Status in Lhasa, 1971–2000

Month	1	2	3	4	5	6	7	8	9	10	11	12
ITEM												
Temperature (°C)												
Mean	−1.6	1.5	5.2	8.4	12	15.9	16	14.7	13	8.7	2.9	−1.2
Mean maximum	7.2	9.2	12.7	15.9	19.9	23.2	22.5	21.4	20	17.0	12.0	8.1
Extreme maximum	18	20.6	25.0	25.0	29.4	29.9	28.8	27.2	24.9	23.0	21.6	17.8
Mean minimum	−9.1	−5.9	−2.1	1.5	5.6	9.8	10.4	9.7	7.8	2.0	−4.2	−8.2
Extreme minimum	−16	−14.5	−10.2	−6.0	−2.4	2.0	4.5	3.3	0.3	−6.3	−10.3	−15.5
Mean precipitation/mm	0.8	1.2	2.9	6.1	27.7	71.2	116	120.6	68.3	8.8	1.3	1.0
Precipitation days/d	0.7	1.0	1.6	4.3	9.9	14.3	19.1	20.0	15.4	4.5	0.7	0.6
Mean wind speed/(m/s)	1.9	2.4	2.4	2.4	2.2	2.0	1.7	1.6	1.6	1.5	1.5	1.6

TABLE 2.7 Monthly Mean Global Solar Radiation in Lhasa

Month	Cumulative Monthly Mean Global Solar Radiation (MJ/m^2)
January	485.27
February	517.40
March	605.50
April	715.78
May	828.90
June	798.79
July	756.51
August	707.50
September	624.97
October	604.39
November	503.20
December	465.69

Golmud River in Qaidam Basin Midsouth Area with a mean elevation of 2780°m. Golmud has a plateau continental climate; it is hot in summer and cold in winter; the temperature climbs back slowly in spring and drops quickly in autumn, varying greatly between day and night; the area has sufficient sunshine, strong solar radiation, scarce precipitation, extensive evaporation, and extremely arid weather; it is not only a sensitive zone for climate change, but also a fragile ecological environment.

The Golmud area has comparatively stable annual solar radiation variations with an annual mean sunshine percentage of 70.2%. Monthly variations indicate that solar radiation is mainly concentrated in the period from April to August, which accounts for over 54% of the total. With intensive solar radiation and long sunshine duration, this region has abundant solar energy resources.

2.5.3.1 Weather Conditions of Golmud [8]

Golmud has a high elevation, few rainy days, long sunshine duration, high radiant intensity, and excellent atmospheric transparency; the mean daily sunshine duration approximates 8.5 h with annual mean sunshine duration of 3096.3 h and annual global solar radiation of 6604.48–7181.1 MJ/m^2; the area has abundant solar energy resources. Statistics on major meteorological elements from the Golmud Meteorological Station are shown in Table 2.8.

TABLE 2.8 Statistics for Major Meteorological Elements at the Golmud Meteorological Station

S/N	Item	Value	Memo
1	Multiyear mean temperature/°C	5.3	
2	Multiyear extreme maximum temperature/°C	35.5	
3	Multiyear extreme minimum temperature/°C	−33.6	
4	Maximum frozen ground depth/cm	105	In November 1997
5	Multiyear maximum snow depth/cm	6	In December 1992
6	Annual mean atmospheric pressure/kPa	72.47	
7	Air mean relative humidity/%	32	
8	Annual evaporation/mm	2504.1	
9	Multiyear hail days/day	0.5	
10	Sandstorm annual mean days/day	13.2	
11	Multiyear mean wind speed/maximum wind speed/[(m/s)/(m/s)]	2.8/22	
12	Regional oxygen content/%	74	Corresponding to sea level

2.5.3.2 *Solar Radiation on Golmud*

Measured solar radiation data for 1971–2007 provided by the Golmud Meteorological Station indicate that over that 36-year period, the annual solar radiation distribution variation in Golmud was basically stable with a respective value interval maintained between 6604.48 and 7181.1 MJ/m^2. The difference between the maximum and minimum value was only 538.4 MJ/m^2; annual mean solar radiation was 6908.17 MJ/m^2. During the decade from 1997 to 2007, mean solar radiation was 6852.64 MJ/m^2. The maximum annual value during the three-decade study period occurred in 1985 when it reached 7181.12 MJ/m^2, whereas the minimum value of 6604.48 MJ/m^2 occurred in 1998. July has the largest monthly global radiation, which has reached 803.64 MJ/m^2, more than two times that of December and January. Monthly global radiation is mainly concentrated in the period from April to August, which accounted for over 54% of global annual radiation. Compared with other areas of China, Golmud is a region with abundant solar radiation.

2.5.3.3 *Golmud Sunshine Duration*

During the 1971–2007 period, the maximum sunshine duration in Golmud occurred in 1985, at about 3323 h; whereas the minimum value occurred in 2007, at about 2918 h. Yearly sunshine duration has maintained a range of 2900 to 3320 h. Especially after 1999, however, solar radiation has remained stable but with a decreasing trend. In Golmud, May has the longest monthly sunshine duration at about 296 h. It is basically consistent with changes in global solar radiation. Based on national solar energy resource distribution, Golmud has a comparatively longer sunshine duration.

2.5.4 The Solar Resource in Dunhuang

Dunhuang city is in the westernmost part of the Hexi Corridor, Gansu province, the coordinates of which fall in the range 92°13′–95°30′ E and 39°40′–41°35′ N. The city area is 31,200 km^2; its elevation falls in the range of 800–1800 m; the elevation of the urban area is 1138 m. The area has higher terrain in the south and north than in the middle and has a basin–plain topography that inclines from southwest to northeast; it is also known as the "Andun Basin" together with Guangzhou County. Dunhuang city is inland and is surrounded by the Gobi Deserts and subject to a typical continental climate; the area has strong solar radiation, sufficient sunshine, abundant thermal, a short frost-free period, limited precipitation, large variability, intensive evaporation, and frequent disasters [9].

2.5.4.1 *Weather Conditions of Dunhuang*

Climatic features of Dunhuang obviously include its arid climate, low precipitation, large evaporation, great temperature difference between day and night, and long sunshine duration. The area has four distinctive seasons: a warm and windy spring, hot summer, cool autumn, and freezing winter. The annual mean temperature is 9.4°C with a maximum monthly mean temperature of 24.9°C (July), whereas the minimum monthly mean temperature is −9.3°C (January); the maximum extreme temperature is 43.6°C, whereas the minimum temperature is −28.5°C; the annual mean frost-free period is 142 days; the area has a typical warm temperate zone and arid climate (one minor type of temperate zone in a continental climate). Being inland, Dunhuang is separated by high mountains and far from the humid ocean climate; it has an extremely arid continental climate with arid weather all year and limited precipitation. It has three major features: (1) sufficient sunshine; (2) aridity, low precipitation, and a downward air flow to the north over Dunhuang that create an arid region with an annual mean precipitation of only 39.9 mm,

TABLE 2.9 Seasonal Weather Factors of Dunhuang

Month	January	April	July	October
INDEX				
Sunshine percentage/%	77	71	70	82
Sunshine duration/h	232.1	282.4	318.9	280.8
Ground mean temperature/°C	−9.9	16.0	31.5	10.4
Mean precipitation/mm	0.5	2.4	15.2	0.8
Relative humidity/%	46	31	43	45
Mean wind speed/(m/s)	2.2	3.3	2.9	2.3

of which the summer season accounts for 63.9% and winter for only 7.5%, while annual evaporation can be as high as 2400 mm; and (3) distinctive seasons, with a winter that is longer than summer as well as huge day and night temperature differences. The annual temperature variation is as high as 34°C. During most of the year, Dunhuang experiences east and northwest winds with a surface mean wind speed of 3 m/s. Hot-dry winds and black sandstorms are major natural disasters.

Dunhuang is located at the junction of Gansu, Qinghai, and Xinjiang, with the Qilian Mountain to the south, Mazong Mountain to the north, and Gobi deserts to the east and west with an oasis area of 1400 km^2, thus accounting for just 4.5% of the total area. The total cultivated land area of the entire city is 1.7747×10^4 hm^2, accounting for 12.68% of total oasis area. The annual mean temperature of the area is 9.5°C; the effective accumulated temperature over 10°C for the year is 3605.9°C, and sunshine duration is 3246.7 h. Values based on the respective seasonal distribution are shown in Table 2.9.

2.5.4.2 *Solar Radiation in Dunhuang*

The Dunhuang area features strong solar radiation and abundant sunshine resources. Meteorological authority observations for multiple years indicate that the annual sunshine duration of Dunhuang city can reach 3246.7 h with a sunshine percentage of 75%, global annual radiation of 6882.27 MJ/m^2, and daily mean radiation of 18.86 MJ/m^2; it is a region of China with abundant solar energy resources.

2.5.5 The Solar Resource in Turpan

Turpan city is located in the eastern Xinjiang Uygur Autonomous Region, south of Bogda Mountain's main peak in the mideastern Tian Shan

Mountain, and is in the center of the Turpan Basin. With an east–west width of 90 km and north–south length of 262 km, the terrain is higher in the south and north than it is in the middle. Turpan's geographic coordinates are 88°29′28″–89°54′33″ E and 42°15′10″–43°35′ N with elevation in the range of −154 to 4000 m and total land area of 13,589 km^2. The area is adjacent to Hami in the east and adjoins Hejing, Heshuo, Weili, and Ruoqiang Counties of the Bayingolin Mongol Autonomous Prefecture in the west and south; it is also separated by Tian Shan Mountain in the north and connected to Urumqi City as well as Qitai, Jimsar, and Mori Counties in the Changji Hui Autonomous Prefecture. The prefectural administrative office of the Turpan area is located in Turpan city, 183 km from Urumqi City, the capital of the Xinjiang Uygur Autonomous Region, which governs Turpan City, Shanshan County, and Toksun County. The area has sufficient sunshine, abundant thermal energy, a long frost-free period, great daily temperature variation, low precipitation, and intensive evaporation; it features a typical continental warm-temperate-desert climate [10].

2.5.5.1 Weather Conditions in Turpan

The Turpan Basin has a typical continental warm temperate zone arid desert climate. As Turpan is located in the basin and surrounded by high mountains, its temperature rises fast and radiates slowly, which has resulted in the formation of five major characteristics for the area, namely long sunshine duration, high temperature, large temperature differences during day and night, low precipitation, and strong wind. The area has sufficient sunshine and abundant thermal energy, but it also features extreme aridity and dry, hot weather; it has thus been given the name "Fire Continent." As the atmospheric pressure in the basin is low, airflow can be attracted; added to the scarce precipitation and frequent strong wind, it has become a well-known "Wind Storehouse" in China. More than 100 days of the year have an force 8 (fresh gale) or greater wind, sometimes exceeding force 12.

The annual mean temperature is 13.5°C with more than 100 days of the year having a temperature over 35°C and 28 days having an average temperature over 40°C. During June–August, the mean maximum temperature is always above 38°C; at noon, the sand surface temperature can reach 82.3°C. During summer, the mean temperature is around 30°C; the extreme maximum temperature in summer is 49.6°C, and the ground surface temperature is over 70°C most of the time. The extreme minimum temperature in winter is −28.7°C; both daily and annual temperature differences are great, effective accumulated temperature over 10°C for the year exceeds 5300°C, and the frost-free period lasts 224 days. Due to the hot and dry weather, Turpan is arid and scarcely experiences rain, with annual mean precipitation of merely 16.7 mm but evaporation as high as 3000 mm (refer to Table 2.10).

TABLE 2.10 Weather Conditions in the Turpan Region

Turpan Region	Turpan	Shanshan	Toksun	Regional Mean
INDEX				
Annual mean temperature/°C	14.5	11.8	14.2	13.5
Annual precipitation/mm	15.5	26.9	7.7	16.7
Annual sunshine duration/h	2938.0	3102.9	2978.1	3006.3
Annual mean wind speed/(m/s)	1.2	1.4	2.9	1.8
Annual mean relative humidity/%	41	43	41	42

2.5.5.2 *Solar Radiation of Turpan*

Solar resources are rich in Turpan. Its annual sunshine duration approximates more than 3000 h, about 1000 h more than other areas in eastern China located at the same latitude; annual mean global radiation is 5938 MJ/m^2, second only to that of the Qinghai–Tibet Plateau.

2.5.6 The Solar Resource in Guizhou

Guizhou is located east of the Yunnan-Guizhou Plateau in southwestern China within the subtropical East Asian continental monsoon region; it lies in a subtropical humid monsoon climate region, serves as the watershed of the Yangtze River and the upper reaches of the Pearl River, and is a mountainous province with a typical karst landscape. Exposed karst area accounts for 61.9% of the total area of the province, and 73% when considering covered area. Mountain and hill areas account for 92.5% of the province's total area, whereas flat ground between mountains accounts for a mere 7.5%. The mean slope of the province is as high as 17.8 degrees; it is the only province in China not supported by a plain. The region has tall mountains and steep slopes, shattered landscape and shallow soil, and features weak erosion resistance [11].

2.5.6.1 *Weather Conditions of Guizhou*

Guizhou is located in the low-latitude plateau mountain area and adjoins Guangxi Hills in the south; the region is not distant from the ocean and thus is sufficiently moisturized; it borders upon the Sichuan Basin, Qinling Mountains, and Daba Mountains to the north, keeping cold air from the north from reaching the region; the terrain of Guizhou is high in the west and low in the east with its peaks rising one above the other with rolling hills that influence the reallocation of thermal, water, and solar resources; Guizhou is within the East Asia monsoon region while being influenced by westerly and subtropical circulation systems. It also has a frequent and intensive airflow intersection from the north and south. In such a specific geographic location and natural environment, which experiences the impacts of solar radiation and atmospheric circulation, the unique climatic characteristics of Guizhou are formed, namely temperature and humid weather and an obvious three-dimensional climate; distinctive seasons; a long frost-free period; no intensive thermal energy during summer and no severe cold in winter; abundant thermal energy, extreme as well as mild cold weather; sufficient precipitation; frequent drought and flood; weak global radiation with widely scattered sunshine; more rain and less sunshine; lower wind speed; and a variety of frequently occurring disasters.

In most areas of Guizhou, the annual mean temperature is 14–18°C, the accumulated temperature over 10°C is 3500–5500°C, and the frost-free period is 260–330 days. During winter months, the mean temperature is 4–9°C, and the extreme minimum temperature is −5 to 7°C; whereas during summer months, the mean temperature is 20–25°C, and the extreme maximum temperature is 32–36°C. Annual precipitation falls in a range of 1100–1300 mm, which is nearly double the amount for the North China Plain. During the summer half-year, the temperature in Guizhou is comparatively high with concentrated precipitation, sufficient sunshine as well as thermal, water, and light synchronization.

2.5.6.2 *Solar Radiation of Guizhou*

Influenced by a stationary front, Guizhou features numerous rainy days, significant cloud cover, great cloud thickness, and a low amount of global solar radiation reaching the ground; in most areas, global annual radiation falls in a range of 3349–4186 MJ/m^2; it is a region with minimal solar global radiation. Scattered radiation in Guizhou accounts for an especially large proportion of global radiation, with Zunyi and Guiyang both having over 60% scattered radiation; Xianning, with more clear days, has a scattered radiation proportion above 50%. Especially in winter, the proportion of scattered radiation exceeds 67%; in specific years, global radiation may consist almost entirely of scattered radiation.

2.5.7 The Solar Resource in Hainan

Hainan is located north of the South China Sea south of 20°N; the Qiongzhou Strait facing the mainland is just 20–30 km across the sea. Being within a tropical climate region, Hainan is not only greatly influenced by the mainland but also significantly regulated by the ocean; atmosphere–ocean thermal and water exchange as well as seasonal variations have direct impacts on temperature and precipitation in Hainan.

Hainan has a tropical marine monsoon climate in which summer is long and winter is not significant; sunshine is sufficient and temperature is high with abundant rainfall; the eastern area is wet, western area is dry, southern area is warm, northern area is cold, and photosynthetic potentials are great. It is warm all year with abundant rainfall and distinctive dry and wet seasons; the wind is normally strong with tropical storms and typhoons occurring frequently; and climatic resources are diversified. Basic characteristics of the region are four indistinctive seasons with no intensive thermal in summer and no severe cold in winter; the annual temperature range is small, and the annual mean temperature is high; the dry and rainy seasons are clearly distinguished from each other with a dry winter and spring and rainy summer and autumn; there are many tropical cyclones; light, thermal, and water resources are abundant; wind, aridity, cold, and similar climatic disasters occur frequently [12,13].

2.5.7.1 *Solar Radiation and Sunshine Duration of Hainan*

The annual global solar radiation in Hainan is 4500–5800 MJ/m^2; variation in global radiation days is not significant. Hainan is located south of the Tropic of Cancer, and various areas of Hainan enjoy long sunshine durations with annual sunshine durations of 1750–2650 h and illumination ratios of 50%–60%. The sunshine duration in the west coastal area can be as high as 2650 h, whereas in the central mountainous areas it may be as low as 1750 h due to more clouds and mist. The seasonal distribution of sunshine durations indicates that summer (June–August) has the longest sunshine duration followed by spring (March) and autumn (September–November); winter has the shortest sunshine duration.

2.5.7.2 *Thermal Characteristics of Hainan*

Annual mean temperatures of various areas in Hainan fall in the range of 22.8–25.0°C, a bit lower in central mountainous areas and higher in the southwest. There is no winter season, and January and February are the coldest months with a mean temperature between 16 and 24°C and a mean extreme minimum temperature of over 5°C at most times. Summer is from the middle 10 days of March to the first 10 days of November; July and August have the highest mean temperatures at 25 and 29°C. Thermal characteristics of various meteorological stations throughout the province during the 1971–2005 period are shown in Table 2.11.

2.5.7.3 *Precipitation Conditions in Hainan*

Most areas of Hainan have sufficient precipitation. The annual mean precipitation for the entire province is above 1500 mm. Influenced by monsoon and terrain, precipitation varies significantly in terms of time and space. Annual precipitation has a circular distribution with an east

TABLE 2.11 Temperature and Precipitation Characteristics From Various Meteorological Stations in Hainan

	Temperature/°C			
Area	**Annual Mean**	**Maximum**	**Minimum**	**Annual Mean Precipitation/mm**
Haikou	24.1	27.9	21.5	1372
Dongfang	25.0	28.6	22.0	798
Lingao	23.7	28.2	20.7	1202
Chengmai	23.9	29.0	20.5	1488
Danzhou	23.5	28.7	20.3	1539
Changjiang	24.6	30.0	21.0	1392
Baisha	23.0	28.8	19.4	1626
Qiongzhong	22.8	28.7	19.1	2050
Ding'an	24.1	28.7	21.1	1480
Tunchang	23.7	28.6	20.6	1753
Qionghai	24.3	28.5	21.4	1735
Wanning	24.7	28.3	22.1	1802
Lingshui	25.0	29.0	22.1	1434
Wuzhishan	22.8	28.4	19.2	1493

area that is larger and obviously wetter than the west; the east-central mountainous area serves as the center of precipitation with annual amounts of about 2000–2400 mm; precipitation in the west, with less rainfall, is 800–1200 mm. Hainan has great humidity throughout the year, and the annual mean water vapor pressure varies from 23 hPa (Qiongzhong) to 26 hPa (Sanya). The central and eastern coastal areas are humid, whereas the southwestern coastal area is semiarid; the other areas are semihumid. Hainan is within the monsoon climate region with dominant wind changes by season. During the winter half-year, regular wind mainly consists of northeast and east winds with a mean wind speed of 2–3 m/s. During summer half-year, southeast and southwest winds blow throughout Hainan; in addition, more typhoons occur in summer and autumn, and moisture and thermal are more sufficient than during the winter half-year with its north wind and provide abundant water vapor resources for precipitation in Hainan. Statistical results for precipitation from various meteorological stations for the 1971–2005 period are shown in Table 2.11.

2.5.7.4 *Typhoons and Thunderstorms*

Hainan is a region with many tropical storms and typhoons. Most tropical storms and typhoons that affect Hainan occur on the surface of the western Pacific (5°–20° N). There are plenty of tropical storms and typhoons, typically 8–9 times a year but as high as 11 times a year. The season is long, starting in May with tropical storms and typhoons that reach their highest levels during the August–October period. About 75% of tropical storms and typhoons land on coastal areas of the Wenchang–Qionghai–Wanning region; no landing records have been found in the western coastal region. Windstorms are more severe on the northeastern coast. Typhoon rain plays a crucial role in annual precipitation, and for most areas in Hainan it accounts for 31%–36% of total annual rainfall and 44%–45% in the west and south. Hainan is the region with the most frequent thunderstorm activity; there are 60–85 thunderstorm days in the southern coast and over 100 days for the other areas. Thunderstorm often occur daily after midday. The thunderstorm period is from March to October each year.

2.5.8 The Solar Resource in Harbin

With coordinates of 45°45′N and 126°46′E, Harbin is north of Northeast China and south of Heilongjiang Province. It is adjacent to the branch hill of the Zhangguangcai Mountains to the southeast and the Xiao Hinggan Mountains to the north; the Songhua River passes through the middle of Harbin. With an elevation of 143 m and total area of 56,579 km^2 (7086 km^2 of urban area), Harbin governs eight districts and ten counties (cities). Compared with other provincial capitals in China, Harbin has the highest latitude as well as the lowest winter temperature. Harbin City is controlled by a continental polar air mass through the winter and influenced mainly by a subtropical marine air mass during summer; spring and autumn are seasons exhibiting winter and summer wind alternation. According to China's division of climate zones, Harbin City has a midtemperate continental monsoon climate. Influenced by the geographic environment, marine and territorial air masses, and monsoon alternation, the seasonal climate of the entire city varies significantly [14].

2.5.8.1 *Basic Climatic Characteristics of Harbin*

Harbin City has four climatic characteristics. The first is a cold dry winter with low precipitation: winter (November–March) in Harbin lasts more than 5 months with cold dry weather. The period with a mean temperature below 10°C starts October 3 and ends April 30, lasting for 210 days. The coldest month is January with a mean temperature of −19.1°C and extreme minimum temperature of −38.1°C. Precipitation is scarce for the entire year; 5 months have precipitation totaling only 35.1 mm and accounting for only 6.6% of total annual precipitation. The second characteristic is a warm humid summer with concentrated precipitation: summer (June−August) in Harbin has plenty of rainy days and concentrated precipitation influenced by a subtropical marine air mass. It has a high temperature season that demonstrates obvious characteristics of simultaneous thermal and moisture. The mean precipitation of Harbin in summer is 350.9 mm, accounting for 66.1% of the annual total; intermittent precipitation is most commonly seen, with about 11.1 rainy days. The third characteristic is a dry spring with changing temperatures and strong winds. During spring (April−May), dominant cold air in the North is less severe, while warm air in the south starts to increase along with a rapid temperature increase. Due to frequent cyclones during this period, the temperature changes constantly; a one-time temperature increase or drop can be as large as 20°C. The fourth characteristic is an autumn with a fast temperature drop and an early first frost. Due to plunging temperatures in autumn, frost may occur; the mean first-frost day is September 21 (refer to Table 2.12).

2.5.8.2 Solar Radiation in Harbin

Harbin is located between the Tropic of Cancer and the Arctic Circle. Solar altitude variation and various corresponding seasonal solar radiation variations are significant. In Harbin, the sun rises at 7:37 a.m. and sets at 4:20 p.m. on the winter solstice; the respective day length is a mere 8 h 43 min with a solar altitude during midday of 20°15′. The sun rises at 4:19 a.m. and sets at 7:53 p.m. on the summer solstice; the respective day length is as long as 15 h 34 min with a solar altitude during midday of 67°45′. During winter and summer, length of day varies significantly and the midday solar elevation on the winter solstice is less than one-third that on the summer solstice.

According to data from the solar radiation observation station of Harbin during 1961−86, the mean annual global solar radiation of Harbin is 4634 MJ/m^2 with a maximum value of 5009 MJ/m^2 (1976) and minimum value of 4173 MJ/m^2 (1978). The maximum value for monthly global radiation occurred in June at 703 MJ/m^2; the minimum value occurred in December at 105 MJ/m^2; the amplitude of variation within the year is 598 MJ/m^2. The daily variation of global radiation increased from February to March with an amplitude as high as 1549 MJ/m^2; it decreased from July to September with a monthly decrease amplitude of 109−113 MJ/m^2.

Among the global solar radiation types in Harbin, direct and scattered radiation components change significantly by season. Due to differences

TABLE 2.12 Monthly Statistical Indices for Harbin, 1959–2006

Index	January	February	March	April	May	June	July	August	September	October	November	December
Mean temperature/°C	−19.1	−15.1	−4.1	6.6	14.6	20.1	22.6	20.9	14.3	5.4	−5.7	−15.3
Minimum daily temperature/°C	−23.9	−20.0	−9.8	0.4	8.0	14.5	18.3	16.4	9.0	0.5	−10.1	−19.8
Maximum daily temperature/°C	−12.6	−7.5	2.1	13.3	21.2	26.1	27.9	26.4	20.8	12.0	−0.0	−9.5
Mean relative humidity/%	74	70	58	51	51	66	77	78	71	65	67	73
Mean of total precipitation/mm	4.0	4.7	11.1	21.2	36.7	83.4	155.0	112.5	61.5	25.6	9.4	5.9
Mean wind speed/(m/s)	3.4	3.5	4.3	5.2	4.9	3.9	3.5	3.2	3.6	4.1	4.3	4.7

in monthly atmospheric circulation and the nature of the underlying surface, the constitution of the two components shows significant monthly variations that are mainly determined by differences in cloud cover and atmospheric transparency for various months.

2.5.8.3 Sunshine Duration in Harbin

The mean sunshine duration of Harbin is 2641 h, which is longer in summer and shorter in winter, with values in spring and autumn falling in between the two; however, the amplitude of variation of sunshine duration for the year is less than that of the possible sunshine duration. As the temperature is low, humidity is low and the cloud cover is low in winter, while the sunshine percentage (percentage of actual sunshine duration to possible sunshine duration) is high; in February, it can be as high as 67%. Although daylight time is short, total sunshine duration during November–March still accounts for about 34% of the yearly total. With plenty of rainy days in summer, the sunshine percentage decreases, to just 53% in July. The actual sunshine duration for July is lower than that of May and June.

The maximum value of annual sunshine duration in Harbin over the years is 2878 h (1978), and the minimum value is 2315 h (1968). The maximum value of monthly sunshine duration is 327 h (August, 1975), and the minimum value is 87 h (January, 1985). The maximum value of monthly sunshine duration is normally not in June when the summer solstice occurs. This mainly results from the arrival of summer winds in June, increased cloud cover, and a reduction in sunshine percentage; the summer half-year's monthly sunshine duration is significantly influenced by cloud cover.

2.6 SOLAR IRRADIANCE PREDICTION METHODS

2.6.1 Estimation Method for Solar Direct Normal Irradiance

As required by optical principles, site selection for concentrating solar power (CSP) plants shall rely on solar DNI resources. Meteorological stations in China don't possess such test data. Nevertheless, we can acquire global irradiation data and sunshine duration on the horizontal surface corresponding to the period of 1994–2003 from domestic meteorological stations for DNI estimation, a method based on global radiation and sunshine duration on the horizontal surface that is described as follows.

When astronomic radiation coming from outside Earth's atmosphere passes through to Earth, due to the influences of scattered radiation, absorption, and reflection by Earth's atmosphere, radiation finally reaching

Earth undergoes significant changes; in addition, due to different atmospheric environments in the calculation area, solar radiation that reaches the surface has significant regional characteristics that are far less significant than the corresponding astronomic radiation. Many scholars have estimated respective local solar radiation resources by establishing mathematical models and empirical relationships. Many meteorological parameters, such as atmospheric turbidity, relative humidity, clearness index, cloud amount, and sunshine duration, have been proven to have impacts on local solar irradiation resources, among which the most crucial influencing factor is sunshine duration.

By utilizing meteorological data (including solar radiation data and sunshine duration) from typical climatic regions of eight typical cities during the period of 1994–2003, the mathematical relationship between the monthly mean clearness index and the sunshine percentage was established to estimate global solar radiation. The results indicate a model with excellent estimation accuracy. In order to further demonstrate the model's precision, comparisons were made between estimation results of the established model and other mathematical models in terms of mean percentage error (MPE), mean bias error (MBE), and root-mean-square error (RMSE). The results proved that the estimation model based on local meteorological data samples is more reliable than the estimation model that has already been published for global applications.

Many scholars have proposed ways to estimate solar radiation resources based on the mathematical relationship between global solar radiation and sunshine percentage on the horizontal surface through linear regression, among which the most representative is the Angstrom–Prescott model. Here, a mathematical relationship applicable for full-site single-month estimation of domestic solar radiation resources estimation is established by utilizing meteorological data of eight typical cities in different climatic regions of China to estimate the domestic solar radiation resources.

Refer to Eq. (2.16) [5] for the mathematical expression of the Angstrom-Prescott model:

$$\frac{\overline{H}}{\overline{H_0}} = a + b\frac{\overline{S}}{\overline{S_0}} \tag{2.16}$$

In which $\overline{H}$ refers to the monthly mean value of daily global radiation on the horizontal surface, $MJ \cdot m^{-2} d^{-1}$; $\overline{H_0}$ refers to the monthly mean value of the daily total astronomic radiation within the calculation area, $MJ \cdot m^{-2} d^{-1}$; $\overline{S}$ refers to the monthly mean value of sunshine duration; $\overline{S_0}$ refers to the maximum value of the monthly mean value of sunshine duration; a and b are coefficients.

Astronomic radiation on the horizontal surface is calculated as follows:

$$H_0 = \frac{24 \times 3600}{\pi} I_0 f \left(\cos\lambda \cos\delta \sin\omega_s + \frac{\pi}{180}\omega_s \sin\lambda \sin\delta \right) \tag{2.17}$$

in which $I_0 = 1367\ \mathrm{W/m^2}$ (solar constant); λ refers to the local geographic latitude; and δ refers to the declination angle.

$$f = 1 + 0.033 \cos\left(360 \frac{n}{365}\right) \tag{2.18}$$

in which n is the chronological order (for example, the chronological order for January 1 is 1).

Declination angle is calculated as follows:

$$\delta = 23.45 \sin\left(360 \frac{248 + n}{365}\right) \tag{2.19}$$

ω_s refers to the sunset hour angle, which is calculated as follows:

$$\omega_s = \cos^{-1}(-\tan\lambda \tan\delta) \tag{2.20}$$

The longest sunshine duration is calculated as follows:

$$S_0 = \frac{2}{15}\omega_s \tag{2.21}$$

in which values of $\overline{H_0}$, $\overline{S_0}$ can be obtained through Eqs. (2.17) and (2.21). Coefficients a and b can be obtained from $\frac{\overline{H}}{\overline{H_0}}$ and $\frac{\overline{S}}{\overline{S_0}}$ samples through linear regression, in which the values of $\overline{H}$ and $\overline{S}$ can be obtained from the meteorological database of eight selected typical cities during the period of 1994–2003.

Coefficients a and b are calculated as follows:

$$b = \frac{n \sum_{i=1}^{n} \frac{\overline{H}}{\overline{H_0}} \frac{\overline{S}}{S_0} - \left(\sum_{i=1}^{n} \frac{\overline{H}}{\overline{H_0}} \right) \left(\sum_{i=1}^{n} \frac{\overline{S}}{S_0} \right)}{n \sum_{i=1}^{n} \left(\frac{\overline{S}}{\overline{S_0}} \right)^2 - \left(\sum_{1=1}^{n} \frac{\overline{S}}{S_0} \right)^2} \tag{2.22}$$

$$a = \frac{1}{n} \sum_{1=1}^{n} \frac{\overline{H}}{\overline{H_0}} - \frac{b}{n} \sum_{1=1}^{n} \frac{\overline{S}}{S_0} \tag{2.23}$$

in which n is the quantity of measuring points.

Global solar radiation and sunshine data of selected sites during the period of 1994–2003 correspond to the typical climatic cities provided by the China Meteorological Administration (CMA), which serves as the main data analysis source in the book. The positions of various sites are shown in Table 2.13.

TABLE 2.13 Geographical Positions of Representative Cities

S/N	Site	Latitude (N)	Longitude (E)	Elevation/m
1	Harbin	45°45′	126°46′	142.3
2	Lanzhou	36°03′	103°53′	1517.2
3	Beijing	39°48′	116°28′	31.3
4	Wuhan	30°37′	114°08′	23.1
5	Kunming	25°01′	102°41′	1892.4
6	Guangzhou	23°10′	113°20′	41
7	Urumqi	43°47′	87°39′	935
8	Lhasa	29°40′	91°08′	3648.9

To calculate deviation levels for the estimation model, the primary comparative parameters appearing in the book are as follows:

RMSE

$$RMSE = \sqrt{\sum_{i=1}^{N} (H_{\text{estimated}} - H_{\text{measured}})^2 / N} \tag{2.24}$$

MABE

$$MABE = \sum_{i=1}^{N} \left| \frac{H_{\text{estimated}} - H_{\text{measured}}}{H_{\text{measured}}} \right| / N \tag{2.25}$$

MBE

$$MBE = \left[\sum_{i=1}^{N} (H_{\text{estimated}} - H_{measured}) \right] / N \tag{2.26}$$

in which $H_{\text{estimated}}$ refers to the estimation value of global solar radiation and H_{measured} refers to the measured value of global solar radiation.

According to Eqs. (2.22) and (2.23), single-month results of regression coefficients a and b can be obtained as shown in Table 2.14 and are used to compare estimation results with actual results for RMSE, MABE, and MBE calculated by Eq. (2.24)–(2.26) using the estimation model of global solar radiation. RMSE mainly reflects the deviation level of the data; MABE mainly reflects the mean status of relative error; and MBE mainly reflects the deviation status of estimation results, which can be either positive or negative; a positive value means the estimation result is higher than the actual result, whereas a negative value means the estimation result is lower than the actual result. Based on the above error analysis, a

TABLE 2.14 Comparison of Estimation Result Errors for Monthly Mean Global Solar Radiation Resources

	Model			
Month	**Model Expressions**	*RMSE*	*MABE*	*MBE*
January	$H/H_0 = 0.157 + 0.680\ S/S_0$	6.18 E-02	1.12 E-01	−3.50 E-03
February	$H/H_0 = 0.120 + 0.721\ S/S_0$	5.43 E-02	9.10 E-02	−3.30 E-03
March	$H/H_0 = 0.116 + 0.709\ S/S_0$	4.80 E-02	8.35 E-02	−1.55 E-09
April	$H/H_0 = 0.111 + 0.678\ S/S_0$	4.33 E-02	7.75 E-02	−1.30 E-03
May	$H/H_0 = 0.132 + 0.624\ S/S_0$	4.70 E-02	8.16 E-02	−5.50 E-03
June	$H/H_0 =0.127 + 0.621\ S/S_0$	4.83 E-02	8.90 E-02	−3.82 E-02
July	$H/H_0 =0.193 + 0.478\ S/S_0$	5.45 E-02	9.35 E-02	−1.40 E-03
August	$H/H_0 = 0.192 + 0.487\ S/S_0$	5.36 E-02	9.38 E-02	−4.20 E-03
September	$H/H_0 = 0.165 + 0.559\ S/S_0$	5.11 E-02	8.30 E-02	−3.80 E-03
October	$H/H_0 = 0.102 + 0.685\ S/S_0$	4.52 E-02	7.66 E-02	−1.80 E-03
November	$H/H_0 = 0.078 + 0.754\ S/S_0$	5.33 E-02	8.77 E-02	−1.50 E-03
December	$H/H_0 = 0.160 + 0.656\ S/S_0$	7.22 E-02	1.23 E-01	−9.01 E-05
Mean value	$H/H_0 = 0.139 + 0.637\ S/S_0$	5.27 E-02	9.10 E-02	−5.40 E-03

conclusion can be made that the smaller the RMSE of the comparison result, the higher the precision.

As shown in Table 2.14, the mean RMSE for various months of the full-site single-month estimation model is 5.27%. Under clear weather conditions, atmospheric transmission coefficients can be reflected through sum of regression coefficients a and b. Monthly variation of sum of regression coefficients corresponding to the estimation model developed in the book is shown in Fig. 2.9.

According to Fig. 2.9, on average in China, the sum of these coefficients gradually decreases with the beginning of the new year, bottoming in July before enjoying a gradual increase thereafter, with atmospheric transparency reaching a maximum in spring and winter and a minimum in summer. This is caused by several factors: in most areas in China, the second and third quarters are the main precipitation seasons and respective weather conditions are comparatively humid; whereas during spring and winter, the weather is comparatively dry; it is not easy for precipitation to form. Humidity is form easily on cloudy days under

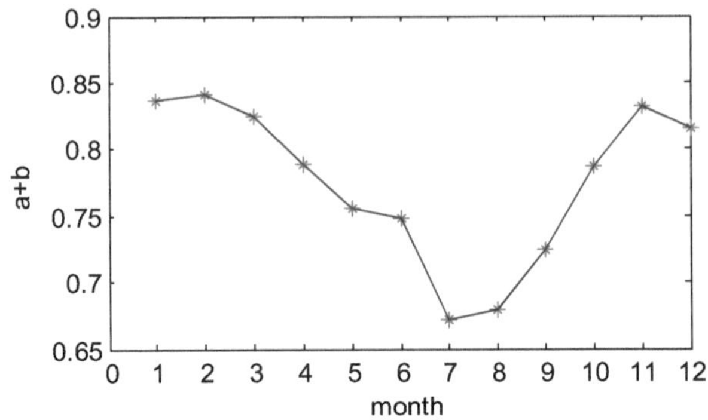

FIGURE 2.9 Monthly variation tendency of sum of coefficients a and b.

various weather conditions, resulting in solar radiation resources being reflected, absorbed, and scattered by numerous clouds. Therefore, the respective atmospheric transparency is low, which leads to minimum atmospheric transparency in June. Weather in spring and winter is dry and frequently influenced by cold air; there are fewer clouds in the sky and thus the respective atmospheric transparency is high. Furthermore, according to Fig. 2.9, the average sum of mean values of regression coefficients a and b is 0.7754, which is quite close to the value of 0.80 published in literature applicable for most domestic areas under a mid-temperate rainforest climate (i.e., weather is normally dry in winter). To a certain extent, it has demonstrated the estimation precision of the model.

By utilizing climatic data of eight representative cities in typical climatic regions during the period of 1994–2003, the Angstrom–Prescott model relationship is obtained through linear regression; it is possible to establish a full-site single-month estimation model. The model has satisfactory estimation precision; when there is no measured radiation value, it is possible to use the model to estimate local solar radiation resources.

2.6.2 Influences of Climate Change on Solar Direct Irradiance

According to "Solar Radiation Status Research in China for the Recent 30 Years," by utilizing ground surface global, direct, and scattered radiation variations at China's 55 solar radiation stations for nearly 30 years from 1961 to 1990, LI Xiaowen et al. [15] of China believed that (for the 1990s) solar radiation and direct radiation had shown a decreased tendency in recent years in most areas in China. Meanwhile, by integrating with and analyzing visibility observation data from China's 60 sites daily at 2:00 p.m. Beijing time that corresponded with the same period, they discovered that visibility in most areas of China showed a decreased tendency; they also believed that increases in atmospheric turbidity and the concentration of atmospheric suspended particles served as possible reasons for decreased direct radiation in some domestic areas. The influences of atmospheric suspended particles and other aerosols on solar radiation are a complex problem. HU Liqin et al. [16] studied the influences of clouds and aerosols on solar radiation absorbed by the atmosphere. During the same period, some scientists pointed out that solar radiation actually absorbed by clouds was much more than the amounts estimated using current theory. Robert D. Cess et al. from the State University of New York at Stony Brook measured solar radiation in five different locations by satellite and concluded that global mean radiation absorbed by clouds was around 25 W/m^2, which is much more than the estimated range of 0–10 W/m^2 given by the theoretical model. Generally speaking, current studies on the influences of clouds and aerosols on solar radiation are still not complete.

In terms of radiation observation, LIU Guangren et al. [17] measured solar radiation gradients by utilizing the 325-m meteorological tower in Beijing and preliminarily achieved certain observation results that indicated that under pollution conditions, radiation attenuation was significant between the ground and a height of 320 m; under perfectly clear weather conditions, solar direct radiation during midday hours varied around 10 W/m^2. Otherwise, under severe pollution conditions on the surface, the relative difference was as high as 140 W/m^2, indicating that under severe pollution conditions on the surface, solar direct radiation consumption was great. With clear weather, total ultraviolet radiation in the sky and on the ground varied insignificantly; in cases of poor atmospheric visibility, total ultraviolet radiation decreased, but the difference between that in the sky and on the ground comparatively increased.

2.7 DISTRIBUTION OF SOLAR DIRECT NORMAL RADIATION RESOURCES IN CHINA

Presently, research methods on solar direct normal radiation mainly include ground station observation, satellite remote sensing, and numerical simulation. Ground station observation has the advantage of time continuousness, but stations are located separately. Normally, it is suggested to adopt the method of geographic space interpolation for compensation, which is accompanied by great interpolation errors, especially for areas where stations are sparsely located. H. Broesamle et al.

calculated solar radiation resources by utilizing satellite cloud images and estimated the solar direct normal radiation resources in North Africa. ZUO Dakang et al. carried out studies on the spatial distribution characteristics of global solar radiation in domestic areas for the first time and drew monthly and yearly global radiation distribution charts for China. Using radiation measurement through satellite remote sensing, FU Bingshan et al. established a statistical model of satellite-measured values as well as solar direct and scattered radiation that corresponded to various heights in the atmosphere under clear weather conditions by utilizing data from radiosonde stations located in southeast coastal areas of China. Satellite remote sensing data are connected in terms of spatial distribution yet separated in terms of time distribution; furthermore, data quality is easily influenced by weather conditions. There exist certain difficulties in inverting the status of ground solar radiation based on current techniques. China's solar resource division indices were first proposed by a domestic scholar named Wang Bingzhong in 1983, and since then many researchers have studied the spatial and time distributions and division of solar resources in partial provinces and cities on the basis of observational data from meteorological stations. A common conclusion of solar radiation research has been proposed; that is, solar and direct radiation in partial domestic areas have shown decreased tendency, which likely is caused by an increased concentration of suspended particles in the atmosphere. Recent research has indicated that from the 1960s to 1980s, China's direct radiation resources as a whole have shown decreased tendency; yet since the 1990s, decreased tendency has stopped, and there has even been some increased tendency.

2.7.1 Distribution of China's Annual Mean Daily Solar Direct Normal Irradiation

China has abundant solar resources. According to rough calculations and estimations, the mean global annual irradiation on horizontal surfaces for the entire country over the past 30 years is 5648 MJ/m^2; in areas accounting for more than two-thirds of the total area, annual global solar irradiation has exceeded 5000 MJ/m^2. China enjoys favorable conditions for solar energy utilization. General characteristics of China's global annual radiation distribution show a trend of the west being higher than the east.

In the west, annual global solar radiation shows a zonal distribution with the south being higher than the north, in which southern Tibet (except for the Shannan Prefecture) and the Qaidam Basin have higher solar radiation; the Tarim and Turpan Basins in Xinjiang have lower amounts.

The two low-value zones exist separately around Tian Shan Mountain and the Shannan Prefecture of Tibet.

In the east, North China has the largest global annual radiation, whereas the southeast and northeast have lower solar radiation; in the area of 20°–40°N, solar radiation decreases with increases in latitude. The main factor influencing solar radiation distribution in East China is cloud cover distribution caused by atmospheric circulation. The area is influenced by moist airflow from the ocean with large amounts of intermediate and low-level clouds, which may weaken the solar radiation reaching the ground, especially in the southeast with sufficient precipitation, even for multiple preliminarily selected low-value minor centers (which are estimated to be related to the complex terrain of the locality; the southeastern area has a mountainous hilly topography that may easily create topographic precipitation); in the northeast, global annual radiation shows significant zonal distribution characteristics. Solar elevation has a dominant role toward the solar radiation distribution of the area. The area features high latitude and low solar altitude; when radiation passes through the atmosphere, the optical path is long and radiation is greatly weakened. Thus the global annual radiation is low.

The southern Qinghai–Tibet Plateau (except for the Shannan Prefecture) is a major high-value center for annual global solar radiation due to its high elevation and the atmosphere's short optical path. Solar radiation loss in the atmosphere is insignificant, and solar short-wavelength radiation reaching the ground is strong; there is also a comparatively high-value zone along the Qaidam Basin, Altun Mountains, and Kunlun Mountains north of the Qinghai–Tibet Plateau; on a national scale, there are two comparative deep low-value centers for annual global solar radiation: the leeward slope in the eastern Qinghai–Tibet Plateau–Sichuan Basin and the Shannan Prefecture of Tibet, in which the Sichuan Basin low-value center is mainly influenced by cloud cover at the junction of the south–north flows of the Qinghai–Tibet Plateau, with frequent weather system activities and rainfall that have greatly weakened the global radiation reaching the ground. The Shannan Prefecture is south of the Himalayas with great elevation variation. When the moist Indian monsoon branch passes through the area, the air mass climbs over the mountains and creates an intensive precipitation. The area is the continental rain pole of the world. The large number of intermediate and low-level clouds brought by rainy weather has weakened the global solar radiation reaching the ground, leading to low values.

The operational observation system of the CMA has 98 radiation observation stations, including 17 first-level stations capable of conducting direct radiation observation. China's meteorological radiation observation stations were categorized as Class A and Class B before 1993; observation factors of Class A stations include global, scattered, and direct radiation, and observation factors of Class B stations only include global radiation. During this period, various observation sites have been adjusted many times. Since 1993; observation stations have been adjusted from Class A and Class B to Level I, II, and III observation stations. Observation factors of a Level-I station include global, net, scattered, direct, and reflective radiation; observation factors of a Level-II station include global and net radiation; and observation factors of a Level-III station include global radiation.

Before 1993, radiation observation instruments applied in radiation observation stations nationwide included thermoelectric type (constantan and manganese steel welding) and inductive surface (regular black paint) radiation meters with a relative error of 10%; during and after 1993, radiation observation stations nationally started to apply domestically developed thermoelectric type (wire-wound constantan and plated copper) and inductive surface (specific optical black paint) automatic telemetering radiation meters with a relative error of 0.5%.

Except for solar irradiation, China's regular meteorological data can be divided into two parts; one is ground observation data including wind speed, wind direction, atmospheric pressure, temperature, dew-point temperature, and sunshine duration, which are covered by China's main meteorological stations (excluding data from meteorological stations in Hong Kong, Macao, and Taiwan), involving 740 stations with a spatial density higher than radiation sites. The other part is sounding data, the observation factors of which mainly include wind speed, wind direction, atmospheric pressure, elevation, temperature, and dew-point temperature.

Actual observation data in China for DNI are far from satisfying the need for a spatial distribution of the solar resource. Thus calculations in the book are conducted using annual mean daily DNI data for China with a "40 km × 40 km" spatial resolution provided by the National Renewable Energy Laboratory (NREL) in the United States based on a daily radiation model (Climatological Solar Radiation Model). The respective data are mainly calculated and deduced based on remote sensing data for the period from January 1, 1985, to December 31, 1991. The model is used for analysis and calculation while considering various factors including clouds, water vapor, trace gases, and aerosol content in the atmosphere. The calculation model for DNI being transferred from the atmosphere to the ground is as follows:

$$DNI = E_0(\tau_R \tau_{Ozoo} \tau_{Gas} \tau_{wv} \tau_{Ae}) \tau_{Cl} \tag{2.27}$$

in which E_0 refers to the solar constant; τ_R refers to the Rayleigh scattering transparency; τ_{Ozoo} refers to the ozone absorption transparency; τ_{Gas} refers to the mixed gas absorption transparency; τ_{wv} refers to the water vapor absorption transparency; τ_{Ae} refers to the aerosol absorption or scattering transparency; and τ_{Cl} refers to the cloud absorption or scattering transparency.

Based on NREL data, China's annual mean daily DNI distribution can be obtained. The data have not been calibrated through ground measurement.

Daily DNIs of "2 kWh/m^2" and "4.4 kWh/m^2" occur most frequently on the ground in China. The maximum value appears in the Tibet Autonomous Region at 9.365 kWh/m^2 while the minimum value appears in the Sichuan Basin at 0.785 kWh/m^2. The annual mean daily DNI in China is 4.033 kWh/m^2.

2.7.2 Influencing Factors for Spatial and Temporal Distributions of Solar Direct Irradiation

Solar direct radiation decreases along with reduced elevation, in that when solar radiation enters the atmosphere from above, the further it travels to the lower level, the greater the content of water vapor, aerosol, and mixed gas it contains and the greater and faster its attenuation is. In the upper part of the troposphere (with a height from the top of 7–17 km) and above, the attenuation of solar direct radiation by the atmosphere is insignificant.

In different places and at different time points, the distribution, constitution, and concentration of water vapor and aerosol in the atmosphere vary from each other, which leads to different absorption and scattering of solar radiation.

2.7.3 Basic Characteristics of China's Solar Resource

Characteristics of China's solar resource are mainly determined by geographic latitude, topography, and atmospheric circulation conditions. Normally, the bigger the solar altitude (lower the latitude), the shorter the path that solar radiation follows when passing through the atmosphere (higher the elevation) and is less weakened by the atmosphere with more solar radiation reaching the ground.

1. Characteristics of latitude and season influences. The influence of latitude on global radiation distribution is mainly manifested in East China (to the east of 105°E), including the zonal tendency of the annual total irradiation isoline. It serves as the general background for global radiation distribution, which has manifested the dominant role of the astronomic factor (astronomic radiation) on global radiation. Along with seasonal changes, the influence level of latitude may change according to variation of the astronomic factor and atmospheric circulation factor effect contrast.

It normally follows the tendency of degradation of annual global irradiation along with latitude increase. In western areas, zonal characteristics of the global radiation distribution may be concealed by the major influences of geographic conditions that have appeared in northern Xinjiang.

2. Features of topographic influences. The influences of topography on global radiation distribution are mainly manifested in elevation differences. In the Qinghai–Tibet Plateau, the mean elevation is above 4000 m and is the strongest influence on annual solar radiation distribution. According to the annual solar radiation distribution chart, the Qinghai–Tibet Plateau has a high-value center mainly caused by weakened absorption and scattering of the atmosphere on solar radiation due to high elevation. The radiation status at Tian Shan Mountain, Qilian Mountain, and other high mountains is similar to that of the plateau. At the eastern margin of the Qinghai–Tibet Plateau, due to rapid elevation change, global radiation isolines are concentrated.
   Fig. 2.10 shows the annual solar radiation distribution by elevation. According to Fig. 2.10, solar radiation distribution variation and elevation changes are subject to an approximate linear relationship. The Qinghai–Tibet Plateau has a lower latitude and greater solar altitude; it has the highest elevation, and the optical path that solar radiation follows through the atmosphere to reach the ground is

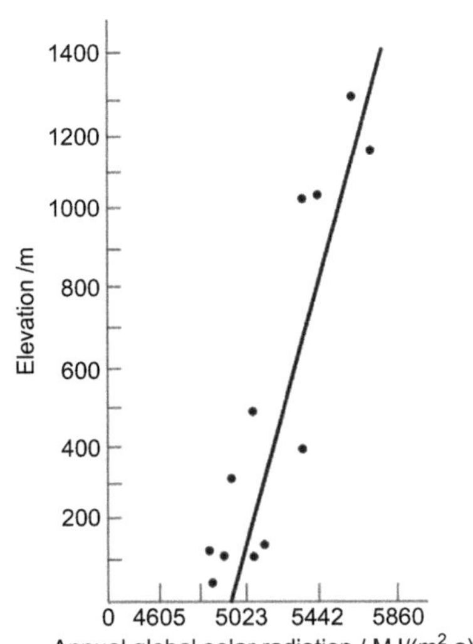

FIGURE 2.10 Distribution of annual solar radiation by elevation.

shorter; on the plateau, the atmospheric density is lower (thin air), the contents of water vapor and solid impurities in the atmosphere are smaller; there are fewer clouds; and the atmospheric transparency is satisfactory. Based on the above, the refraction, scattering, and absorbing effects of solar radiation are greatly weakened, which intensifies solar radiation; summer has more clear days with longer sunshine duration than other areas. Thus the Qinghai–Tibet Plateau is the region in China with the largest annual global solar radiation, and also the region in China with intensive solar radiation during summer. However, due to the high elevation of the Qinghai–Tibet Plateau, the air is thin and thus there are fewer clouds in the sky and less atmospheric counter-radiation. On the other hand, the thermal insulation function of the atmosphere is quite poor and cannot preserve thermal from ground radiation well; in addition, the wind speed is higher on the plateau, making it even less conductive to thermal storage and preservation. Even in summer, mean temperatures in most areas of the Qinghai–Tibet Plateau are quite low; it is the region in China with the lowest mean summer temperature.

Due to isolated topographic conditions in the Sichuan Basin, a low-value center for global radiation is created. The Qinghai–Tibet Plateau and Sichuan Basin are typical in terms of the reflecting influences of topography on radiation distribution. Topographic influences of other areas are comparatively less significant due to low mountain heights or the small scale of horizontal extension of mountains, etc.

3. Features of atmospheric circulation influences. The influences of atmospheric circulation on global radiation distribution are mainly reflected through cloud status development. The actual global radiation distribution is the result of the integrated influences of latitude, topography, and atmospheric circulation conditions. The influences of the first two are comparatively stable, whereas the influences of atmospheric circulation conditions feature the greatest variability. Summer global radiation is most significantly reflected by subtropical anticyclone in the middle and lower reaches of the Yangtze River and its southern areas as well as the rainband in North China. The influence of rainy weather on the global radiation of Qinghai–Tibet Plateau is also quite strong.

   The influences of atmospheric circulation conditions on the annual variation of global radiation in various areas are also considerable, mainly resulting in displacement months in certain areas when maximum and minimum values of global radiation occur.

2.7.4 Regionalization of China's Solar Resource

To facilitate the exploration and utilization of the solar resource in China, China's solar resource has a three-level division distribution in which the indices of the first-level division are the levels of global annual radiation on the horizontal surface [kWh/($m^2 \cdot a$)]. The distribution of China's solar resource according to first-level division indices is shown in Table 2.15. There are four categories.

The number of days each month when sunshine duration is above 6 h is the index for the second-level division. The ratio of the maximum and minimum values for the number of days each month when sunshine duration is above 6 h can be used to measure the scale of annual variations in the local solar resource; the smaller of the ratio, the more stable the annual variation of the solar resource and the more conductive it is for utilization of the solar resource. In addition, the season when the maximum and minimum values occur has manifested a feature for local solar resource distribution.

The characteristic value of solar energy daily variation is used as the index of the third-level division. It is specified as the annual mean sunshine duration at 9–10 local true solar time to represent sunshine conditions in the morning, 11–13 for midday, and 14–15 for afternoon; a longer annual mean sunshine duration means that the specific period is conductive to solar energy utilization. The third-level division index has indicated optimum or unfavorable hours in a day for solar energy utilization.

2.7.5 Measurement of Solar Direct Normal Irradiation

Using primary GIS analysis on solar radiation data resources in China published by the NREL, the mean solar direct normal radiation distribution in China on a multiyear climate basis with a "40 km × 40 km" resolution can be obtained.

1. Solar radiation in China generally shows the characteristics of a distribution with the west being higher than the east bounded by midwest Inner Mongolia–Ningxia–northwest Gansu–west Sichuan–northwest Yunnan; most areas to the west of the boundary

TABLE 2.15 Classification of the Solar Resource Based on Annual Global Irradiation

Category	S/N	Index [kWh/($m^2 \cdot a$)]
Zone with abundant resources	I	≥1740
Zone with comparatively abundant resources	II	1400–1740
Zone with comparatively poor resources	III	1160–1400
Zone with poor resources	IV	<1160

have annual total direct solar irradiation above 1400 kWh/m^2, showing zonal distribution characteristics of the south being higher than the north; areas to the east of the boundary have annual total direct solar irradiation of less than 1400 kWh/m^2; North China has the largest amount.

The national annual mean solar direct normal irradiation is 1472 kWh/m^2. The area southwest of Tibet is an extremely high-value zone, while the extremely low-value zone is located in the Shannan Prefecture of Tibet, Sichuan Basin, and Chongqing–Guizhou–Hunan region.

2. Based on the geographic statistics of solar radiation resources in various provinces (cities, districts), annual total direct solar irradiation on an average climate condition in various provinces (cities, regions) can be obtained. Tibet, Qinghai, Xinjiang, Gansu, Ningxia, and Inner Mongolia are the provinces (regions) in China with the most abundant solar radiation resources, while various provinces (cities, regions) in North China and Yunnan have intermediate radiation resources, and Chongqing, Guizhou, and Hunan have the poorest radiation resources.
3. During a clear day, the global irradiance on the horizontal surface is less than the DNI; during a cloudy day, the global irradiance on the horizontal surface is more than the DNI. Therefore, in northwestern areas of China with fewer cloudy days, DNI values are higher than the global irradiance on the horizontal surface.

Solar meteorological stations that are currently the most commonly used include APL in the United States, Kipp & Zonen in the Netherlands, and the RSP-3G irradiation meter in Germany, which are all available for measuring solar direct normal, global, scattered, direct, and other radiation (refer to Figs. 2.11 and 2.12 and Table 2.16). Data collection frequency

FIGURE 2.11 Kipp & Zonen irradiance meters at an operational station.

FIGURE 2.12 RSP-3G irradiance meter.

TABLE 2.16 Parameter Indices of CMP21 Pyranometer in the Kipp & Zonen System

Item	Index
Spectral coverage	285–2800 nm
Sensitivity	7–14 μV/(W m^2)
Response time	5s
Directional error (80 degrees, 1000 W m^2)	<10W m^2
Sensitivity upon temperature (−20°C to +50°C)	±1%
Suitable temperature range	−40°C to +80°C
Maximum solar irradiance	4000 W m^2
Field of view	180°C

is available for adjustment; normally, as for the perennial observation, it can be adjusted to 10 min per time.

The main configuration of a reference irradiation station is shown in Table 2.17. Fig. 2.13 indicates the mean DNI on a 10-min basis measured at Badaling during the period from 10:00 a.m. on November 29, 2011, to 9:30 a.m. on January 5, 2012. According to the figure, during December Badaling had plenty of clear days.

Figs. 2.14 and 2.15 separately display the relationships between DNI and global irradiance on the horizontal surface, as well as DNI and diffuse irradiance on the horizontal surface. According to Fig. 2.14, during a clear day the global irradiance on the horizontal surface is less than the DNI

TABLE 2.17 Main Configuration of a Solar Irradiance Testing Station

Name	Quantity
Data collecting instrument	1
Base with batteries	1
Pyrheliometer ISO primary standard	1
Solar position tracker	1
Case Heating plate	1
Pyranometer ISO secondary standard	1
Scattered radiation meter, ISO secondary standard	1
Artificial ventilating hood	1
Atmospheric long wavelength radiation meter	1
Reflection radiation meter	1
Pallet-mounted with a reflection radiation meter	1
Ground long wavelength radiation meter	1
Pallet-mounted with a ground long wavelength radiation meter	1
Communication unit	1

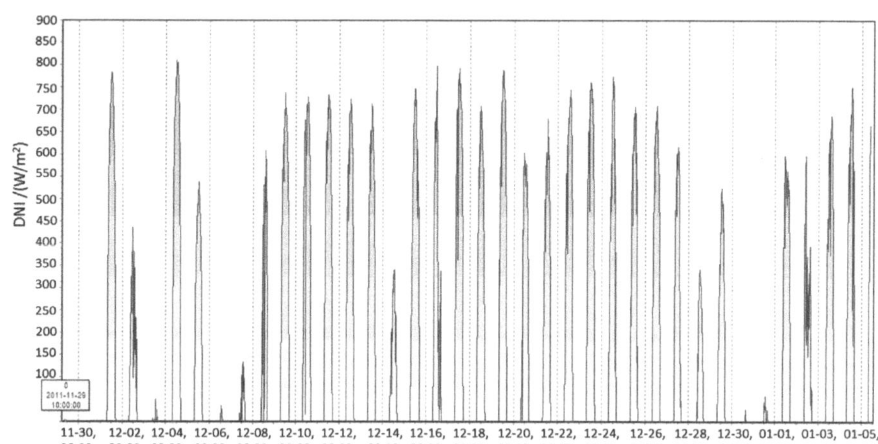

FIGURE 2.13 Measured values of direct normal irradiance for month of December, 2011 in Badaling (statistical period: 10 min).

and diffuse irradiance is insignificant, yet on a cloudy day as shown in Fig. 2.15, the global irradiance on the horizontal surface is more than the DNI and diffuse irradiance on the horizontal surface is equivalent to global irradiance.

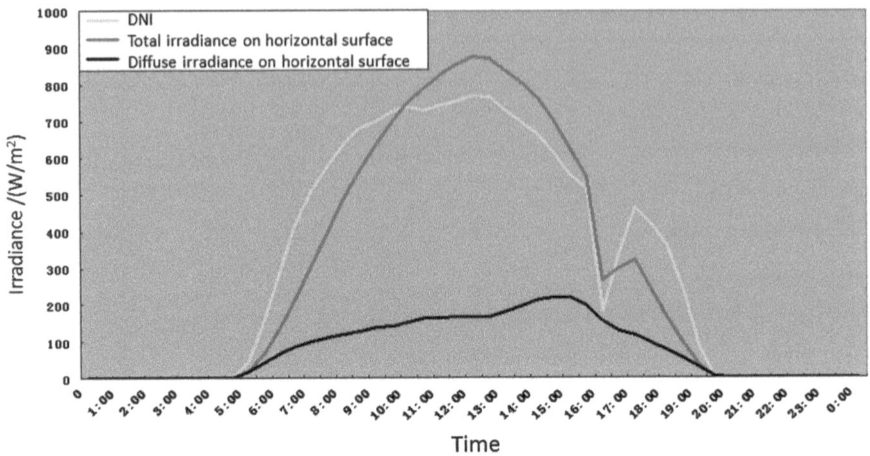

FIGURE 2.14 Relationship between global/diffuse irradiance on the horizontal surface and direct normal irradiance during a clear day [15].

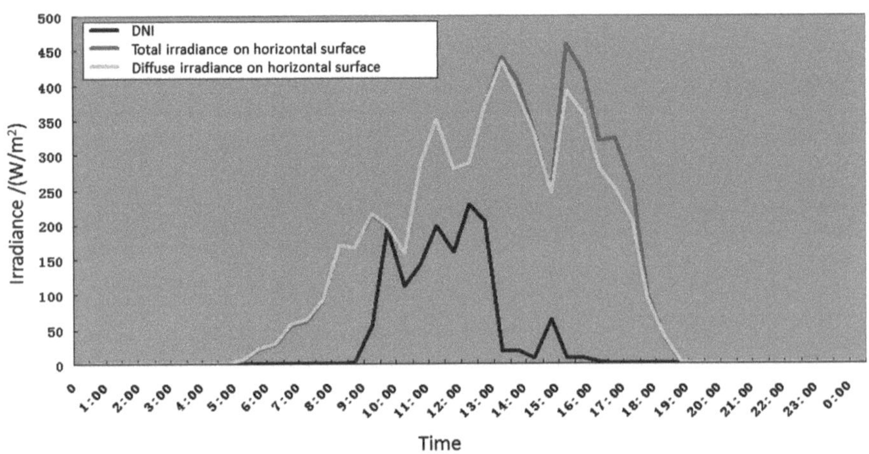

FIGURE 2.15 Relationship between global/diffuse irradiance on the horizontal surface and direct normal irradiance during a cloudy day [15].

2.8 VARIOUS SPECIAL CLIMATE CONDITIONS IN THE PLANT AREA

During site selection, except for DNI determination, the following parameters shall also be provided to the design unit and the equipment manufacturer for reference in equipment design and type selection. The mean monthly parameter values since meteorological records began, shall be recorded in Tables 2.18 and 2.19.

TABLE 2.18 Mean Value Statistics of Monthly Meteorological Factors

Month	**1**	**2**	**3**	**4**	**5**	**6**	**7**	**8**	**9**	**10**	**11**	**12**	**Annual mean**
Item													
Mean temperature/°C													
Mean relative humidity/%													
Mean precipitation/mm													
Mean evaporation/mm													
Mean wind speed/(m/s)													
Mean sunshine duration/h													

TABLE 2.19 Annual Extreme Value Statistics for Meteorological Factors of Meteorological Station

Item	Value/°C	Appearance Date	Item	Value/m	Appearance Date
Extreme maximum temperature			Maximum frozen ground depth		
Extreme minimum temperature			Maximum snow accumulation depth		

2.8.1 Ambient Air Temperature

1. Multiyear mean temperature, including daily mean temperature, the month of annual maximum temperature, the mean temperature range, and the month of minimum temperature shall be recorded.
2. The extreme maximum temperature, which is the maximum temperature of the locality since meteorological records began.
3. Extreme minimum temperature refers to the appearance date, month, and year of the extreme minimum temperature.
4. The typical day temperature variation diagram by month shows the data used to understand the relationship between sunshine and temperature, which is of great significance in the calculation of collector heat losses.

2.8.2 Wind Speed

Although there are differences in the evaluated total amount of current land wind energy resources, the corresponding wind energy resource distribution in China is basically consistent; that is, the locations with abundant resources and the respective distribution features are basically the same. Major causes for the huge difference in total amounts are the evaluation methods, data sources, and levels applied by these studies, which differ from each other; the Second and Third National Wind Energy Resources General Surveys were based on observation data obtained from a meteorological station 10 m above the ground through statistical analysis, whereas numerical simulation results were obtained on the basis of a 50-m height; when the value was simply estimated on the basis of a 10-m height according to the land wind shear exponent, land wind forces corresponding to heights of 50 and 70 m above ground can be measured.

Areas with stronger wind are mainly distributed on the southeast coast and neighboring islands, Inner Mongolia, Xinjiang, and the Hexi Corridor of Gansu, as well as North China and partial areas in the Qinghai–Tibet Plateau. Furthermore, partial areas in Central China also have strong winds.

Wind in China is significantly influenced by region and season; northern areas have stronger wind than southern areas; winter and spring are under the influences of Siberian high pressure and exhibit wind forces that are stronger than those of summer and autumn.

1. Wind direction variations for various months can be produced to facilitate concentrator load design by filling out Table 2.20.
   From Table 2.20, a wind direction diagram can be drawn.
2. Maximum wind speed and wind direction data from neighboring meteorological stations by year, as well as maximum wind speed, mean wind speed, and annual dominant wind direction on a 10-m height basis for a period of 50 years, are recorded.

2.8.3 Precipitation Parameters

1. Annual mean precipitation is the mean height of liquid or solid (after melting) water per square meter every year falling from the sky to the ground, which can be obtained from the local meteorological station.
   Precipitation within unit time is referred to as the precipitation intensity. Normally the unit time is 10 min, 1 h, or 1 day.
2. Annual mean evaporation is the amount of water that spreads into the air through evaporation within a year, which is normally scaled by evaporated water layer thickness (mm).

2.8.4 Disastrous Weather Phenomena and Respective Parameters

1. Hail. Annual mean hailing days, diameter of hailstone, maximum diameter, and maximum annual hailing days.
2. Sandstorm. Annual mean frequency of sandstorm, which can be measured by days and used as a reference for solar shade calculation and concentrator design. Furthermore, the travel rate of sandstorms shall also be provided.

TABLE 2.20 Wind Direction Frequency (%)

Wind Direction	N	NNE	NE	ENE	E	ESE	SE	SSE	S	SSW	SW	WSW	W	WNW	NW	NNW	C
Time																	
Annual																	
Summer																	
Winter																	

3. Sand blow. Monthly mean sand-blowing days, the minimum sand-blowing month, and the no-sand-blowing month by season and year. These data can be used for references in mirror surface cleaning.
4. Rainfall. Annual mean precipitation is the mean value obtained by dividing the sum of multiyear precipitations at a certain location with the number of years or the mean value of annual precipitation measured in various observation sites at a certain location, which also serves as a major measurement index for a specific local climate.
5. Snowfall. Annual mean snowfall is the mean value obtained by dividing the sum of multiyear snowfall at a certain location with the number of years or the mean value of annual snowfall measured in various observation sites at a certain location, which also serves as a major measurement index for a specific local climate.
6. Thunderstorms. Thunderstorms can be used as a reference for the lightning protection design of the concentrator. The mean number of thunderstorm days in a year obtained through multiple years of observation is referred to as the mean number of thunderstorm days in a year, in which a day with at least one instance of audible thunder is recorded as one thunderstorm day.
7. Overcast conditions. The number of longest continuous overcast days and dates of occurrence have significant reference values for thermal storage system design. In Changsha of Hunan, during 1440 h from January to February 2012, sunshine duration was only 38.6 h. The city encountered its most severe continuous overcast condition in 60 years, with more than 50 overcast days. Such a location is obviously not suitable for establishing an CSP plant.
8. Freeze. Moist soil freezes when the temperature drops to 0°C. This is called frozen ground in meteorology. The maximum frozen ground depth refers to the maximum value of frozen ground depth over a year, which can be used as a reference for foundation and thermal pipeline design. In the urban area of Beijing, the maximum frozen ground depth approximates 85 cm and exceeds 100 cm in Badaling, is about 103 cm in Lanzhou, and is 200 cm in Harbin. An alternate concept is standard frozen ground depth, which refers to the mean value of frozen ground depth over years.
9. Snow accumulation. On flat and broad ground, snow accumulation depth refers to the vertical depth of the snow layer; maximum snow accumulation depth refers to the maximum depth of snow accumulation at a certain location over years, which can be used as a reference in concentrator load design. If the concentrator is equipped with snow-avoidance control, it is not necessary to consider the respective load during heliostat and parabolic trough concentrator design. Figs. 2.16 and 2.17 indicate the condition of the concentrator after a heavy snow on November 5, 2012.

FIGURE 2.16 Surface of heliostat after Blizzard (Badaling, Beijing).

FIGURE 2.17 Parabolic dish concentrator and solar furnace after Blizzard (Badaling, Beijing).

2.8.5 Designed Wind Speed and Ambient Air Temperature

Wind and ambient air temperature are major indices that influence effective power generating hours and the efficiency of CSP generation and the reliability of equipment operation. Currently, working wind speed during CSP generation concentrator design normally has a value of 14 m/s (force 6 wind). The working ambient air temperature of a power plant built in the Gobi deserts of China falls in the range of −30 to 40°C.

As western areas of China have greater wind speeds, determining the maximum working wind speed of a concentrator involves a major factor influencing primary investment and system running time. Along with increases in maximum working wind speed, concentrator costs increase;

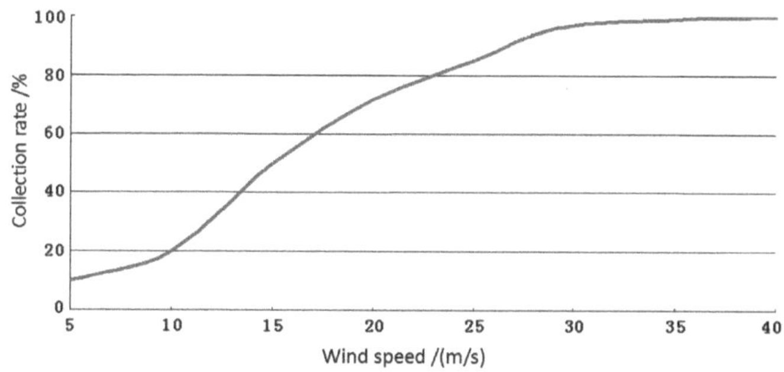

FIGURE 2.18 Relationship between maximum working wind speed of concentrator and solar energy collection rate.

however, this also extends system running time, and thus more power will be generated by the power plant. In this case, price shall be calculated based on local wind speed conditions.

As shown in Fig. 2.18, the longitudinal coordinate is the fraction (also known as the collection rate) of time available for solar utilization. The product of this fraction and annual global solar radiation is equivalent to the actual available solar resource while considering wind speed. For example, assuming the annual total solar DNI at a certain location equals 2100 kWh/m^2:

1. When the maximum working wind speed of the concentrator is 5 m/s as shown in the curve, the proportion = 10%.
   The annual total available solar direct normal irradiation of the power plant equals 2100 $kWh/m^2 \times 10\%$, namely 210 kWh/m^2.
2. When the maximum working wind speed of the concentrator is 15 m/s as shown in the curve, the proportion = 50%.
   The annual total available solar direct normal irradiation of the power plant equals 2100 $kWh/m^2 \times 50\%$, namely 1050 kWh/m^2.
3. When the maximum working wind speed of the concentrator is 22 m/s as shown in the curve, the proportion = 75%.
   The annual total available solar direct normal irradiation of the power plant equals 2100 $kWh/m^2 \times 75\%$, namely 1575 kWh/m^2.

2.9 MEASURING INSTRUMENT

2.9.1 Global Solar Radiation Meter

1. Structure and constitution of the instrument. An global solar radiation meter (also known as pyranometer) is an instrument for measuring annual solar radiation and consists of a solar irradiance sensor,

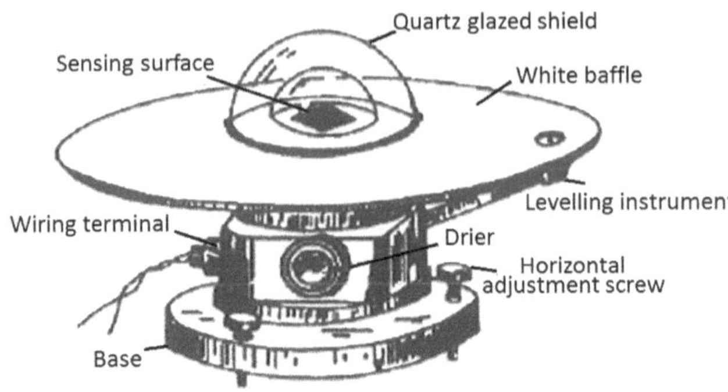

FIGURE 2.19 Constitution of the pyranometer.

a sealed glazed shield (quartz glazed shield), a level, and electronic and mechanical accessories (refer to Fig. 2.19).

The sensor consists of a sensing surface and thermopile. The blackened sensing surface is normally circular or square. The thermopile is made of constantan or plated-copper constantan, whereas the black-and-white sensing surface is made of black-and-white metal sheets that measure the lower-level thermopile thermoelectromotive force by utilizing the different absorption rates of black-and-white metal sheets and converting them into irradiance. The sensitivity of the instrument is 7–14 μV m^2/W with response time ≤ 60 s (99% response). The annual stability is superior to 5%. As for cosine response indices, solar altitudes of 10 and 30 degrees result in cosine response errors that are less than 10% and 5% separately. The all-black glazed shield is made of hemispherical double-layered quartz glass, which not only is windproof but also can penetrate short-wavelength radiation within a range of 0.29–3.05 μm with a constant transparency of nearly 0.9. A double-layered shield is used to prevent the outer shield from being influenced by infrared radiation and to eliminate measurement errors.

Accessories include the machine body, dryer, white baffle, base, level, and wiring terminal. In addition, a metal cover protects the glazed shield (also called the protective shield). The dryer is filled with the drying agent (silica gel) and connected with the glazed shield to keep dried air inside the shield. The while baffle is used to prevent solar radiation from heating the lower section of the machine body as well as prevent the sensing surface from being influenced by radiation below the horizontal surface of the instrument. The base is designed with fixing screw holes for instrument installation and three screws for adjusting the sensing surface on the horizontal level.

2. Installation and maintenance of the instrument first involves firmly mounting the solar radiation meter on a specific stand. The stand is a piece of metal pipe or wooden post fixed on the top with a metal sheet or wooden plate slightly larger than the base of the global radiation meter. The stand is more than 1.50 m from the ground to avoid the influences of ground radiation. The lower section of the stand shall be firmly buried in the ground and prevent sinking or deforming for a long period even while under severe shock vibration (such as a heavy wind), and shall not change the horizontality of the instrument.
   During installation, the white baffle of the global radiation meter shall be dismantled and the global radiation meter shall be mounted on the stand while keeping the wiring terminal of the instrument facing northward. The instrument shall be fixed on the stand with three screws (rustless materials are best); if metal plates are used for the stand, three holes shall be drilled in advance and the instrument shall be fixed by bolts. Then the three screws on the base can be adjusted by utilizing the attached leveling instrument to keep the sensing surface of the global radiation meter horizontal before finally mounting the white baffle.
   After the instrument is installed, the wiring terminal and recorder shall be connected by wires (please pay attention to the positive and negative electrodes during connection). A wiring terminal normally has three lead-out lines, with one piece connected to the machine body and the shielding layer of the cables to avoid interference and inductive lightning strikes.
3. The instrument is used for observing global radiation. Before sunrise, the metal cover shall be opened so that the radiation meter starts to function, and the recorder automatically displays instantaneous values and cumulative total of global radiation. After sunset, observation shall be stopped and the instrument shall be covered. In cases of no rainfall or when no possible damage can occur to the instrument during the night, the global radiation meter and the regular meter may remain uncovered.

Opening and closing of the metal cover shall be handled with special care. As the quartz glazed shield is expensive, heavy, and fragile, the metal cover shall be closed with slight force without touching the glazed shield. In winter, water drips and other coagulations on the glazed shield and its surrounding areas shall be wiped out before closing the metal cover to avoid freezing. Once the metal cover becomes frozen it becomes difficult to remove, so it is suggested that frozen parts are melted using a blower or by adopting other methods before removing the metal cover; the operator shall be careful and strive to not cause any damage to the glazed shield.

The global radiation meter shall be inspected and serviced at least once in the morning and once in the afternoon:

1. Determine whether the instrument is level and whether the sensing surface and glazed shield are intact.
2. Determine whether the instrument is clean. If any dust, frost, mist, snow, or rain falls onto the glazed shield, it shall be completely cleaned in a timely manner using a lens brush or a chamois leather rag, mainly focusing on not scratching or abrading the glass.
3. The glazed shield shall be kept away from water; there shall be no vapor condensate inside the shield. When inspecting whether the silica gel in the dryer has been dampened (changing from blue to red or white), the dampened silica gel shall be replaced in a timely manner and can be reused after baking in an oven until the color turns back to blue.
4. A global radiation meter has excellent waterproof performance. When used for a short period or under limited precipitation, the instrument may be uncovered. However, once there is heavy rain (snow, hail, etc.), rainfall, or snowfall for a long period, to protect the instrument the observer shall close the cover in a timely manner based on specific circumstances and open the cover once the rain stops.

In the event of strong thunderstorms or other adverse weather, the cover shall be closed and patrol shall be intensified; when problems are discovered, they shall be handled in time. As the global radiation meter is applied outdoors, it must be qualified through a verification certificate issued by the measuring department before application. Normally, the verification period is 2 years.

2.9.2 Solar Direct Normal Irradiance Meter

1. The structure and constitution of the instrument (pyrheliometer) consists of the solar position tracker, solar radiation sensing element, solar radiation collimated light tube and sighting device as well as the connector socket and incoming light protective shield. The sensing element consists of a sensing surface and thermopile. When the sensing surface receives solar radiation, the thermopile produces thermoelectromotive force in proportion to the received radiation.
   Performance indices:
   Response time (95% response) < 20 s
   Allowable range of sensitivity: 7–14 μV m^2/W
   Zero offset (response to ambient air temperature variation of 5 K/h): ± 4 W/m^2.
   Annual stability: $\pm 1.0\%$
   Temperature response (with an interval of 50K): $\pm 2\%$
   Nonlinearity: $\pm 0.5\%$

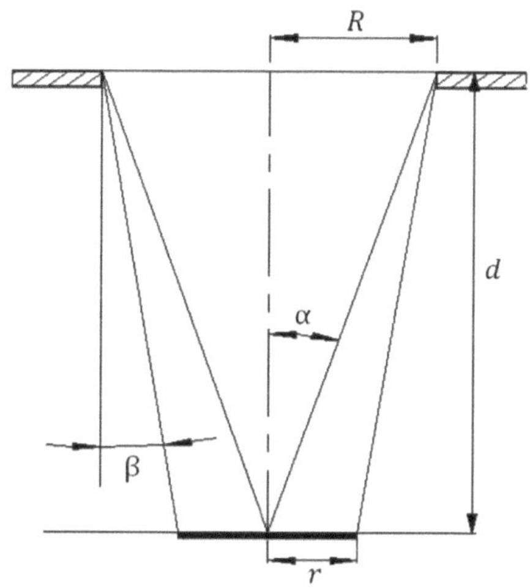

FIGURE 2.20 Geometrical dimensions of incoming light tubes of a pyrheliometer.

Tilt response: ±0.5%
Environmental conditions:
Temperature: −40 to +50°C
Relative humidity: 0%–100%

2. Installation and maintenance of the instrument. A collimated light tube is a metal cylinder that consists of an interior and an exterior tube. There is an annular diaphragm inside the tube that has been boiled black; the sealed opening is available for the transmission of direct solar radiation with a wavelength of 0.3–3.0 μm. There is an aperture at the front and an optical target at the rear of the exterior tube; the line that connects both of them is parallel to the axis of the light tube. The aperture of the collimated light tube is defined by semiopen angle α and oblique angle β (refer to Fig. 2.20). Normally, for a working-level solar direct normal radiation meter, $\alpha = 2.5°$, $\beta = 1°$.

2.9.3 Atmospheric Transmittance Meter

Energy attenuation of radiation in the atmosphere is normally indicated using atmospheric transmittance. Atmospheric transmittance can be defined as a multifactor function related to the distance of detecting path, angle of view, wavelength of radiation, atmospheric pressure, air temperature, atmospheric composition, etc. Atmospheric transmittance attenuation is caused by the absorption and scattering of atmospheric molecules and suspended aerosol particles in the atmosphere; atmospheric absorption attenuation is mainly caused by the intensive radiation absorption of H_2O, CO_2, and O_3 in the atmosphere; and atmospheric scattering attenuation is mainly caused by the scattering effect of atmospheric molecules and suspended aerosol particles in the atmosphere (such as mist and raindrops) on optical waves. The visible light band (0.38–0.77 μm) is influenced by the atmospheric window in which atmospheric absorption is insignificant. Radiation attenuation within this band is mainly caused by the scattering effect of atmospheric molecules and aerosol particles.

1. The structure and constitution of the instrument consists of the optical transmitter-receiver-integrated machine, electrical box, and software. It is used to measure the transmittance of light under various circumstances, road visibility under different climatic conditions, and the tunnel smoke concentration inspection index.
2. Installation and maintenance of the instrument is quite simple; it only needs to be placed at the location to be measured.
   Considering that the testing environment might be adverse, some instruments have been equipped with internal memories and some even with a data-processing function for indicating either measured values at different time points or mean value within a certain period.

Data display: Light transmission rate (%), smoke concentration (L/m), visibility (m).

Testing error: ±2%, light transmission rate.

Ambient air temperature: −50 to 65°C; ambient humidity: 0%−100% (relative humidity); whole-machine mass: about 10 kg.

2.10 GLOBAL DIRECT NORMAL IRRADIANCE DISTRIBUTIONS

2.10.1 Site Adaptation of Satellite-Based Direct Normal Irradiance

Satellite-derived solar radiation data can be used where ground-based measurements are not available. Satellites measure reflected radiation from Earth's surface in several wavelength bands. Known albedo values by location, complex model, and algorithm can be used to determine global, diffuse, and direct-beam irradiance components. Site adaptation of satellite-based DNI is necessary for bankable DNI data in CSP projects. One of the best–known data sets is provided by NASA (http://eosweb.larc.nasa.gov/sse/). Another popular satellite-based DNI provider is Solargis (http://solargis.com). Some introductions of Solargis is discussed in the webpage (https://doi.org/10.1063/1.4949234).

2.10.2 Global and District Direct Normal Solar Irradiation Distributions

Many Sunbelt countries within 40 degrees of the equator are rich in solar resources, and the long-term average annual total direct normal solar irradiation is mainly distributed in five regions of the Sunbelt. These regions are the Mediterranean and Northern Africa, South Africa, China and India, Latin America, and Australia. Fig. 2.21 shows the recent global map of satellite-based annual DNI by Solargis.

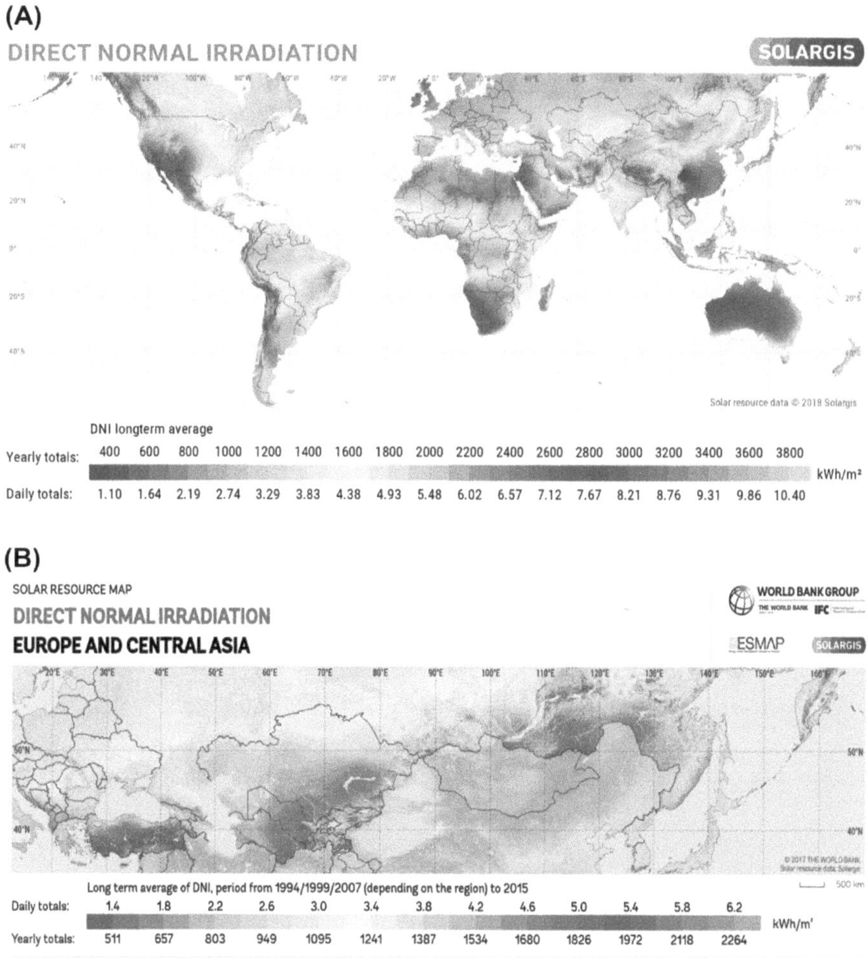

FIGURE 2.21 Global and district annual direct normal irradiation (DNI) distribution: (A) global annual DNI, (B) Europe and Central Asia annual DNI, (C) Latin America and Caribbean annual DNI, (D) Middle East and North Africa annual DNI, (E) South Asia annual DNI. *From SolarGis company website (https://solargis.com/maps-and-gis-data/download/)* © *2017 The World Bank, Solar resource data: Solargis*

(C)

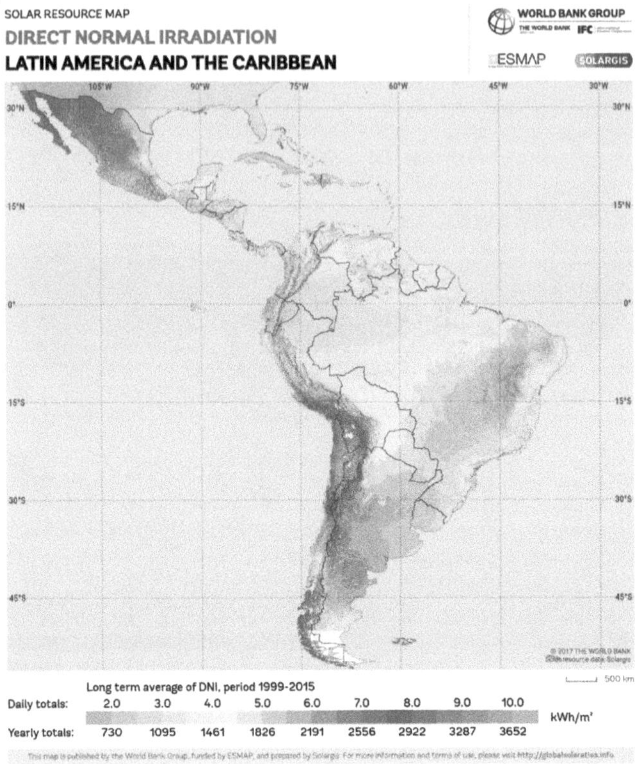

(D)

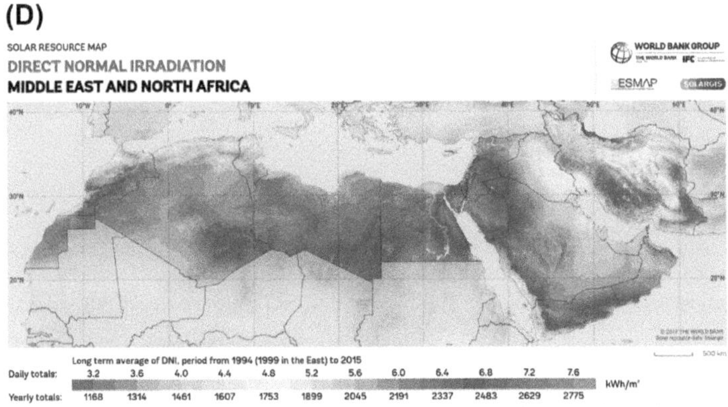

FIGURE 2.21 cont'd.

(E)

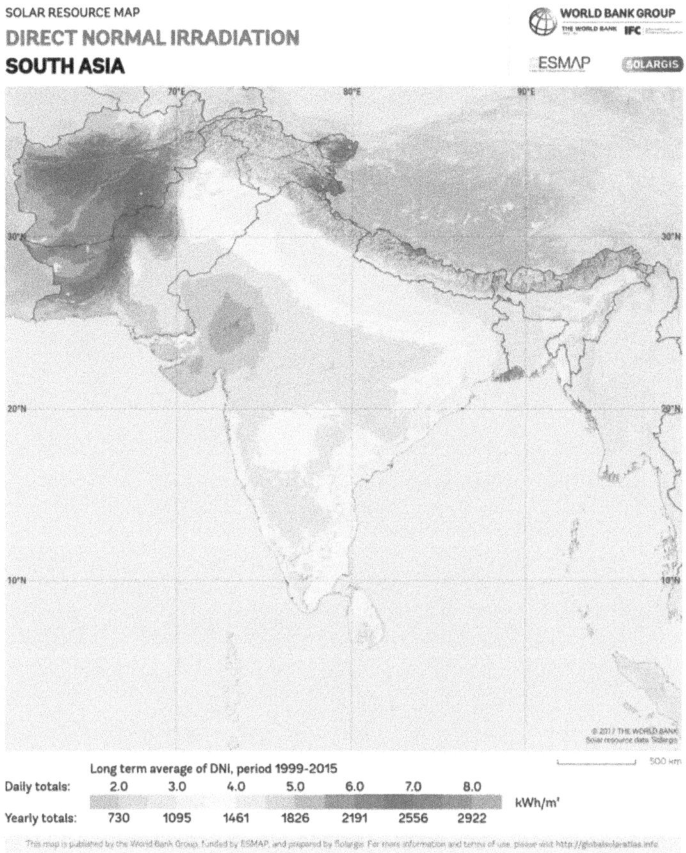

FIGURE 2.21 cont'd.

[A] Suri M., Cebecauer T., Requirements and standards for bankable DNI data products in CSP projects. In: Proceedings of the SolarPACES conference, Granada, Spain, September, 20–23 2011.

[B] Cebecauer T., Suri M., Gueymard C., Uncertainty sources in satellite-derived Direct Normal Irradiance: How can prediction accuracy be improved globally? In: Proceedings of the SolarPACES conference, Granada, Spain, September, 20–23 2011.

[C] Tomas Cebecauer and Marcel Suri, Site-adaptation of satellite-based DNI and GHI time series: overview and Solargis approach, SolarPaces 2015. In: AIP conference proceedings 1734; 2016, 150002. https://doi.org/10.1063/1.4949234.

CHAPTER

3

General Design of a Solar Thermal Power Plant

3.1 POWER PLANT DESIGN POINT

3.1.1 Significance of Design Point

Design point is a primary parameter in concentrating solar power (CSP, also known as solar thermal power) plant design and can be referred to when defining the area of the concentration field, thermal receiver capacity, thermal storage capacity, the rated capacity of the power generator unit, and the power plant's annual power output, as well as other key parameters such as the efficiencies of various equipment. The definition of design point is explained in detail in the chapter 1.2.3 of this book.

3.1.2 Calculation Examples of Applying the Design Point

Design point is related to the longitude and latitude of the power plant, annual mean ambient air temperature, local annual mean wind speed, annual mean solar direct normal irradiance (DNI), and other natural conditions. Normally, CSP plant design requires taking the solar time midday on typical days, such as the spring equinox, autumnal equinox, or summer solstice, during a typical year as the time mark for the design point. The Badaling CSP plant in China has a power generation capacity of 1 MW, which means that its power generation capacity at the design point is 1 MW.

Energy balances for various system units at the design point are shown in Table 3.1.

Design of Solar Thermal Power Plants
https://doi.org/10.1016/B978-0-12-815613-1.00003-1

TABLE 3.1 Energy Balances at Design Point for the Badaling Power Tower Plant

S/N	Item	Input Power (kW)	Loss (kW)		Residual (kW)
1	Solar DNI Input in the heliostat field at the design point	10,000			
2	Concentration process: Shading and blocking, cosine effect, specular reflectance, atmospheric transmission loss, receiver intercept factor	10,000	3500		6500
3	Thermal receiver process: Reflection, convection, radiation, thermal conduction	6500	1300		5200
4	Thermal storage process: Thermal dissipation loss, thermal charging and discharging loss				(Not used)
5	Transmission process	5200	90		5110
6	Stream turbine working: Steam thermodynamic cycle loss, thermal dissipation	5110	4045		1065
7	Self-dissipation loss: Consumption and loss of auxiliary system (rated operational conditions)	1065	Steam turbine 20 Concentration field and communication 20 DCS 5 Cooling tower pump 60 Others (high pressure feed pump, industrial water pump, feed booster pump) 50	155	910
8	Power plant net output				910

3.2 HELIOSTAT FIELD EFFICIENCY ANALYSIS FOR POWER PLANTS

Under conditions corresponding to the design point, the energy flow of various system parts shall be balanced and calculated by applying the calculation method for the Yanqing (Badaling) 1 MW power plant as shown in Table 3.1. According to Table 3.1, solar concentration losses in the heliostat field rank second-largest among all losses and are caused primarily by cosine and interception errors. Cosine loss can be determined from the relationship between solar position and the concentrating field. Software currently available for calculating concentrating field efficiency for a solar tower power plant includes WinDELSOL, heliostat optical code (HOC) [18,19], and Fiat Lux.

The optical efficiency of the solar tower concentrator is important to the thermal performance of the entire solar tower collector. The aperture plane of a cavity receiver, and the (inner or external) absorbing surface of any central receiver, are key interfaces of energy flux. So it is necessary to simulate and analyze the concentrated time-changing solar flux density distributions on the flat or curved receiving surface of the collector with the main optical errors considered. The transient concentrated solar flux on the receiving surface is the superimposition of the flux density distributions of all normal working heliostats in the field.

3.2.1 Brief Introduction to Heliostat Optical Code for Solar Towers

HOC is a solar tower optics code developed by the Institute of Electrical Engineering, Chinese Academy of Sciences (IEE-CAS) that can be used for the layout and optimization design of heliostat fields and for optical performance analysis of partial and full heliostat fields on receiver apertures. HOC was developed with MATLAB and can be used for both cavity and external cylinder receivers. HOC is based on new backward ray tracing (BRT) and shading-and-blocking methods to simulate the transient concentrated solar flux on the receiving surface for a solar tower power plant [18,19].

3.2.2 Algorithmic Principles of Heliostat Optical Code

HOC uses the BRT method combined with the lumped effective solar cone to simulate the flux density map on the receiving surface. For BRT, bundles of rays are launched at the receiving-surface points of interest, strike directly on the valid cell centers among the uniformly sampled mirror cell centers in the mirror surface of the heliostats, and are then directed to the effective solar cone around the incident sun beam direction after reflection. All optical errors are convoluted into the effective solar cone. The brightness distribution of the effective solar cone is here assumed to be circular and Gaussian. The mirror curvature can be adequately formulated using a certain number of local normal vectors at the mirror cell centers of the heliostat.

The principle of the BRT method coupled with the effective sun-shape (solar cone brightness distribution) is schematically illustrated by Fig. 3.1. In Fig. 3.1, (O; North-East-Height) is the global coordinate system, $\vec{S}_c$ is the unit solar vector, $\vec{w}$ is the target surface normal at the target surface center T_0, (u_j, v_j, w_j) are the coordinates of point T_j in the target surface coordinate system $\left[\mathbf{T_0};\ \vec{u}-\vec{v}-\vec{w}\right]$, $\vec{p}_j$ is the unit normal vector of target surface at point T_j, (x_i, y_i, z_i) are the coordinates of the center point M_i of the ith mirror surface element in the mirror surface coordinate system $\left[\mathbf{M_0};\ \vec{x}-\vec{y}-\vec{z}\right]$, $\vec{n}_i$ is the mirror surface normal at the ith mirror point M_i, $-\vec{r}_i$ is the unit vector of the backward ray from target point T_j to mirror point M_i, $\vec{S}_i$ is the reflection unit vector for $-\vec{r}_i$ after being reflected at point M_i, and the radial brightness distribution of the effective solar cone can be denoted as $\phi_{eff}(\cdot)$ in kW/(m^2·sr) integrated with $\phi_{eff}(\cdot)$ over the full solid angle range equal to the value of DNI in kW/m^2. Following the denotations in Fig. 3.1, the concentrated solar flux density at the target surface point $T_j(u_j, v_j, w_j)$ is $F(u_j, v_j, w_j)$. Here, $\Delta x_i \Delta y_i$ and $\rho_{m,i}$ are the area and specular reflectance of the ith mirror surface element, L_i is the distance from target point $T_j(u_j, v_j, w_j)$ to mirror surface

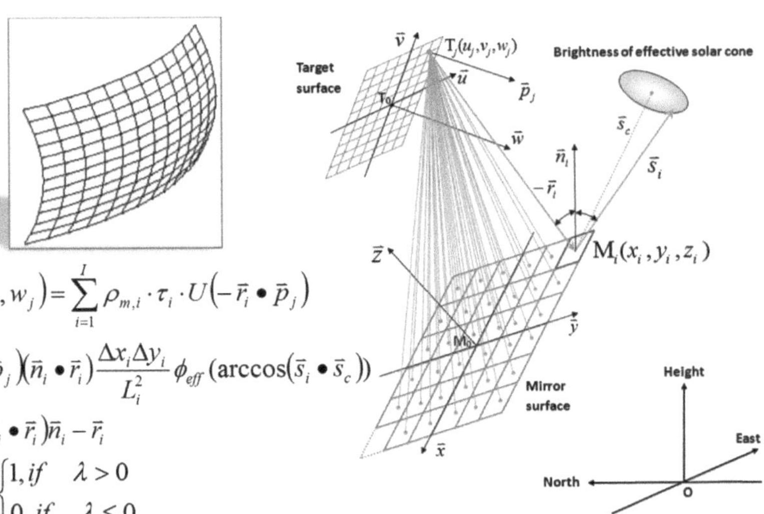

$$
\begin{cases}
F(u_j,v_j,w_j)=\sum_{i=1}^{I}\rho_{m,i}\cdot\tau_i\cdot U(-\vec{r}_i\bullet\vec{p}_j) \\
\cdot(-\vec{r}_i\bullet\vec{p}_j)(\vec{n}_i\bullet\vec{r}_i)\dfrac{\Delta x_i\Delta y_i}{L_i^2}\phi_{eff}(\arccos(\vec{s}_i\bullet\vec{s}_c)) \\
\vec{s}_i=2(\vec{n}_i\bullet\vec{r}_i)\vec{n}_i-\vec{r}_i \\
U(\lambda)=\begin{cases}1, if & \lambda>0\\ 0, if & \lambda\le 0\end{cases}
\end{cases}
$$

FIGURE 3.1 Principle of backward ray tracing to simulate the concentrated flux densities on the target surface.

point $M_i(x_i, y_i, z_i)$, and τ_i is the absorptivity (absorption ratio) of the receiving surface. In Fig. 3.1, the grid element of the mirror surface at the lower corner marked with dotted lines is invalid, possibly due to shading and blocking by neighboring heliostats or the solar tower.

Fig. 3.1 schematically shows the BRT method's ray trace principle on the mirror surface plane and the target plane for simulating concentrated solar flux density at target surface point $T_j(u_j, v_j, w_j)$, but the equation shown in Fig. 3.1 is simply the general expression for the BRT method, with mirror surface normal $\vec{n}_i$ changing with mirror point positions and the target surface normal $\vec{p}_j$ changing with target surface positions.

Since the shading-and-blocking computation is the most complicated, time-consuming, and important optical computing section for the concentrated solar flux map on the receiving-surface for a field of heliostats, we also developed a new shading-and-blocking method to compute the shading-and-blocking factors (efficiencies) of all heliostats given a time or sun position in the sky. The shading-and-blocking mirror region of a heliostat by neighboring heliostats and solar tower shading on the heliostat mirror are all computed using a flat-ground-plane platform; i.e., the mirror contours and the envelope cylinder of the tower are projected onto the horizontal ground plane along the sunbeam incident direction or in the directions of the reflection. If the shading projection of a sampled mirror point of the current heliostat is inside the shade cast by a neighboring heliostat or in the shade cast by the tower, that mirror point should be shaded from the incident sun beam.

Fig. 3.2 illustrates the new model for identifying the invalid shading-and-blocking mirror region of a heliostat by neighboring heliostats and also the solar tower. The shading-and-blocking mirror region of a heliostat by neighboring heliostats and also the solar tower shading on the heliostat mirror are all computed on a flat-ground-plane platform; i.e., projecting the mirror contours and envelope cylinder of the tower onto the horizontal ground plane along the sunbeam incident direction or along reflection directions. In Fig. 3.2, the shading-and-blocking projections on the horizontal ground plane of the heliostat and solar tower contours are drawn with dashed lines. Fig. 3.2A shows the orientations of all normal-working heliostats around the circle cylinder solar tower marked in red given a solar vector, whereas the current heliostat, and the incident and reflected solar rays by the heliostat surface center, are marked with green lines. In this imaginary case, the field ground slopes from north to south. Fig. 3.2B shows the shading projections for all heliostat contours (including the current heliostat marked in green) and the solar tower cylinder contour together with heliostat plan positions. Fig. 3.2C shows the reflection-blocking projections for all heliostat contours (including the current heliostat marked in green) together with the heliostat plan positions. Fig. 3.2D is the partial enlarged drawing of Fig. 3.2B, in which the

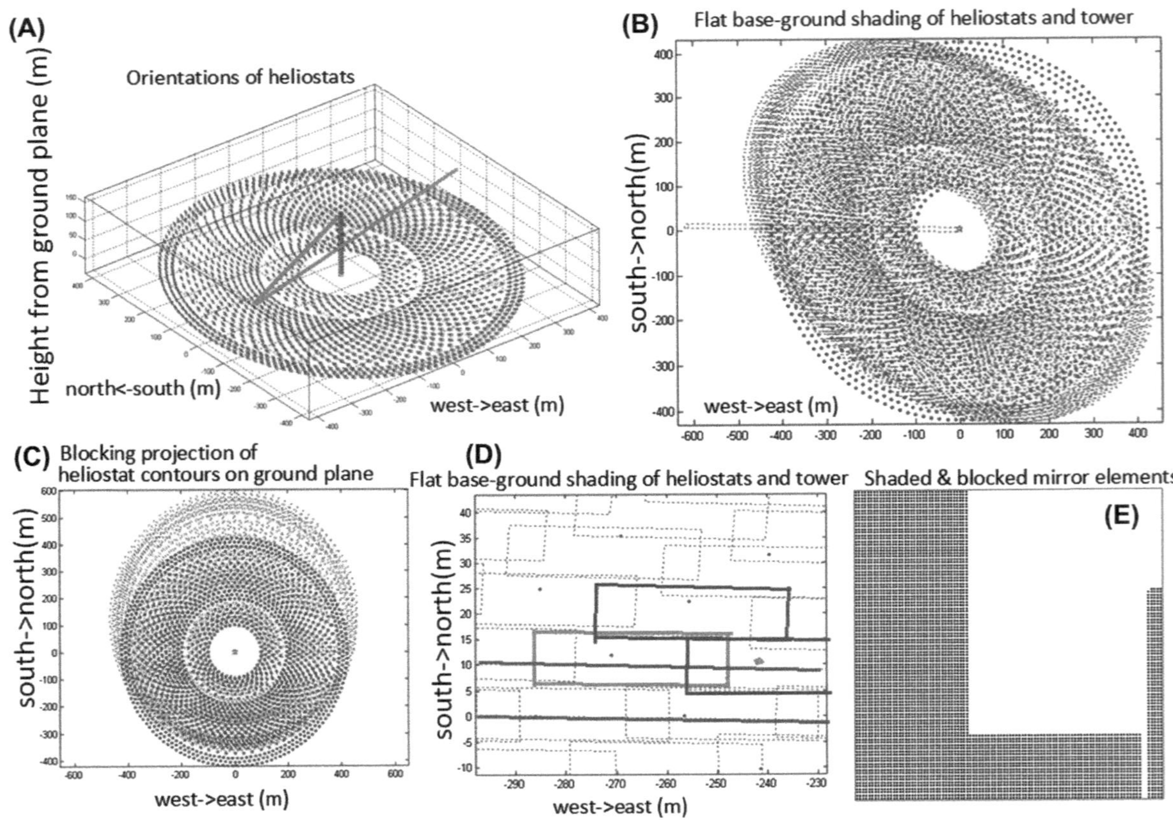

FIGURE 3.2 Illustration of the flat-ground-plane projection model for identifying a Heliostat's Shading-and-Blocking mirror region caused by neighboring heliostats and the solar tower shading region of the heliostat mirror.

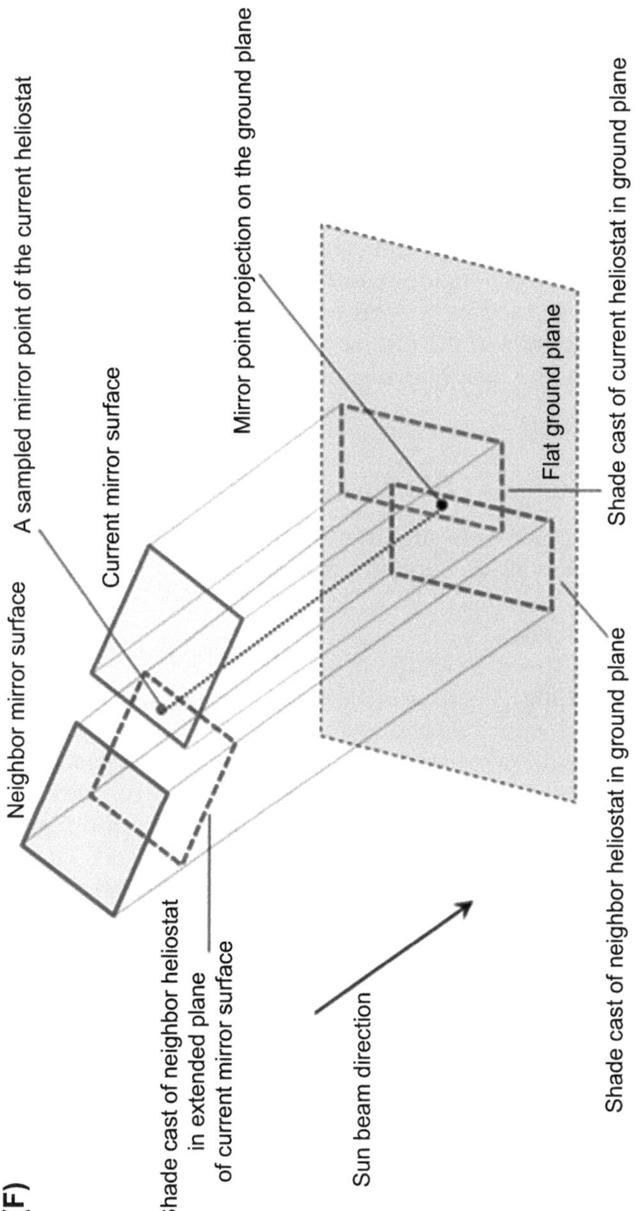

FIGURE 3.2 cont'd.

shadow of the current heliostat contour on the horizontal ground plane overlaps with the shadows from the two neighboring heliostats and the shadow of the solar tower. The current heliostat and its neighboring heliostats are marked by green dots and lines, and the tower is marked by red dots and lines. Fig. 3.2E marks the shaded mirror region of the current heliostat by neighboring heliostats and the solar tower.

Fig. 3.2F shows the geometric topological invariance of a mirror point of the current heliostat inside the shade cast by a neighboring heliostat under shading projection from the mirror surface plane to the flat ground plane. If the shading projection of a sampled mirror point of the current heliostat is shaded by a neighboring heliostat or the tower, the mirror point should be shaded from the incident sun beam, i.e., if a mirror point of the current heliostat is shaded by a neighboring heliostat, the shading projection of the mirror point is in the intersection of the shade regions of the current heliostat and the neighboring heliostat in the flat ground plane, as Fig. 3.2F shows.

3.2.3 Two Specific Versions of Heliostat Optical Code

HOC based on the BRT and new shading-and-blocking methods for a solar tower power plant was developed using MATLAB, as Fig. 3.3 shows. The following two versions of HOC are for the cavity receiver and cylinder receiver, respectively.

Fig. 3.3A–D show the specific version of HOC for the DAHAN solar tower power plant in Beijing. There are 100 azimuth-elevation tracking heliostats in the solar field, and the total tower is 118 m high. The mirror surface of the heliostats is 10 m wide and 10 m long; it is composed of 8 rows $\times$ 8 columns of square mirror facets, and each mirror facet is 1.25 m $\times$ 1.25 m. The target center is 74.5 m higher than the ground plane.

Fig. 3.3E–G show the cylinder central receiver solar field layout (CCRSFL) version of HOC. A CCRSFL adopts the traditional radial-staggered pattern layout for a heliostat field. In CCRSFL, given the parameters of tower height, heliostat size, and receiver size, the heliostats are initially arranged as a compact circular field with sufficient total mirror area; then the heliostat field grows gradually by increasing the radial spaces between adjacent heliostat rings during the optical optimization process. The optimizing objective function collects the maximum annual solar energy from the receiver aperture. After first-stage optimization of solar field layout, the solar field is trimmed by selecting the needed number of heliostats that have higher annual optical efficiencies. The trimmed heliostat field is optimized further through a second stage similar to the first. CCRSFL can efficiently simulate the concentrated solar flux map of heliostats on the receiving surface using only a pair of cross-

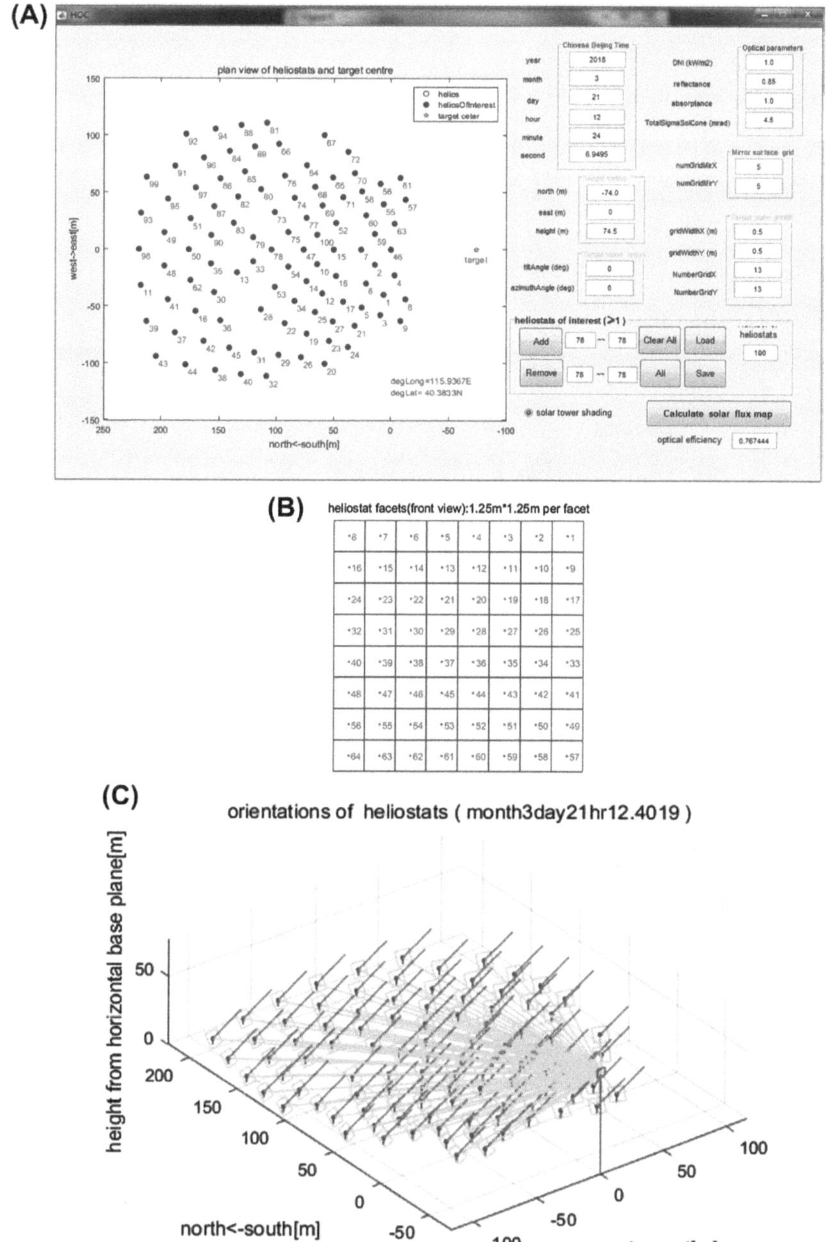

FIGURE 3.3 Two versions of heliostat optical code (HOC) for the cavity receiver and the cylinder receiver respectively: (A) interface of the specific version of HOC for the DAHAN solar tower power plant in Beijing with data loaded for all 100 heliostats, (B) arrangement of mirror facets for a 100-m^2 DAHAN heliostat, (C) light path drawing of all DAHAN heliostats at the solar noon of the spring equinox, (D) simulation of the concentrated solar flux image on the horizontally north facing target plane by all HAHAN heliostats at the solar noon of spring equinox, (E) interface of cylinder central receiver solar field layout version of HOC with field data loaded, (F) simulated solar flux density on the cylinder receiver surface from all heliostat fields with respect to given parameter settings, and (G) efficient HOC simulation of the concentrated solar flux map of heliostats on the receiving surface using only a pair of cross-flux-density profiles to reconstruct a flux density map of the elliptical Gaussian type.

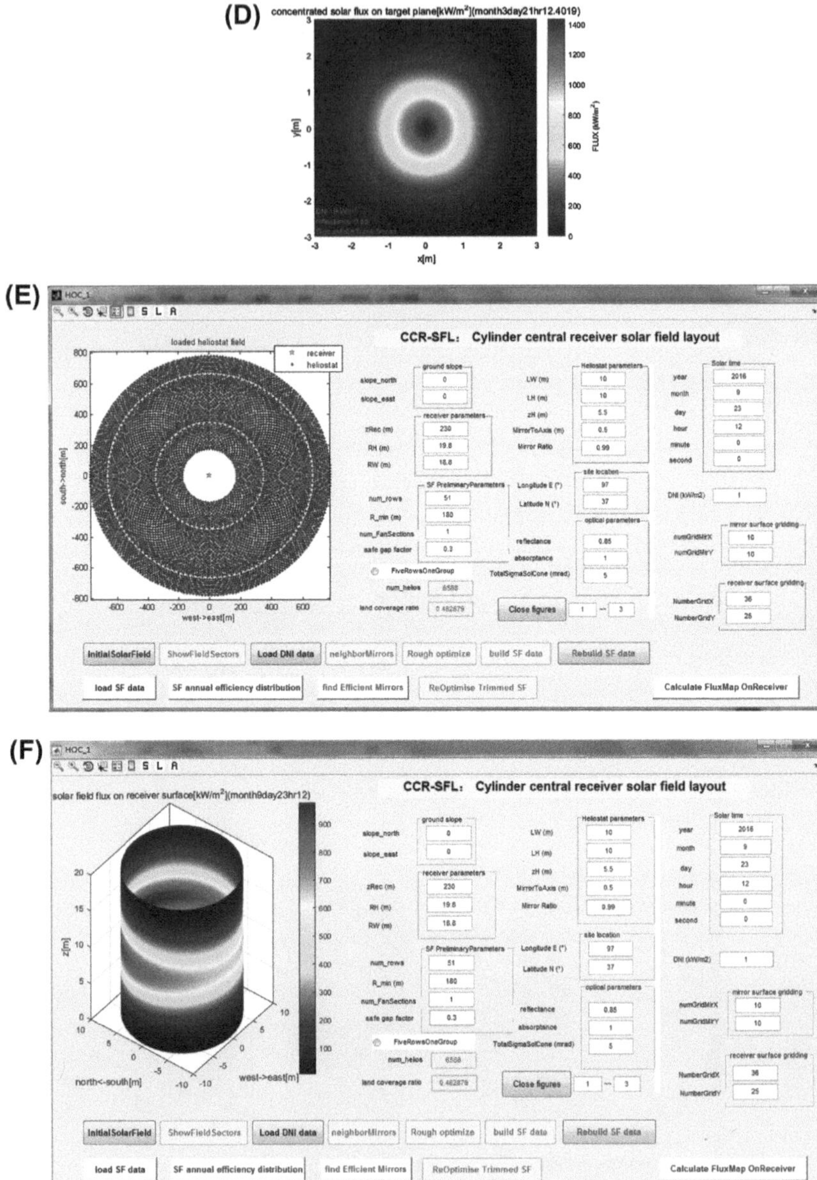

FIGURE 3.3 cont'd.

(G)

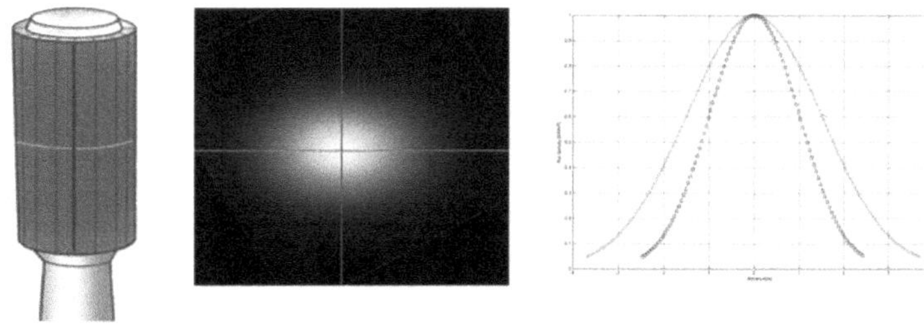

$$f(x,y)=I_0 \exp\left\{-\left[\frac{(x-x_0)^2}{2\sigma_x^2}+\frac{(y-y_0)^2}{2\sigma_y^2}\right]\right\}=\frac{1}{I_0}\times\left[I_0 \exp\left(-\frac{(x-x_0)^2}{2\sigma_x^2}\right)\right]\times\left[I_0 \exp\left(-\frac{(y-y_0)^2}{2\sigma_y^2}\right)\right]$$

$$f(x,y)=\frac{f(x,y_0)\times f(x_0,y)}{f(x_0,y_0)}=\frac{1}{f(x_0,y_0)}\times f_1(x)\times f_2(y)$$

FIGURE 3.3 cont'd.

flux-density profiles to reconstruct a flux density map not limited to the elliptical Gaussian type, as Fig. 3.3G shows. Fig. 3.3E and F show a typical calculation example of cylinder receiver flux density distribution at the solar noon of the autumnal equinox.

Fig. 3.3F shows the mathematical principle of flux map reconstruction using a pair of cross-flux-density profiles along the X and Y axial directions. $f(x, y)$ is the flux density function over the X−Y receiving surface, where (x_0, y_0) is the coordinate pair of the peak value, and σx and σy are the standard deviations along the X and Y axes. $I_0 = f(x_0, y_0)$ represents the peak flux value at point (x_0, y_0). $f(x, y_0)$ and $f(x_0, y)$ are the flux density profiles passing through point (x_0, y_0) along the X and Y axes, respectively. Thus HOC can efficiently simulate the concentrated solar flux map $f(x, y)$ on the receiving surface with just the flux density values of the X and Y flux-density profiles and then using them to reconstruct the flux density map.

3.2.4 Values of Specular Reflectance

The value of specular reflectance has a direct impact on energy calculations. Normally, all values are taken at the design point when mirrors are clean and fall in a range of 92%–94% [20].

However, when calculating annual energy for a power plant in an area with significant sand dust, the effect of dust accumulation on the mirror must be considered. Generally speaking, in a season with a large amount of dust, heliostat reflectance drops by 0.8% per day [20]. When functioning by facing upward to the sky, the reflectance of a parabolic trough

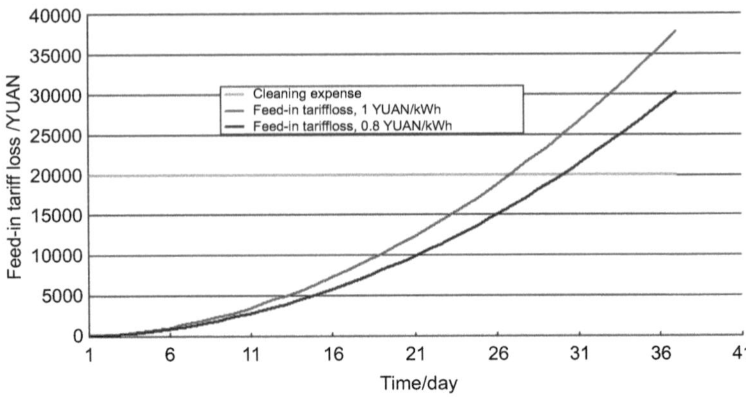

FIGURE 3.4 Defining the mirror cleaning time.

reflective surface may drop even faster. In the case of cleaning the mirror twice a month, it is more appropriate to take a value of about 83% as annual mean reflectance.

The ratio of cleaning times, energy loss, and paid cleaning expense shall be calculated. Currently, the cleaning expense for each square meter of mirror is about 2 yuan. For a heliostat concentration field of 10,000 square meters, the respective cleaning expense would be 20,000 yuan. The daily loss of reflectance of the mirror is 0.8%. When the expense of power loss exceeds the concentration field cleaning expense, the cleaning process shall start. Fig. 3.4 indicates the cleaning time defined by the intersection of the mirror cleaning expense and economic loss caused by power attenuation from dust accumulation loss.

Assuming the feed-in tariff of an CSP plant is 1 yuan/kWh, the cleaning time indicated in Fig. 3.4 approximates 27 days. Obviously this value changes according to the cleaning costs, daily dust accumulation status, and feed-in tariff of the power plant. After the mirrors are cleaned, reflectance bounces back to 94% from 72.4%.

Assuming the feed-in tariff of the CSP plant is 0.8 yuan/kWh, the cleaning time indicated in Fig. 3.4 approximates 30 days.

3.2.5 Atmospheric Transmittance Analysis

Atmospheric transmittance is related to the degree of local sand blow. Due to the short focal distance of a parabolic trough collector, it is not necessary to consider transmittance. However, for a large-scale power tower plant, this factor must be considered.

During transmission of optical waves in the atmosphere, absorption and scattering of atmospheric gas molecules and aerosols may result in energy attenuation of the solar beam, and the nonuniform air refractive index may result in the variation of amplitude and phase of the optical wave. When the power of the optical wave is large enough and lasts for an extremely short time, the nonlinearity effect may also influence the characteristics of solar beams.

The integrated effects of absorption and scattering results in attenuation of the radiant intensity of transmitted light. Partial energy of incident light is converted into other forms of energy through absorption (such as thermal), and partial energy deviates from its original direction due to scattering (namely spatial reallocation of radiation energy).

Monochromatic radiation with intensity is assumed to pass through a thin atmospheric layer with thickness dl. Without considering nonlinearity effects, optical intensity attenuation dI is proportional to I, namely

$$\frac{dI}{I} = \frac{I' - I}{I} = -\beta dl \tag{3.1}$$

Atmospheric transmittance can be obtained through integral calculus

$$T = \frac{I}{I_0} = \exp\left(- \int_0^L \beta dl \right) \tag{3.2}$$

$$T = \exp(-\beta L) \tag{3.3}$$

in which β is the atmospheric extinction coefficient, l/km, and L refers to the transmission distance of solar radiation, km.

This is Lambert's law for describing atmospheric attenuation, which manifests the exponential law of decreased optical intensity with increases in transmission distance.

As extinction coefficient β describes the influences of the two independent physical processes of absorption and scattering on the radiant intensity of transmitted light, β can be manifested as

$$\beta = k_m + \sigma_m + k_a + \sigma_a \tag{3.4}$$

in which k_m and σ_m are molecular absorption and scattering analysis coefficients separately, and k_a and σ_a are absorption and scattering coefficients of aerosols in the atmosphere.

Research on atmospheric attenuation can be summarized as studies of the four basic attenuation parameters mentioned above. When applying, the commonly used unit for β is either L/km or dB/km, and the conversion relationship is

$$\beta(\text{dB/km}) = 4.343 \times \beta(1/\text{km}) \tag{3.5}$$

1. Molecule absorption. Atmospheric molecules polarize under the influence of the optical wave electric field and perform forced vibrations on the frequency of incident radiation. Thus in order to overcome the internal resistance of atmospheric molecules, energy is consumed and takes the form of absorption of atmospheric molecules.

   Molecular absorption characteristics strongly rely on the frequency of radiation wavelength. The inherent absorption frequency of a molecule is determined by its internal motion modality. The internal motion of a polar molecule normally consists of the electronic motion within the molecule, vibration of atoms that compose the molecule, and rotation of molecules surrounding the center of mass. The respective resonant absorption frequencies correspond to the ultraviolet, visible light, near-infrared, intermediate-infrared, and far-infrared regions of the optical wave.

   Although N_2 and O_2 molecules take up the largest share of the atmosphere (about 90%), their visible light and infrared regions show hardly any absorption, and then only show significant absorption toward far-infrared and microwave bands. Thus within the visible light and near-infrared regions, the effects of absorption are normally not considered.

   Except for the above molecules, the atmosphere also contains He, Ar, Xe, O_3, Ne, and so on. All these molecules have considerable absorption spectral lines in the visible light and near-infrared regions; however, due to their insignificant amount in the atmosphere, the respective effect of absorption is normally also not considered. Only when the remaining attenuation factors are already quite weak above the sky is the absorption effect considered. For a tower power plant, the influences of molecular absorption are not considered.

   H_2O and CO_2 molecules also play a certain role in the absorption of solar radiation. In special cases, the H_2O molecule has a broad vibration–rotation and sole-rotation structure within the near-infrared region, and thus the H_2O molecule acts as the most important absorption molecule in visible light and near-infrared regions and serves as the main cause for optical attenuation of the atmosphere on clear days. The central wavelengths of some major absorption spectral lines of H_2O molecules are as follows: 0.72, 0.82, 0.93, 0.94, 1.13, 1.38, 1.46, 1.87, 2.66, 3.15, 6.26, 11.7, 12.6, 13.5, and 14.3 μm.

   For certain specific wavelengths, the atmosphere may present extremely intensive absorptions that optical wave can barely pass through. Solar radiation wavelength bands applied by CSP generation are mainly below 3 μm and greatly overlap with the above parts. Thus in the power tower plant with concentrators located far away from the receiver, influences of atmospheric humidity on radiation transmission shall be considered. It is suggested not to build up large-scale power tower plants in areas with high atmospheric humidity.

2. Molecule scattering. There is always the statistical deviation of partial density from the mean density in the atmosphere, which is known as the density fluctuation. It ruins the optical uniformity of the atmosphere; partial solar radiation light will transmit toward other directions, which leads to the scattering of the solar radiation in various directions.

   In visible light and near-infrared wave regions, radiation wavelength is always much greater than molecular dimensions. Under such conditions, scattering is referred to as the Rayleigh scattering. The intensity of Rayleigh scattering light is inversely proportional to the biquadratic value of wavelength λ. The empirical equation of Rayleigh scattering coefficient is:

$$\sigma_m = 0.827\, NA^3/\lambda^4 \tag{3.6}$$

   in which N refers to the quantity of molecules within a unit volume, cm^{-3}; A refers to the molecular scattering section, cm^2; λ refers to the length of the optical wave, cm. The bigger of wavelength λ, the weaker of the scattering effect; the smaller of λ, the stronger of the scattering effect.

   When encountering atmospheric molecules or aerosols particles, the optical wave will interact with them and once again emit the light (called wavelet) of the same frequency and less intensity with the incident light in all directions, which is referred to as the optical scattering. The wavelet is called the scattering radiation; atmospheric molecules and aerosol particles that receive the original incident light and emit wavelets are called scattering particle. When scales of scattering particles are much smaller than the wavelength of incident light (for example, scattering effect of atmospheric molecules on visible light), it is called molecular scattering or Rayleigh scattering, in which scattering lights have been distributed on an even and symmetric basis.

   As molecular scattering is inversely proportional to the biquadratic value of wavelength, the bigger of the λ, the weaker of the scattering effect; and the smaller of the λ, the stronger of the scattering effect, scattering effect of visible light is stronger than that of infrared light, and scattering effect of blue light is stronger than that of red light. In a clear sky, other particles are insignificant in content; thus Rayleigh scattering has been playing a dominant role.

In addition, as the scattering effect of blue light is the strongest, a clear sky is normally blue-colored.

3. Attenuation of atmospheric aerosols. Definition of atmospheric aerosols: There is a large amount of solid and liquid particles in the atmosphere with the size between 0.001 and 10 μm. They are mostly dusts, soot deposits, water droplets, sized salts, organic microbes, etc. As they are suspended in a gel state in the atmosphere, these particles are normally called atmospheric aerosols. As a major factor for influencing climate change, atmospheric aerosols have aroused widespread attention from the world scientific circle, which mainly include six categories of seven types of aerosol particles. Sand dust aerosols or so-called mineral aerosols are the main components of troposphere aerosols. China's sand dust aerosols mainly come from deserts in Xinjiang, Gansu, Inner Mongolia, Loess Plateau and other arid and semiarid areas. Recently, frequent sandstorms in North China have aroused broad attention from both home and abroad. Sandstorm has turned into a major environmental problem of the Earth. Sizes of atmospheric aerosol particles are distributed on a very complex basis, and are greatly influenced by climate variation as well. Maximum values of concentration and dimension (diameter) of atmospheric aerosol particles corresponding to different types of weather have been listed in Table 3.2.

   Optical wave attenuation of atmospheric aerosols includes aerosol scattering and absorption (refer to Fig. 3.5). Sizes of the

TABLE 3.2 Physical Parameters of Haze, Cloud, and Precipitation

Weather Type	N/cm^{-3}	$a_{max}/\mu m$	Type of Atmospheric Aerosols
Haze, M	100	3	Maritime or coastal aerosols
Haze, L	100	2	Territorial aerosols
Haze, H	100	0.6	High-altitude or stratosphere aerosols
Rain, M	100	3000	Drizzle or moderate rain
Rain, L	100	2000	Heavy rain
Hail, H	10	6000	Hail containing a large number of small particles
Cumulus cloud, C.1	100	15	Cumulus cloud or stratus cloud, mist

N refers to particle concentration, a_{max} represents the maximum diameter of particle.

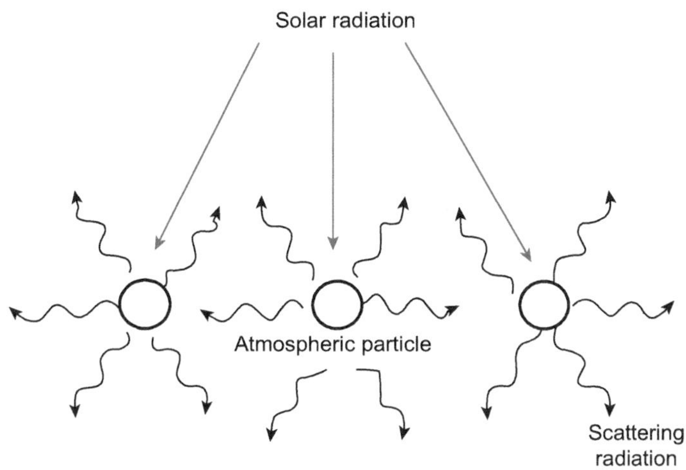

FIGURE 3.5 Atmospheric aerosol attenuation.

sand dust particles correspond to visible light and near-infrared light, which shall be handled by applying the Mie-Debye scattering theory. The Mie-Debye scattering, on the other hand, mainly relies on various characteristics, like the size of the scattering particle, density distribution and refractive index; the relationship between it and the wavelength is much less intensive than the relationship with Rayleigh scattering (which can be approximately deemed as irrelevant to wavelength).

According to the IPCC report, the mean life of sand dusts is about 4 days, the mean column-perpendicular integral content is about 32.2 mg/m^2, and the mass extinction coefficient when wavelength is 0.550 μm is 0.7 m^2/g. Based on the research analysis results, particles within a range of 0.1−1.0 μm are the main light extinction particles. NIU Shengjie et al. entered into desert sources (Tengger Desert, Badain Jaran Desert, Mu Us Desert) during April to May every year from 1996 to 1999, and conducted systematic observations toward the sand dust weather; by utilizing a plane to observe atmosphere aerosols in desert areas, they also systematically analyzed the microstructure of sand dust aerosols in Helan Mountains. ZHANG Wenyu et al. conducted the ground multiwave band solar radiation observations in Shapotou Station in Tengger Desert during April to September 2001. According to their research, optical thickness of local atmospheric aerosols varied significantly under different weather conditions. On April 6, 2000, during the catastrophic sandstorm, according to the analysis on sand dust particles, concentration of coarse particles ($d > 2$ μm) during the sandstorm was more than 20 times of that after the sandstorm; whereas the concentration of fine particles ($d < 2$ μm) is seven times that after the sandstorm. Optical thickness, extinction efficiency factor, scattering efficiency factor, and absorption efficiency factor of aerosols are major physical parameters to describe the conditions of atmospheric aerosols, as well as the crucial factors to evaluate the energy flow transmission of atmospheric concentration.

Extinction efficiency factor Qe refers to the ratio of the light extinction section to geometric section of the respective particle. According to the Mie-Debye scattering theory, extinction efficiency factor Qe has the important features of fluctuating along with the attenuation of sand dust particle radius r and approaching 2. According to Fig. 3.6, position of the primary main peak of Qe falls in a range of 0.1−1.0 μm. Along with the increase of wavelength λ, position of the primary main peak moves toward the direction of the increase of r. According to the measured data from both home and abroad, it is exactly within the above-mentioned main peak area, in which aerosol spectrum normally has sharp peak values. However, along with the increase of particle radius r, variation amplitude of extinction efficiency factor gradually decreases; when $r > 5.0$ μm, the extinction efficiency factor Qe gradually loses its sensitivity against the particle size r, and gradually approaches 2, that is, solar energy eliminated by large particles from the incident beam is exactly two times of the optical energy prevented by its section.

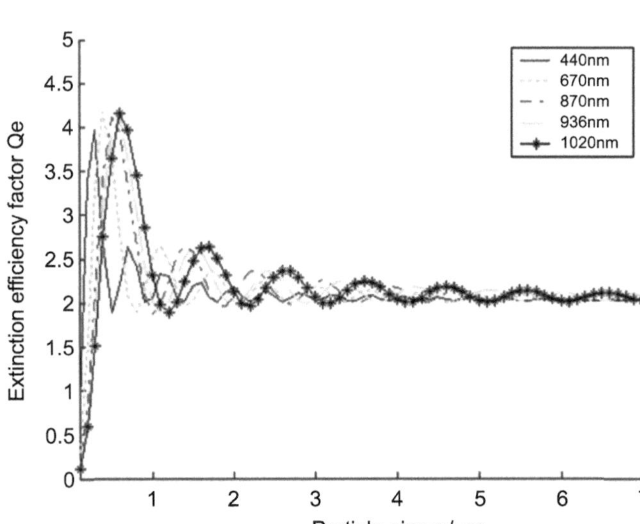

FIGURE 3.6 Variation of extinction efficiency factor Qe over particle size r.

Scattering effect factor Qs refers to the ratio of scattering section to the geometric section of the respective particle. Fig. 3.7 shows the variation rule of scattering efficiency factor Qs over particle size *r*. Position of the primary peak value of Qs also falls in a range of 0.1−1.0 μm. Along with the increase of wavelength λ, position of the primary peak value moves toward the direction of the increase of *r*. When the particle size *r* is very small, Qs is much smaller than 1, that is, the scattering ability of particles are much smaller than any energy projected onto its geometric section. Along with the increase of particle radius *r*, Qs also rapidly increases, and gradually approaches the maximum value of 4. When $r > 4.0$ μm, variation amplitude of Qs gradually decreases and is slowly weakened in the form of damped oscillation while gradually approaching 1, that is, for large particles, scattering efficiency factor approaches 1. Such variation is caused by the absorption nature of particles.

Absorption efficiency factor Qa refers to the ratio of absorption section to the geometric section of the respective particle. According to Fig. 3.8, based on the variation of absorption efficiency factor Qa over particle size *r*, Qa gradually increases along with the increase of *r* and approaches 1; in addition, absorption features of particles vary according to different values of wave band λ, absorption ability at the wave band of 440 nm is obviously higher than those at other wave bands. Meanwhile, by integrating with Figs. 3.6 and 3.7, along with the growth of absorption, vibration of Qe and Qs curves gradually decreases until vanishing in the end.

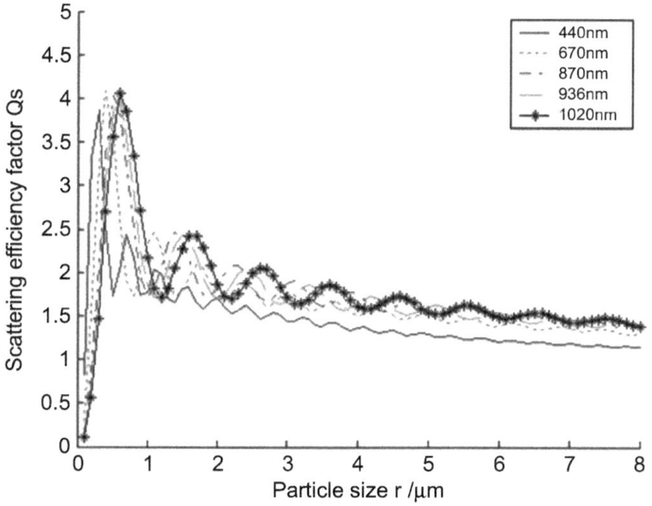

FIGURE 3.7 Variation of scattering efficiency factor Qs over particle size *r*.

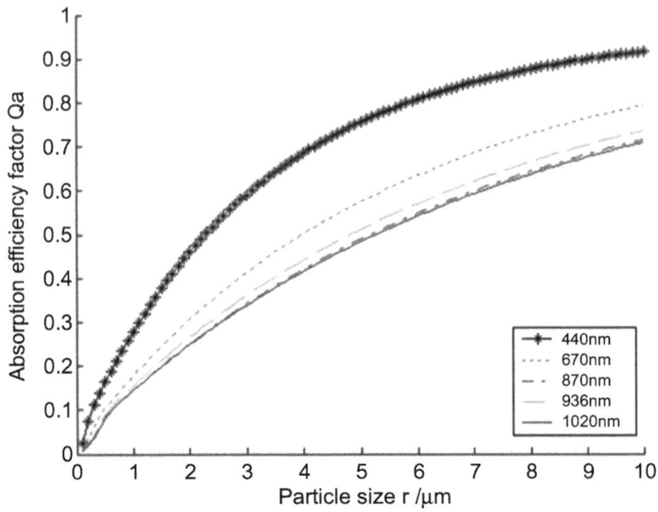

FIGURE 3.8 Variation of absorption efficiency factor Qa over particle size *r*.

4. Extinction coefficient. Aerosol extinction coefficient is used to describe the degree of solar direct radiation being weakened by the atmosphere. Value of the coefficient is related to the category of aerosols and wave band of solar direct radiation; yet in general, the coefficient is comparatively stable.

   Fig. 3.9 offers contributions of differently sized particles to extinction coefficient, according to which, the overall extinction coefficient of all particles is proportional to the area below the curve,

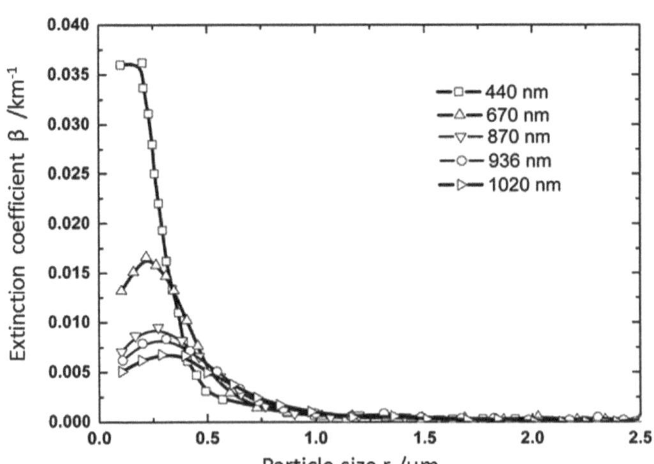

FIGURE 3.9 Variation of extinction coefficient β over particle size *r*.

and particles with the size of 0.1−0.4 μm have the largest contributions to the total extinction coefficient; peak value of extinction appears when the particle size approximates 0.2 μm, $\beta(440) = 0.36/\text{km}$. Along with the increase of particle size, extinction coefficient decreases rapidly; when the particle size reaches up to 1.0 μm, extinction coefficient has already approached 0. Then, based on the variation of extinction coefficient over wavelength, aerosol particle has the best extinction ability when the wavelength is 440 nm; along with the increase of wavelength, extinction ability of the particle gradually decreases; comparing with other wavelengths, the extinction coefficient decreases with a higher speed when wavelength is 440 nm. Yet at the same time, it is also discovered that the relationship between extinction coefficient and wavelength is not onefold, which can be concluded from 0.4 to 1.0 μm section of the particle size curve. In this section of the particle size curve, curves of various wavelengths intersect with each other; there is no obvious law to follow. Many studies have indicated that variation of extinction coefficient over wavelength can normally be manifested as follows:

$$\beta = \frac{A}{r^{\gamma}} \tag{3.7}$$

in which A is a constant; index number γ may vary from 4 (suitable for Rayleigh scattering and extremely small-sized particles) to 0 (suitable for the scattering effect of mist on visible light or near-infrared light).

However, only for extremely large and extremely small particles, can such dependency relationship be manifested by a simple function within a comparatively broad wavelength range. Sand dust particles for a power plant fall in between these two extreme values. Based on the entire wavelength and particle size ranges, variation of extinction coefficient still follows a certain rule.

To sum up, extinction coefficient β in Eq. (3.4) is 0.02. For a 10 MW power plant, the last row of heliostats is located about 0.8 km away from the receiver tower. By substituting extinction coefficient and light transmission length $L = 0.8$ km into Eq. (3.3), atmospheric transmittance of heliostat from the last row can be obtained:

$$T = \exp(-\beta L) = \exp(-0.02 \times 0.8) = 0.984 \tag{3.8}$$

The mean value is

$$T = \exp(-\beta L) = \exp(-0.02 \times 0.4) = 0.99$$

For a 100 MW-level power tower plant, area of the concentration field is about 1 million square meters. The last row of heliostats is 2 km away from the tower.

$$T = \exp(-\beta L) = \exp(-0.02 \times 2) = 0.96$$

3.2.6 Heat Losses of Power Tower Cavity Receiver

Receiver is the device in the CSP generation system that converts solar energy into thermal energy. Receiver of the power tower generation system is normally mounted at the top of the receiver tower. There are mainly two types of receivers, cavity type and cylinder type receivers. The cavity receiver is used to absorb the solar radiation collected by the concentration field arranged in northern sectors (Fig. 3.10A and C), whereas the cylinder receiver is used to receive solar radiation collected by the concentration field arranged in circles surrounding the receiver tower (Fig. 3.10B). Absorber of the cavity receiver is mounted within a cavity structure insulating thermal from outside; opening area of the cavity is less than the surface area of the absorber, which is able to effectively reduce heat loss of the absorber. A cylinder receiver is made by welding a large amount of straight tubes that are vertically parallel to each other on the instrument in order to create strip receiver plates, and put together the receiver plates in order to create an approximate cylinder. Bottom and top of the cylinder have been designed with heat-transfer fluid header for providing cold and hot transfer fluid to the receiver and collecting the thermal transfer fluid. According to different heat-transfer fluid inside the absorber, receivers for the power tower generation system can be categorized into synthetic oil receiver, water/steam receiver, molten-salt receiver, air receiver, liquid metal receiver, solid particle receiver, etc. According to different features of solar radiation absorption process inside the receiver, receivers for the power tower generation system can be categorized into volumetric receiver and tubular receiver. The volumetric receiver's absorber is normally made of honeycomb ceramics, foam ceramics, metal gauze, foam metal and other

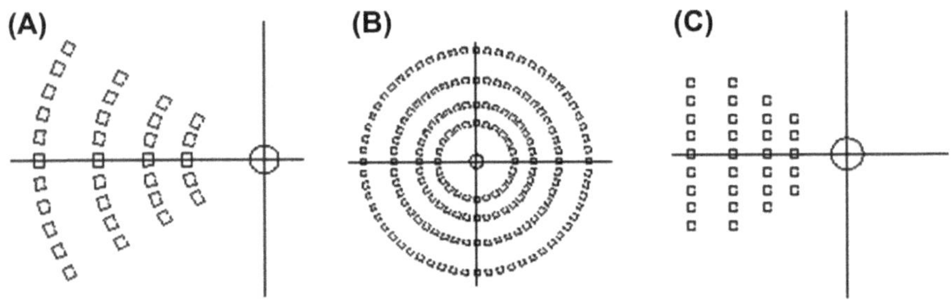

FIGURE 3.10 Various kinds of concentration fields and receivers.

porous materials. Solar radiation is transmitted and absorbed within the entire volume of absorber, which is frequently used together with the system that uses heat-transfer fluid as the gas. The tubular receiver consists of several pieces of tubes. Solar radiation will be absorbed on the surface of the absorber tube; interior wall of the tube exchanges thermal energy with the heat-transfer fluid flowing through the tube by the convection heat transfer, the heat-transfer medium of which is liquid in most of the cases.

For a cavity receiver, the heat loss P_{LOSS} can be calculated as follows [21]

$$P_{LOSS} = P_{REFCAV} + P_{RAD} + P_{CONV} + P_{COND} \tag{3.9}$$

in which P_{REFCAV} refers to the reflective radiation loss of the cavity receiver; P_{RAD} refers to the radiation heat loss from the surface of absorber inside the cavity receiver to the outside through receiver aperture; P_{CONV} refers to the thermal convection from the surface of absorber to the outside through receiver aperture; P_{COND} refers to the conductive heat loss from the surface of absorber to the outside heat loss.

The area of the cavity receiver is shown in Fig. 3.11. A_1 is the aperture area of the receiver (unit: m^2) and A_2 is the area of inner absorbing surface of the receiver.

1. Reflective radiation loss P_{REFCAV} U_{marov} (1983) equation, under the premise of the thermal absorbent surface being a gray body, the receiver's equivalent absorptance α_{eff} is

$$\alpha_{eff} = \frac{\alpha_w}{1 - (1-\alpha_w)\left(1 - \frac{A_1}{A_2}\right)} \tag{3.10}$$

   in which a_w is the solar absorptance on the surface of the absorber of the receiver, which can be measured through experiments and usually depend on the temperature.

   Receiver's equivalent reflectance P_{COV} is

$$\begin{aligned} \rho_{COV} &= 1 - \alpha_{eff} \\ &= 1 = \frac{\alpha_w}{1 - (1-\alpha_w)\left(1 - \frac{A_1}{A_2}\right)} \end{aligned} \tag{3.11}$$

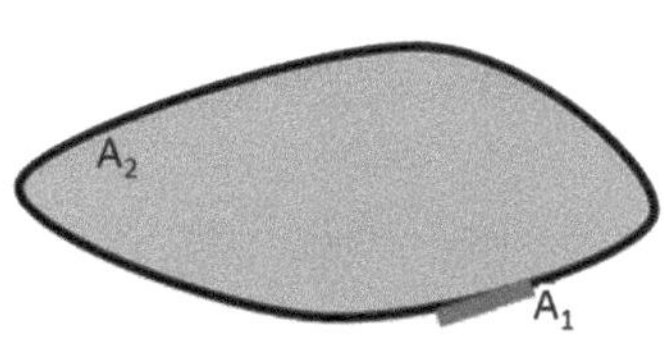

FIGURE 3.11 Schematic diagram of the area of the cavity receiver.

P_{AP} refers to the radiation flux through the receiver aperture after solar radiation is reflected by the concentration field.

$$P_{REFCAV} = \rho_{COV} P_{AP} = \left(1 - \frac{\alpha_w}{1 - (1 - \alpha_w)\left(1 - \frac{A_1}{A_2}\right)}\right) P_{AP} \tag{3.12}$$

P_{AP} can be obtained through some heliostat field simulation software, and corrected through experimental measurements.

In Eq. (3.12), a_w is a constant that changes along with (A_1/A_2). We assume $a_w = -0.85$, and substitute it into Eq. (3.11)

$$\rho_{COV} = 1 - \frac{0.85}{1 - 0.15\left(1 - \frac{A_1}{A_2}\right)} \tag{3.13}$$

Fig. 3.12 displays the variation rule of reflectance, according to which, equivalent reflectance of cavity receiver is normally around 0.05.

In case that the area ratio (A_1/A_2) is a constant and a_w changes accordingly, we assume $A_1/A_2 = -0.2$, and substitute it into Eq. (3.11)

$$\rho_{COV} = 1 - \frac{\alpha_w}{1 - (1 - \alpha_w)(1 - 0.2)} = 1 - \frac{\alpha_w}{1 - 0.8(1 - \alpha_w)} \tag{3.14}$$

According to Fig. 3.13, along with the increase of absorptance of absorber, equivalent reflectance decreases rapidly. Thus it is useful to increase the thermal efficiency of the receiver by

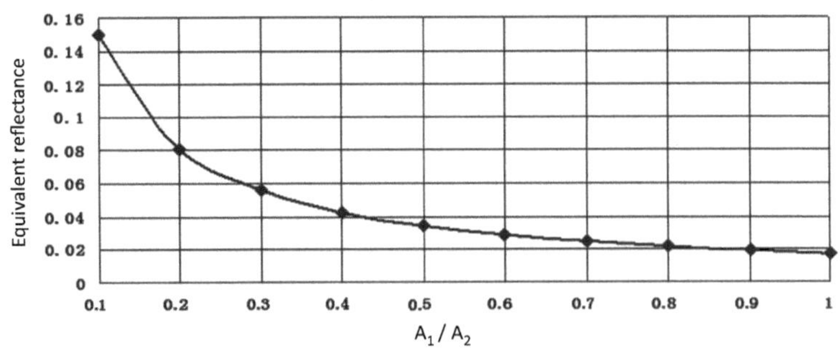

FIGURE 3.12 Equivalent reflectance of cavity receiver ($A_w = 0.85$).

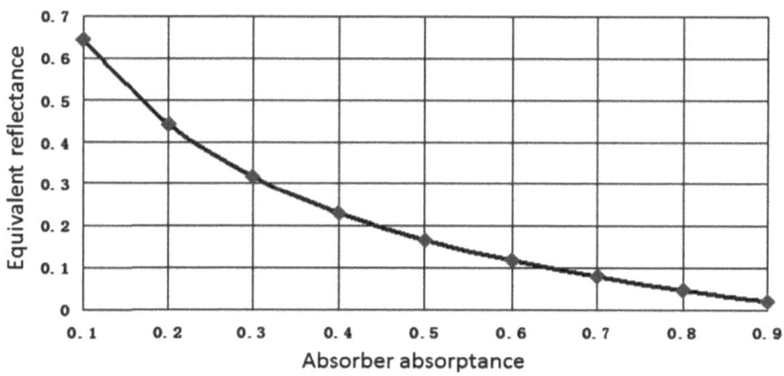

FIGURE 3.13 Equivalent reflectance of cavity receiver ($A_1/A_2 = 0.2$).

selecting coatings with higher absorptance or performing surface preparation.

2. Radiation heat loss PRAD. Radiation heat loss is the thermal radiation from the receiver to the outside through the aperture of receiver.

$$P_{RAD} = \varepsilon_{AP}\sigma\left(T_w^4 - T_g^4\right)A_1 \tag{3.15}$$

in which ε_{AP} refers to the equivalent thermal emittance of receiver; σ is a Stefan-Boltzman constant; T_w refers to the mean temperature of absorber surface; T_g refers to the ground surface temperature.

In Eq. (3.15), as the aperture of cavity receiver tilts downward, it is the radiation heat exchange between the absorber surface and local ground surface that actually occurs (refer to Fig. 3.14).

ε_{AP} can be calculated by applying the U$_{marov}$ (1983) formula

$$\varepsilon_{AP} = \frac{\varepsilon_w}{1 - (1 - \varepsilon_w)\left(1 - \frac{A_2}{A_1}\right)} \tag{3.16}$$

in which ε_w refers to the solar emittance of absorber surface, $\varepsilon_w = 1 - a_w$. We assume $A_2/A_1 = 5$, and substitute it into Eq. (3.16)

$$\varepsilon_{AP} = \frac{\varepsilon_w}{1 + 4(1 - \varepsilon_w)} \tag{3.17}$$

Variation of equivalent emittance along with the absorptance of absorber is shown in Fig. 3.15, according to which, equivalent emittance varies greatly with the absorptance. Absorptance treatment to the absorber surface is an effective approach to reduce radiation heat loss.

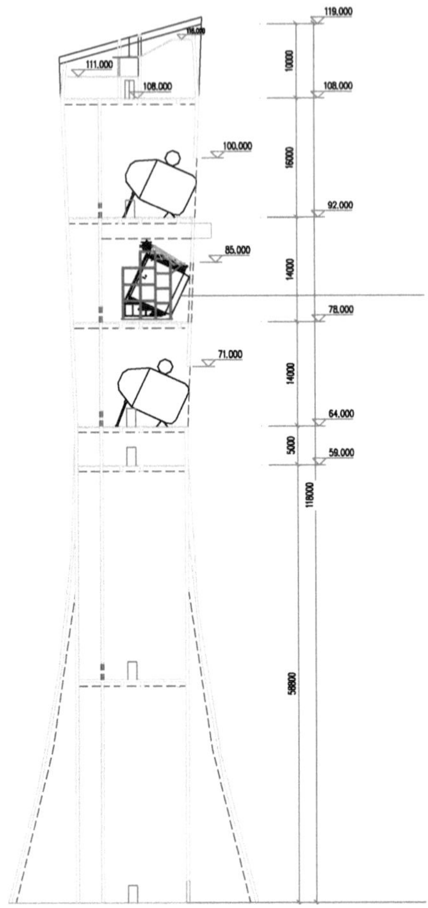

FIGURE 3.14 Radiation heat transfer between the absorber surface and local ground surface.

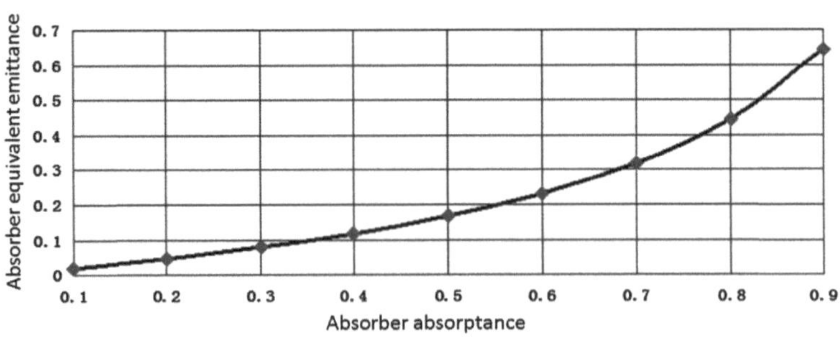

FIGURE 3.15 Equivalent emittance of cavity receiver ($A_2/A_1 = 5$).

After measuring the absorber surface temperature T_w, by substituting Eq. (3.16) into Eq. (3.15), formula for calculating radiation heat loss of receiver can be easily obtained.

$$P_{\mathrm{RAD}} = \frac{\varepsilon_w \sigma \left(T_w^4 - T_g^4\right) A_1}{1 - (1 - \varepsilon_w)\left(1 - \frac{A_2}{A_1}\right)} \tag{3.18}$$

3. Convective heat loss P_{conv}. Convective heat losses of receiver include natural convective heat loss and forced convective heat loss. The forced convective heat loss will not be explained in this section at this moment, only the method for calculating natural convective heat loss will be discussed. Driven by buoyance, natural convection is related to the shape of the absorber, the installation dip of receiver (Fig. 3.16), as well as the surface temperature of absorber and the respective distribution. It is extremely difficult to achieve precise conclusions only based on theoretical analysis. In this section, the experimental formula Sieber Kraabel model is applied to calculate the Nusselt number

$$Nu = 0.088 Gr^{\frac{1}{2}} \left(\frac{T_w}{T_a}\right)^{0.18} (\cos\theta)^{2.17} \left(\frac{d_{\mathrm{AP}}}{L}\right)^{8} \tag{3.19}$$

$$s = 1.12 - \frac{0.982 d_{\mathrm{AP}}}{L} \tag{3.20}$$

in which θ refers to the receiver dip angle (refer to Fig. 3.16); T_a refers to the ambient air temperature; T_w refers to the absorber mean temperature; d_{AP} refers to the diameter of aperture area; L is the characteristic length, which is equivalent to the depth of receiver

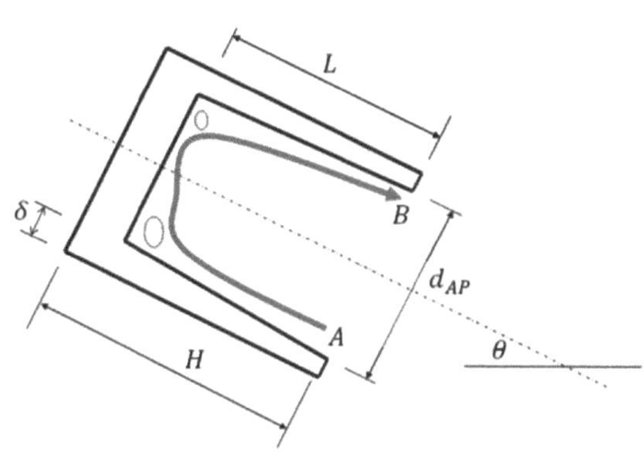

FIGURE 3.16 Cavity receiver.

aperture (refer to Fig. 3.16); G_r is the Grashof number; Nu is the Nusselt number.

In case that the receiver is placed horizontally, $\theta = 0$, the respective Nusselt number is Nu_0.

Nusselt numbers corresponding to other dip angles and the respective ratios are:

$$\frac{Nu}{Nu_0} = \cos^{2.47}\theta \tag{3.21}$$

Fig. 3.17 shows the variation rule of Nusselt number, according to which, the bigger of the receiver dip angle, the less of the convective heat loss. Receiver dip angle is normally determined by the concentration field arrangement, and is related to the reflection directions of heliostats. Considering solar altitudes at different time points, the calculation process is complex. This section mainly discusses the convective heat loss. This angle is assumed to be a given value that will not be explained in any detail.

According to the rules of natural convection, the approximate flow pattern inside the absorber is shown in Fig. 3.16, according to which, in case of designing blocks at corner A and B, it will effectively restrain natural convection.

$$\begin{aligned} Nu &= \frac{\alpha_{AP} L}{\lambda} \\ \alpha_{AP} &= \frac{\lambda}{L} Nu \\ &= 0.088 \frac{\lambda}{L} Gr^{\frac{1}{3}} \left(\frac{T_w}{T_a}\right)^{0.18} \left(\cos^{2.47}\theta\right) \left(\frac{d_{AP}}{L}\right)^{s} \end{aligned} \tag{3.22}$$

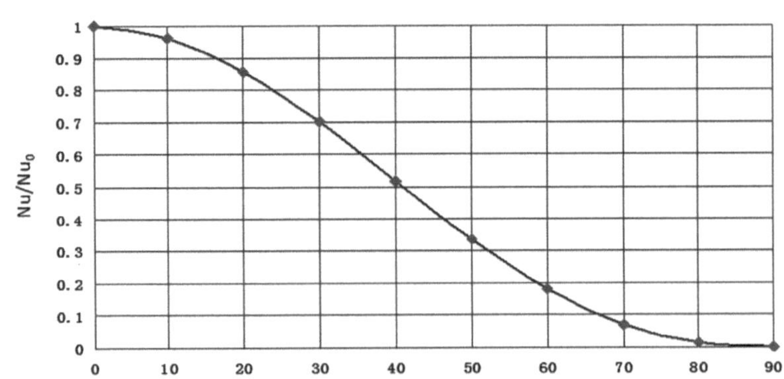

FIGURE 3.17 Variation of cavity receiver Nu over θ.

Overflowed energy through receiver aperture due to natural convection is

$$P_{\text{CONV}} = \alpha_{\text{AP}}(T_w - T_a)A_1 \tag{3.23}$$

We substitute Eq. (3.22) into Eq. (3.23),

$$P_{\text{CONV}} = \alpha_{\text{AP}}(T_w - T_a)A_1$$

$$= 0.088Gr^{\frac{1}{3}}\left(\frac{T_w}{T_a}\right)^{0.18}(\cos^{2.47}\theta)\left(\frac{d_{\text{AP}}}{L}\right)^{\left(1.12-0.982\frac{\lambda_{\text{AP}}}{L}\right)}\left(\frac{\lambda}{L}\right) \tag{3.24}$$

$$(T_w - T_a)A_1$$

in which λ is the air thermal conductivity under ambient air temperature, W/(m·K).

4. Conductive heat loss P_{COND}. Conductive heat loss of receiver depends on the thermal radiation through the receiver wall surface and thermal-insulating materials, values of which are mainly determined by its thermal-insulating performance.

$$P_{\text{COND}} = \frac{T_w - T_a}{R} \tag{3.25}$$

Provided that absorber surface temperature T_w and ambient air temperature T_a have been obtained, heat loss is transmitted through absorber to the thermal-insulating material, and further through thermal-insulating material to receiver exterior wall surface; the exterior wall surface exchanges thermal through natural convection and dissipates it into the atmosphere. Thermal resistance of this process has been calculated as follows:

$$\frac{1}{R} \approx \frac{2\pi kH}{\ln\left(\frac{r_{\text{AP}} + \delta}{r_{\text{AP}}}\right)} + h_{wb}\pi(r_{\text{AP}} + \delta)H \tag{3.26}$$

in which h_{wb} refers to the nature convective heat-transfer coefficient of receiver exterior wall surface, W/(m^2·°C); δ refers to the thickness of thermal insulating material, m; r_{AP} refers to the radius of receiver aperture, m; and H refers to the axial length of receiver, m.

Due to the nonuniform internal temperature of receiver and the installation angle of receiver, Nusselt distribution for heat exchange between receiver exterior surface and the air is also complex. Here, the horizontally placed long cylinder is used to simplify the calculation of Nusselt number through natural convection thermal radiation. The mean Nusselt number of receiver exterior wall surface is calculated through Eq. (3.27) [52]

$$Nu_{wb} = \left\{ 0.6 + \frac{0.387 Ra_{wb}^{\frac{1}{6}}}{\left[1 + \left(\frac{0.559}{PR}\right)^{\frac{9}{10}}\right]^{\frac{8}{27}}} \right\}^2, \quad 10^{-5} < Ra_{wb} < 10^{12} \tag{3.27}$$

$$\begin{aligned} Ra_{\mathrm{wb}} &= \frac{g\beta(T_{wb} - T_a)(r_{\mathrm{AP}} + \delta)^3}{va} \\ &= \frac{g(T_{wb} - T_a)(r_{\mathrm{AP}} + \delta)^3}{vaT_a} \end{aligned} \tag{3.27a}$$

$$h_{wb} = \frac{\lambda Nu_{wb}}{H} = \frac{\lambda}{H} \left\{ 0.6 + \frac{0.387 Ra_{wb}^{\frac{1}{6}}}{\left[1 + \left(\frac{0.559}{P_r}\right)^{\frac{9}{16}}\right]^{\frac{8}{27}}} \right\}^2 \tag{3.28}$$

in which β is the volumetric expansion coefficient of the air, $1/°C$, as $\beta = 1/T_a$; v is the kinematic viscosity of the air, m^2/s; a is temperature diffusion coefficient, m^2/s; P_r is the Prandtl number of the air; T_{wb} is the mean temperature of absorber exterior wall surface, °C.

We substitute Eqs. (3.26–3.28) into (3.25)

$$\begin{aligned} P_{\mathrm{COND}} &= \left[\frac{2\pi k H}{\ln\left(\frac{r_{\mathrm{AP}} + \delta}{r_{\mathrm{AP}}}\right)} + \pi(r_{\mathrm{AP}} + \delta) H h_{wb} \right] (T_w - T_a) \\ &= \left[\frac{2\pi k H}{\ln\left(\frac{r_{\mathrm{AP}} + \delta}{r_{\mathrm{AP}}}\right)} + \delta\lambda \left\{ 0.6 + \frac{0.387 Ra_{wb}^{\frac{1}{6}}}{\left[1 + \left(\frac{0.559}{P_r}\right)^{\frac{9}{16}}\right]^{\frac{8}{27}}} \right\}^2 (r_{\mathrm{AP}} + \delta) \right] \\ &\quad (T_w - T_a) \end{aligned} \tag{3.29}$$

In the case that the reference temperature is the ambient air temperature, $P_r = 0.71$, we substitute it into Eq. (3.29)

$$P_{\mathrm{COND}} = \left[\frac{2\pi k H}{\ln\left(\frac{r_{\mathrm{AP}} + \delta}{r_{\mathrm{AP}}}\right)} + \pi\lambda\left(0.6 + 0.32 Ra_{wb}^{\frac{1}{6}}\right)^2 (r_{\mathrm{AP}} + \delta) \right] (T_w - T_a) \tag{3.30}$$

When Eq. (3.30) is applied, it is necessary to know the mean temperature T_{wb} of receiver exterior wall surface, which is calculated as follows:

$$P_{\mathrm{COND}} = \left[\frac{2\pi kH}{\ln\left(\frac{r_{\mathrm{AP}}+\delta}{r_{\mathrm{AP}}}\right)}\right](T_w - T_{wb}) \tag{3.31}$$

Through substituting Eq. (3.27a) into (3.30), and according to Eq. (3.31) = Eq. (3.30) for energy balance, then T_{wb} can be calculated and substituted into Eq. (3.31) in order to obtain P_{COND}.

In actual projects, while considering the safety of receiver, according to the requirement of design norm, receiver surface temperature shall not exceed 80°C; otherwise, it may result in fire hazards of cables and other inflammable equipment on receiver exterior surface. When calculation is conducted through Eq. (3.31), it is also possible to substitute $T_{wb} = 80°\mathrm{C}$ into the equation, so that

$$P_{\mathrm{COND}} = \left[\frac{2\pi kH}{\ln\left(\frac{r_{\mathrm{AP}}+\delta}{r_{\mathrm{AP}}}\right)}\right](T_w - 80) \tag{3.31a}$$

In case of considering the influence of wind, it is also possible to directly calculate the forced convective heat-transfer coefficient of receiver exterior surface under forced convection conditions.

5. Receiver heat loss. By substituting the above into Eq. (3.9), heat loss can be calculated as follows:

$$\begin{aligned}
P_{\mathrm{LOSS}} &= P_{\mathrm{REFCAV}} + P_{\mathrm{RAD}} + P_{\mathrm{CONV}} + P_{\mathrm{COND}} \\
&= \left[1 - \frac{\alpha_w}{1-(1-\alpha_w)\left(1-\frac{A_1}{A_2}\right)}\right]P_{\mathrm{AP}} + \frac{\varepsilon_w\sigma\left(T_w^4 - T_g^4\right)A_1}{1-(1-\varepsilon_w)\left(1-\frac{A_2}{A_1}\right)} \\
&\quad +0.088Gr^{\frac{1}{3}}\left(\frac{T_w}{T_a}\right)^{0.18}\left(\cos^{2.47}\theta\right)\left(\frac{d_{\mathrm{AP}}}{L}\right)^{\left(1.12-0.982\frac{d_{AP}}{L}\right)} \\
&\quad \left(\frac{\lambda}{L}\right)(T_w - T_a)A_1 + \left[\frac{2\pi kH}{\ln\left(\frac{r_{\mathrm{AP}}+\delta}{r_{\mathrm{AP}}}\right)}\right. \\
&\quad \left. +\pi\lambda\left(0.6+0.32Ra_{wb}^{\frac{1}{5}}\right)^2(r_{\mathrm{AP}}+\delta)\right](T_w - T_a)
\end{aligned} \tag{3.32}$$

Or, the engineering simplified equation can be applied:

$$
\begin{aligned}
P_{\text{LOSS}} &= P_{\text{REFCAV}} + P_{\text{RAD}} + P_{\text{CONV}} + P_{\text{COND}} \\
&= \left[1 - \frac{\alpha_w}{1-(1-\alpha_w)\left(1-\frac{A_1}{A_2}\right)}\right] P_{\text{AP}} + \frac{\varepsilon_w \sigma \left(T_w^4 - T_g^4\right) A_1}{1-(1-\varepsilon_w)\left(1-\frac{A_2}{A_1}\right)} \\
&\quad + 0.088 Gr^{\frac{1}{3}} \left(\frac{T_w}{T_a}\right)^{0.18} \left(\cos^{2.47}\theta\right) \left(\frac{d_{\text{AP}}}{L}\right)^{\left(1.12-0.982\frac{d_{\text{AP}}}{L}\right)} \\
&\quad \left(\frac{\lambda}{L}\right)(T_w - T_a)A_1 + \left[\frac{2\pi k H}{\ln\left(\frac{r_{\text{AP}}+\delta}{r_{\text{AP}}}\right)}\right](T_w - 80)
\end{aligned}
\tag{3.33}
$$

In order to facilitate the understandings of readers, a calculation example is given as follows.

Assume

$$
\begin{gathered}
T_g = T_a = 20^{\circ}\text{C},\ T_w = 400^{\circ}\text{C},\ \alpha_w = 0.9,\ \varepsilon_w = 0.85,\ k = 0.048\ \text{W}/(\text{m}\cdot{}^{\circ}\text{C}), \\
\lambda = 0.033\ \text{W}/(\text{m}\cdot{}^{\circ}\text{C}),\ \delta = 0.3\ \text{m},\ d_{AP} = 5\ \text{m},\ r_{AP} = 2.5\ \text{m},\ L = 5\ \text{m}, \\
H = L + \delta = 5.3\ \text{m},\ \theta = 20^{\circ},\quad v = 22.8 \times 10^{-6}\ \text{m}^2/\text{s},\quad a = 32.8 \times 10^{-6}\ \text{m}^2/\text{s}, \\
A_1 = 25\ \text{m}^2, A_2 = 100\ \text{m}^2,\ P_{AP} = 6500\ \text{kW};\ \sigma = 5.6686 \times 10^{-8}\ \text{W}/(\text{m}^2\cdot\text{K}^4)
\end{gathered}
$$

Grashof number is:

$$
Gr = \frac{g\beta(T_w - T_a)L^3}{va} = \frac{9.81 \times (400-20) \times 5^3}{22.8 \times 10^{-6} \times 32.8 \times 10^{-6} \times 293} = 2.1 \times 10^{12}
$$

When interior of the cavity is in a turbulent state, various parameters are substituted into Eq. (3.33)

$$
\begin{aligned}
P_{\text{LOSS}} &= \left[1 - \frac{0.9}{1-(1-0.9)\left(1-\frac{25}{100}\right)}\right] \times 6500 \\
&\quad + \frac{0.85 \times 5.6686 \times 10^{-8} \times \left(673^4 - 293^4\right) \times 25}{1-(1-0.85)\left(1-\frac{100}{25}\right)} + 0.088 \\
&\quad \times \left(2.1 \times 10^{12}\right)^{\frac{1}{3}} \times \left(\frac{400}{20}\right)^{0.18} \left(\cos^{2.47}20^{\circ}\right) \left(\frac{5}{5}\right)^{\left(1.12-0.982\times\frac{5}{5}\right)} \\
&\quad \left(\frac{0.033}{5}\right)(400-20) \times 25 + \left[\frac{2\pi \times 0.048 \times 5.3}{\ln\left(\frac{2.5+0.3}{2.5}\right)}\right] \times (400-80) \\
&= 176 + 183 + 103 + 4.5 = 466.5(\text{kW})
\end{aligned}
$$

Proportions of various factors in receiver heat loss are shown in Fig. 3.18, according to which, reflection and radiation heat losses have accounted for a comparatively large proportion. Therefore, the solar selective coating on the interior surface of absorber is of great significance. Conductive heat loss is insignificant, which can be almost neglected.

Along with the increase of the dimension of aperture of receiver, intercept factor of receiver increases as well; meanwhile, the heat loss also starts to grow. Fig. 3.19 reveals the variation of cavity receiver heat loss at the design point along with the variation of the dimension of aperture of receiver. At off-design points, it is necessary to calculate the dimension of facula in the concentration field by applying the concentration field calculation software, and then calculate the intercept factor of receiver before substituting it into P_{AP} in Eq. (3.33).

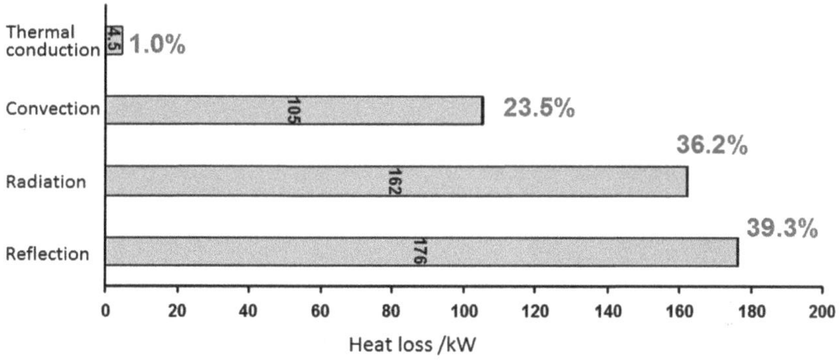

FIGURE 3.18 Heat losses in various parts of cavity receiver.

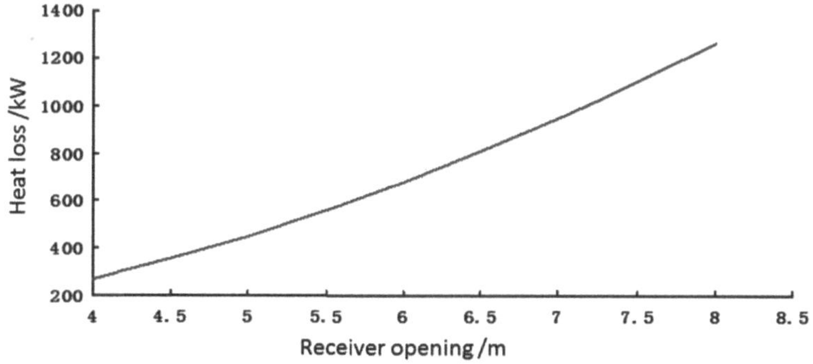

FIGURE 3.19 Variation of heat loss of cavity receiver at the design point along with the dimension variation of aperture of receiver.

The above heat loss has been calculated without considering the influences of wind speed and direction. For a commercial power plant, the receiver is normally mounted inside a sealed absorption space, in which the influence of wind on conductive heat loss can be neglected; yet its influences on convection loss shall be considered.

3.3 THERMAL PERFORMANCE OF PARABOLIC TROUGH COLLECTOR

Calculation of the efficiency of parabolic trough collector is comparatively complex, which is related to the solar irradiance, axial layout of concentrator, optical performance of concentrator, working temperature of heat-transfer medium, ambient air temperature, wind speed, and concentration field features; thus it is very difficult to ensure the precision of calculation. Normally, an efficiency calculation formula shall be offered by the equipment manufacturer.

3.3.1 Parabolic Trough Receiver Tube Heat Loss Parameters

As variable properties of vacuum, transparent glass tubes and coatings against temperature variation are involved, the heat-transfer theory calculation on evacuated tube heat loss is comparatively difficult, which is normally obtained through experimental measurements. The test data of heat loss coefficient of the evacuated tube from SCHOTT of Germany is shown below, which is taken as an example to demonstrate the approximate range of heat loss coefficient of the evacuated tube.

Table 3.3 has listed data corresponding to Fig. 3.20. The unit of heat loss coefficient is the heat loss power along unit length of axial line of the evacuated tube: W/m.

In Fig. 3.20, the lateral axis is the difference of evacuated tube temperature and ambient air temperature, whereas the vertical axis is the heat loss coefficient of the evacuated tube (W/m). When the temperature difference is 293°C, the respective heat loss coefficient is about 113 W/m; when the temperature difference is 393°C, the respective heat loss coefficient is about 257 W/m. Presently, there has been no China state standard on the testing of the evacuated tube heat loss coefficient. When using various unit measured values, it is necessary to carefully check the testing approaches and conditions described in the testing report. The heat loss coefficient is related to the thermal charging mode, materials of heater, steady state conditions during temperature measurement (ambient air temperature, heater temperature), position of

TABLE 3.3 Measured Values of Heat Loss Coefficient of SCHOTT PTR70 Evacuated Tube [22]

Testing	Evacuated Tube Temperature/°C	Mean Glass Temperature/°C	Mean Ambient Air Temperature/°C	Evacuated Tube and Ambient Air Temperature Difference/°C	Heat Loss Coefficient HL/(W/m)
1	100	26	23	77	15
2	153	30	23	130	23
3	213	35	23	190	43
4	246	38	24	222	59
5	317	50	24	293	113
6	346	55	23	323	141
7	390	65	24	366	204
8	418	73	25	393	257
9	453	82	23	430	333
10	458	84	24	434	348
11	506	99	24	482	495

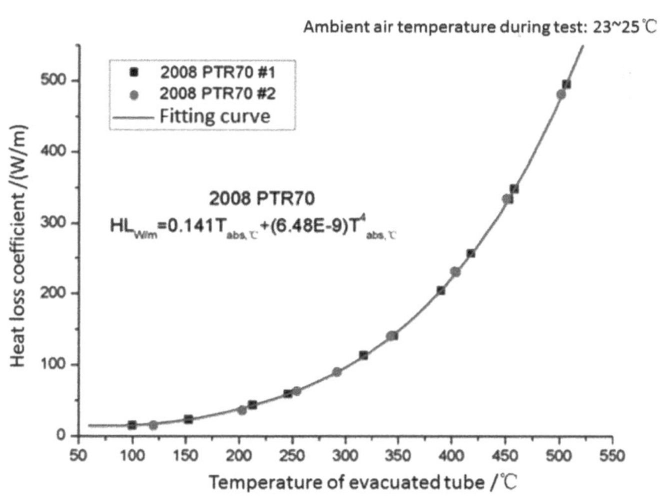

FIGURE 3.20 Heat loss coefficient of PTR70 evacuated tube from schott of germany [22]. *Data Source: Technical Report NREL/TP-550-45633, May 2009.*

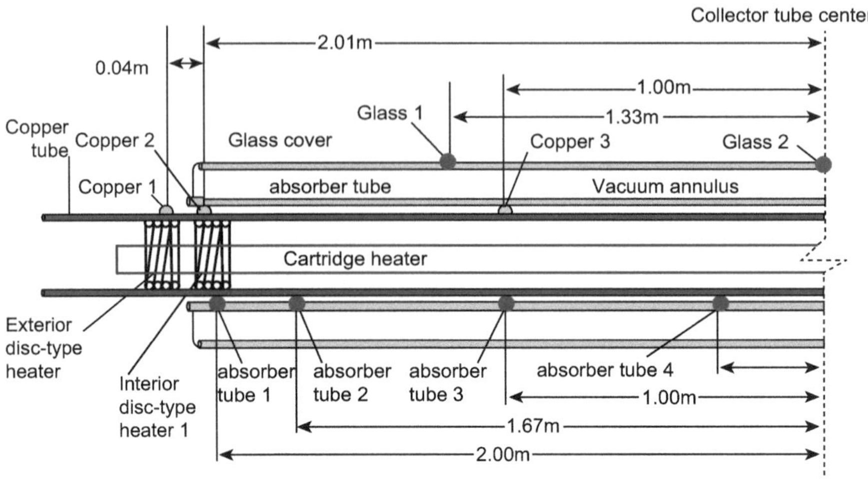

FIGURE 3.21 Positions of temperature measuring points on the test platform of heat loss coefficient of NREL evacuated tube [22].

temperature measuring point and sampling frequency, as well as the temperature and color of the interior wall.

Fig. 3.21 has indicated positions of temperature measuring points on the evacuated tube test platform of the National Renewable Energy Laboratory (NREL), which is part of the Department of Energy (DOE)[22]; the heater material is copper.

Standards currently suitable for testing the thermal performance of parabolic trough solar collectors include the American standard ANSI/ ASHRAE 93 "Methods of Testing to Determine the Thermal Performance of Solar Collectors" [23] and European standard EN 12975-2 "Thermal Solar Systems and Components—Solar Collectors: Part 2: Test Methods" [24]. Although these two standards have already been compared by certain literature, they have focused on low-temperature thermal utilization of solar energy of flat-plate-type and vacuum-tube-type solar collectors. Working temperatures of these two types of solar collectors are normally less than 80°C; nevertheless, working temperature of the parabolic trough solar collector falls in a range of 100–400°C.

3.3.2 Current Status of Measurement Methods for Parabolic Trough Collector Thermal Performance

3.3.2.1 Current Overseas Research Status of Thermal Performance of Parabolic Trough Collectors

Ever since 1970s, commercial products of concentrating solar collectors began to be developed, which made the DOE in the United States and relevant industries to realize that it was highly necessary to conduct systematic and standardized testing and evaluation toward collectors; it would help to enable potential users evaluate such technology through unified tests. In 1973, with the help from the DOE, Sandia National Laboratories of America (SNLA) located in Albuquerque carried out studies on the testing of tracking parabolic trough collectors for the first time. In 1975, SNLA's mid-temperature solar system testing facility started to function, which included the parabolic trough collector module testing platform and system testing platform.

In 1977, American Society of Heating, Refrigerating and Air Conditioning Engineers (ASHRAE) published the standard test method ASHRAE 93-77, which provided guiding principles for testing non-tracking and tracking solar collectors. Henceforth, it was developed synchronously with the construction of tracking solar collector testing platform. Presided by the American Society for Testing Materials (ASTM), many investors participated in the development and evaluation of solar collectors. They applied themselves in developing the test method for tracking solar collectors, and finally published the standard test method ASTM E905 in 1983. However, as tracking solar collectors was normally used in the large-scale array, it was necessary to understand the performance of the entire system before testing a single collector module. These systems included pipelines and other system balance members, as well as unsteady state conditions. In order to acquire such data, DOE initiated a series of projects. The first project involved a large amount of field tests. These tests had focused on numerous large-scale parabolic trough thermal collecting systems on industrial sites. The second project was called the Modular Industrial Solar Retrofit project, which focused on advanced parabolic trough thermal collecting system for industrial steam and carried out development and testing works.

SNLA's parabolic trough collector module test platform includes three test stations (each test station has an independent fluid loop), a data collection system and a parabolic trough collector with an aperture area up to 45 m^2. Based on the heat-transfer fluid in the loop, different testing temperatures can be defined. For example, loop 1 applies Therminol 66 synthetic oil, the maximum operating temperature of which is 315°C; loop 2 applies Syltherm 800 synthetic oil, the maximum operating temperature of which is 425°C (refer to Fig. 3.22). Biaxial rotation test platform is able to make the aperture of receiver of parabolic trough collector face a random direction within a specific test period. In addition, the meteorological station collects all necessary data about natural conditions. The laboratory also published some important testing research reports related to the performance of parabolic trough collectors, such as the 30 MWe SEGS Power Plant Simulation Report [25], Test Report for LS-2 Parabolic

FIGURE 3.22 Parabolic trough collector module test platform of Aandia National Laboratories of America.

Trough Collector Applied in SEGS Power Plant, Test Report for Parabolic Trough Collector Mounted with Schott Vacuum Receiver, and Parabolic Trough Collector Steady State Test Report for Industrial Applications [26].

Early tracking solar collector tests carried out by SNLA and the subsequent Solar Energy Research Institute of America were based on the general technologies in ASHRAE 93-77. However, this method was not completely applicable for the tracking solar collector, because there were no massive explanations, analysis and precise testing technologies specifically described in this standard to be related to the tracking solar collector testing. Being aware of the incompleteness of this method, the solar energy commission of ASTM established a subordinate professional commission, which carried out a series of studies on standards from the aspect of economic test methods. The target of this standard was to define the annual energy output of the appointed tracking solar collector at a specific location. This professional commission operated by following the principle of voluntariness and unanimity through consultation, members of which included manufacturers, users and other representatives from industries, governments, colleges, and universities as well as testing laboratories. Solar Energy Research Institute of America drafted the original standards for this professional commission; the ultimate achievement was the ASTM standard test method E905-87: "Standard Test Method for Determining Thermal Performance of Tracking Concentrating Solar Collectors," the newest version of which was the updated version in 2007.

According to the testing requirements of concentrating solar collectors, this standard has mainly considered the respective massive technical problems, which include:

Influences of tracking/driving system and surface precision of reflector on thermal performance of collectors; Appropriate selection of standardized factors for incident solar radiation; Research on quasi steady state test conditions for solar collectors with high concentration ratios; As the high temperature synthetic oil applied in testing is lack of sufficiently precise specific thermal power parameters, the specific thermal power of synthetic oil shall be determined through calorimetry; Testing and analysis on vertical incidence and angular incidence on aperture of receiver of collector; In the case that heat-transfer fluid within the collector does not flow, sunshine may cause damage to the collector; therefore, requirements on pretreatment of solar collector shall be cancelled; For large-scale solar collectors, most of the solar simulators may introduce interferences and uncertainty; therefore, it is specified to perform the test outdoors under the clear weather.

This published standard shall be applied in axial or biaxial concentrating solar collectors; influences of solar scattering radiation are negligible, only influences of direct radiation, as well as determination of optical responses of collectors toward different solar incidence angles and thermal performance of vertically incident solar radiation under different operating temperatures shall be considered. Methods in this standard are requested to achieve quasi steady state conditions, measure certain environmental parameters, and determine the product of inlet and outlet temperature difference of heat-transfer fluid within the collector and thermal capacity of heat-transfer fluid. The test method has provided experiment and calculation procedures in order to determine such parameters as response time, incidence angle correction factor, range of near-vertical incidence angle, and the thermal gain rate corresponding to the near-vertical incidence angle.

According to the definition, response time refers to the time required for the temperature increase of heat-transfer fluid within a specified collector after the step change of solar radiation, which has determined the time necessary for achieving quasi steady state conditions. Thermal performance of a collector corresponding to a random incidence angle is obtained through the calculation of the incident angular modifier (IAM) and thermal performance of collector under near-vertical incidence. Measurement of IAM is carried out when the collector heat loss is at the minimum level; therefore, inlet temperature of heat-transfer fluid during the measurement is equivalent to or close to the ambient air temperature. In case that the tested collector is mounted on a biaxial tracking test platform, the thermal performance test is able to achieve the condition of vertical incidence of solar radiation through the aperture of receiver of collector all day long. In case that the single axis platform is used for testing, or a linear concentrating collector uses its original tracking/driving device, the range of near-vertical incidence must be determined. The testing standard has also provided the method to obtain this angular range, within which the decrease of thermal performance does not exceed 2%. The testing standard has strict requirements on the variation of measurement parameters. These parameters include heat-transfer fluid inlet temperature of collector, temperature increase of heat-transfer fluid after passing through the collector, product of heat-transfer fluid flow rate and specific heat capacity, ambient air temperature and solar DNI.

During the period from 1974 to 1980, a few of tracking solar collector testing research institutions were established in America. Although they were designed and completed before ASTM testing standard being published, basic principle of ASHRAE 93-77 reserved in ASTM standards served as the design foundation for these test platforms; in addition, many test platform designers also participated in the solar energy subcommittee of ASTM. It was important these research institutions focused on the design of linear concentrating tracking parabolic trough collectors. It had reflected that when these works were performed, linear concentrating parabolic trough solar collectors had already been close to commercialization and manufactured by some American companies. It was also noticeable that each of these research institutions had biaxial rotation test platforms; by controlling the tracking direction, they could achieve the vertical incidence of solar radiation on aperture of receiver of collector. Thus thermal performance testing for collectors could be performed in most times of the day.

In 2010, with the support from the NREL published the performance testing instruction for large-scale parabolic trough solar systems, and offered basic principles of two test methods, namely short-term steady state test and multiday continuous test method. The instruction aimed at creating the official PTC52 concentrating solar power generation performance testing standard of the American Society of Mechanical Engineers; according to the plan, the standard included other parabolic trough concentrating solar power generation technologies. However, it normally costs years of time to complete the preparation and approval works related to an official performance testing standard.

As a matter of fact, before carrying out this research, the laboratory had already carried out loop test on SEGS parabolic trough collectors (refer to Fig. 3.23). Furthermore, for technologies related to the performance of parabolic trough collectors, the laboratory had carried out a series of research works, and published the respective research reports (e.g., outdoor measurement of optical performance of the vacuum-tube-type parabolic trough evacuated tube, pipeline modeling of parabolic trough thermal collection system, rapid analysis on parabolic trough collector

FIGURE 3.23 Testing loop of SEGS parabolic trough collector.

field, heat-transfer analysis and modeling of parabolic trough evacuated tube, heat loss analysis on evacuated tube type parabolic trough evacuated tube and indoor heat loss test method, research on parabolic trough evacuated tube based on site conditions, performance model of parabolic trough solar power plant, simulation model of parabolic trough solar power plant, testing of parabolic trough collector reflective surface, solar advisory model established for parabolic trough solar power plant, technology and performance evaluation of parabolic trough solar power plant, and wind tunnel test method of parabolic trough collector [27]).

Deutsches Zentrum für Luft-und Raumfahrt (DLR) has also carried out field testing research on parabolic trough collectors or the collector field. In order to satisfy field installation requirements, testing based on field conditions requires a set of movable equipment and instruments, which mainly include the clamp-on sensor that is able to measure temperature, flow rate and inclination angle, mobile meteorological station, and data collection system. Such testing is able to accomplish the following tasks: Collector field performance evaluation (depending on field and power plant running conditions), efficiency of collector/array/loop, incidence angle influencing factor, heat loss, annual performance prediction, etc.

Besides, in 2009, by referring to the biaxial test platform of SNLA, DLR started to design the Kontas parabolic trough collector rotation test platform (refer to Fig. 3.24A). This research center also carried out studies on technologies related to parabolic trough collectors, which mainly included influences of measuring equipment on the uncertainty of parabolic trough collector performance testing, parabolic trough collector testing in the REACT project, solar flux density testing in parabolic trough

(A)

(B)

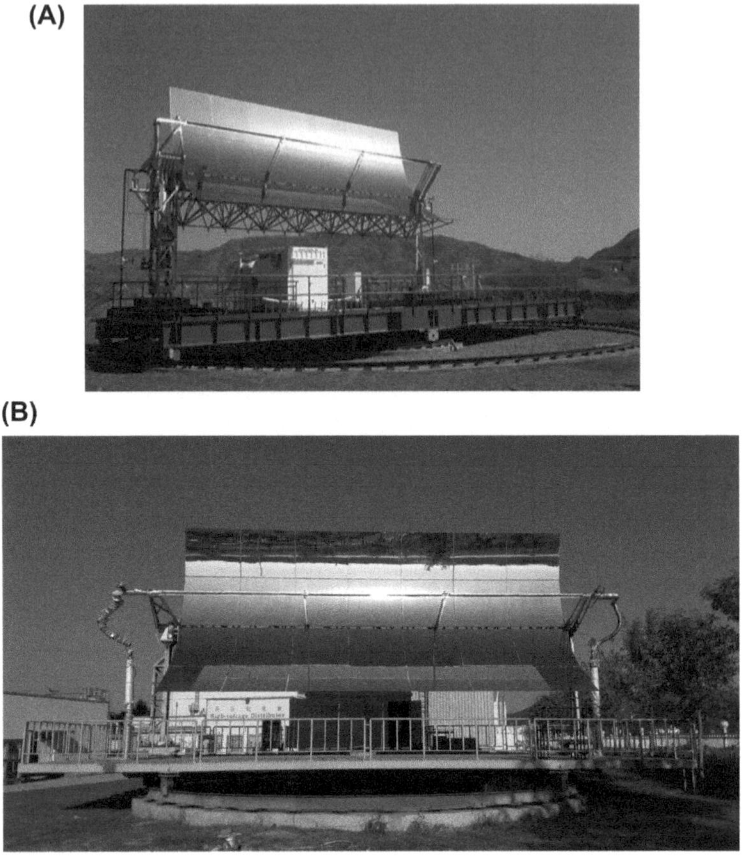

FIGURE 3.24 (A) Kontas test platform in Spain. (B) Institute of Electrical Engineering, Chinese Academy of Sciences test platform in Beijing, China.

collector concentration field, optical efficiency test platform of indoor and outdoor parabolic trough vacuum evacuated tube, transient simulation model for parabolic trough solar power plant and influences of transient variation on energy output, as well as transient thermography method for heat loss testing of parabolic trough evacuated tube.

Fig. 3.24B shows the test platform at IEE-CAS.

Except for the above-mentioned four prominent international research institutions, many scholars in the world have also carried out in-depth studies on parabolic trough collectors. For example, parabolic trough collector efficiency testing by applying the quasi steady state test method in the European standard EN 12975-2; application of American standard ASHRAE 93-2003 and European standard EN 12975-2 in parabolic trough collector performance testing and the respective comparisons; experimental model established on the basis of parabolic trough collector performance analysis; research on performance values of parabolic trough evacuated tube by comparing configurations of nonvacuum and vacuum parabolic trough evacuated tube; prediction on the truncation factor of parabolic trough collector. Furthermore, simulation and control models of parabolic trough power plant have also been established by fully considering the thermal performance of parabolic trough collectors (e.g., a linear prediction control model developed for 30 MW parabolic trough solar power plant; the dynamic simulation model for a parabolic trough solar power plant with thermal storage facilities; a concentration field analytic model that considers heat loss of nonlinear parabolic trough collectors; a PATTO model that is capable of predicting the overall performance of parabolic trough power plant under abnormal working conditions; model computer program SimulCET for parabolic trough power plant based on experiences and physical derivation).

It is a remarkable fact that technical standards of CSP generation are of great significance for the accelerated reduction of costs. Therefore, CSP generation and thermochemistry organization (SolarPACES) subordinate to International Energy Agency (IEA) is now carrying out the international project TASK I CSP generation system, which includes the development of testing procedures and standards for the parabolic trough collector field. Although many works have been conducted in order to realize standardization, thermal performance test method of parabolic trough collector still needs to be improved. EN 12975-2 standard that is applicable for concentrating solar collectors has already been an integral part of ISO 9806, which is now being revised. Under the framework of SolarPACES, a workgroup for CSP generation standards was founded in 2011. A standard must be improved while orienting toward a common framework, which shall also be intensified in terms of the following aspects, namely qualification, certification, testing procedures, components and system endurance test, entrusting procedures, model-based results, concentration field modeling, etc.

3.3.2.2 Brief Introduction to the ASHRAE 93 Steady State Test Method

The newest version of ASHRAE 93 was published in 2010 [23], which specifies that in order to be consistent with the international state standard ISO 9806-1 "Test Methods for Solar Collectors—Part 1: Thermal Performance of Glazed Liquid Working Medium Heating Collectors Including Pressure Drop" [28], partial testing process and certain requirements on measurement parameters in the previous version have been adjusted in this revised standard.

This testing standard offers the specific measurement operation process and calculation procedures, and separately determines thermal efficiencies at the aperture of receiver of parabolic trough collector under near-vertical incidence of solar radiation corresponding to different inlet temperatures of heat-transfer fluid within the collector, and optical responses of parabolic trough collector corresponding to different solar incidence angles. In order to make sure that calculation results are compared in accordance with the unified standard, thermal efficiency on the basis of the aperture area of parabolic trough collectors in ASHRAE 93 has been applied in this section for the regression of steady state test model as well as the respective prediction and calculation, refer to Eq. (3.34).

$$\eta_a = F_R\left[(\tau\alpha)_e\rho\gamma - \frac{A_r}{A_a}U_L\frac{(T_{fi} - T_a)}{G_{bp}}\right] = \frac{\dot{m}c_f(T_{fo} - T_{fi})}{A_aG_{bp}} \tag{3.34}$$

in which F_R refers to the heat remove factor of collector; $(\tau\alpha)_e$ refers to the normal absorption and transmission factor of collector; A_a refers to the area of aperture of receiver of collector; A_r refers to the area of absorber of collector; γ refers to azimuth angle coefficient correction factor; ρ refers to the altitude angle coefficient correction factor; U_L refers to the heat loss coefficient; T_{fi} refers to the inlet temperature of heat-transfer fluid of the collector; T_a refers to the ambient air temperature; G_{bp} refers to the solar DNI; $\dot{m}$ refers to the mass flow of heat-transfer fluid; c_f refers to the specific thermal capacity of heat-transfer fluid; and T_{fo} refers to the outlet temperature of collector.

For thermal efficiency testing at the aperture of receiver of parabolic trough collector under solar near-vertical incidence, within the working temperature range of parabolic trough collector, inlet temperatures of heat-transfer fluid from at least four evenly separated collectors shall be determined, and inlet temperature of heat-transfer fluid from one collector shall approach the ambient air temperature; under the maximum thermal collection, inlet temperature of heat-transfer fluid is the maximum working temperature of parabolic trough collector. For inlet temperature of heat-transfer fluid of each collector, the steady state test model needs to acquire at least four independent data points. Therefore, the total data points shall be not less than 16 points. Only in this case, can regression for steady state test model parameters be conducted.

It is necessary to achieve the strict steady state condition when conducting the collector performance testing in order to ensure the effectiveness of test data, which requires to conduct the test under a clear weather, requires solar DNI to be larger than 800 W/m^2 within the entire test period, and requires the volume flow of heat-transfer fluid passing through the parabolic trough collector to be set to the same value, etc.;

TABLE 3.4 Requirements on Measurement Parameters Within the Steady State Test Period

Parameter	Value	Parameter	Value
G_{bp}	$\pm 32\ W/m^2$	Inlet temperature of heat-transfer fluid within the collector	±2% of set value
Ambient air temperature	±1.5°C	Outlet temperature of heat-transfer fluid within the collector	±0.05°C per minute
Volume flow of heat-transfer fluid	±2% of set value	Mean ambient wind speed	2~4 m/s

and these measurement parameters shall also satisfy the allowable deviation specified in Table 3.4.

Parabolic trough solar collectors operate while tracking the movement of the Sun in the axial direction. However, the rotation test platform is capable of enabling the near-vertical incidence of solar radiation at the aperture of receiver of parabolic trough collector within the test period, which means, in order to make sure that the steady state test model is able to conduct the long-term prediction on the thermal performance of parabolic trough collectors, IAM shall be completed in another independent testing process in order to determine the variation of thermal efficiencies of parabolic trough collectors under different incidence angles. For testing the incidence angle correction factor, the inlet temperature of heat-transfer fluid within the parabolic trough collector shall be within ±1°C of the ambient air temperature. In case of not being able to satisfy such requirement, IAM $K_{\tau\alpha}$ shall be calculated in accordance with Eq. (3.35).

$$K_{\tau\alpha} = \frac{\eta_a + F_R U_L (T_{fi} - T_a)/G_{bp}}{F_R\left[(\tau\alpha)_e \rho\gamma\right]_n} \tag{3.35}$$

in which η_a refers to the intercept efficiency of collector.

The steady state test method has been the most widely recognized performance test method all over the world for parabolic trough solar collectors. Many tests have been completed in accordance with this method (e.g., the test for LS-2 parabolic trough collectors of the first commercial CSP plant (SEGS) in the world, and the test for LS-2 parabolic trough collectors mounted with PTR70 vacuum evacuated tubes, the test for parabolic trough collectors applied in the industrial field and the test for a type of fiberglass-reinforced parabolic trough collectors).

Nevertheless, for a large-scale parabolic trough solar collector in actual operation, the respective length has determined that it cannot be placed on a biaxial rotation tracking test platform, which means, the condition of near-vertical incident of solar radiation on the daylight surface of parabolic trough collector is basically unfulfillable; in addition, testing for the IAM also cannot be conducted. In the case that the parabolic trough collector is arranged horizontally along the north-south axis, condition of near-vertical incident of solar radiation on the daylight surface of parabolic trough collector only occurs once in the morning and once at night every day; yet at this moment, it happens to depend on fluctuation of increase or decrease of solar DNI, even under such circumstances, it cannot be ensured that it may occur at any given time throughout the year; instead, it depends on the latitude of the test site. In the case that the parabolic trough collector is arranged horizontally along the west-east axis, only around midday can incidence of solar radiation along the normal direction near the daylight surface of a parabolic trough collector occur.

Besides, analytic results have indicated that when parabolic trough collectors are subject to the typical intermittent heat-transfer fluid temperature variation in the heating process under cloudy conditions, the steady state test model is ineffective. Therefore, the steady state test duration may be extended by these adverse natural environment and operating conditions, especially for test sites with less favorable natural environmental conditions.

3.3.2.3 *Brief Introduction to the EN 12975-2 Quasi-Dynamic Test Method*

In order to adapt to more extensive natural environmental conditions, except for the steady test method, the European standard EN 12975-2 also provided a quasi-dynamic test method for solar collector thermal performance [29]. The respective quasi dynamic test model is established on the basis of minimum error analysis of solar collector output power. Furthermore, it has also been integrated with the ambient wind speed, sky temperature, the IAM for scattering irradiance, etc., which are shown in Eq. (3.36).

$$\begin{aligned}\frac{\dot{Q}}{A} = {} & F'(\tau\alpha)_{en}K_{\theta b}(\theta)G_{bp} + F'(\tau\alpha)_{en}K_{\theta d}G_d - c_6uG^* - c_1(T_m - T_a) \\ & -c_2(T_m - T_a)^2 - c_3u(T_m - T_a) + c_4\left[E_L - \sigma(T_a + 273.15)^4\right] \\ & -c_5dT_m/d\tau \end{aligned} \tag{3.36}$$

in which $\dot{Q}$ refers to the output power of solar collector; $F'(\tau\alpha)_{en}$ refers to the efficiency factor of collector; $k_{\theta b}(\theta)$ serves as the IAM for DNI, yet the specific function expression of the IAM provided by this standard is only applicable for flat-type solar collectors instead of parabolic trough solar collectors. $k_{\theta d}(\theta)$ refers to the collector's IAM; G^* refers to the hemispherical irradiance of aperture of receiver surface of collector; c_1, c_2, c_3, c_4, c_5 are regression coefficients; T_m refers to the mean temperature of fluid within the collector; E_L refers to the heat loss; σ is the Stefan–Boltzmann constant, $\sigma = 5.67 \times 10^{-8}$ W/(m^2 K^4); τ refers to time; u refers to ambient wind speed. Therefore, for parabolic trough solar collector under higher working temperatures and more complex optical effects, there has been no literature to demonstrate that this function expression remains to be fully convenient and effective.

By using the mathematical tool of multiple linear regression (MLR), the two IAMs, $k_{\theta b}(\theta)$ and $k_{\theta d}(\theta)$ in the quasi dynamic test model are able to be obtained together with other parameters of the model on a simultaneous basis, instead of requiring an independent IAM testing process like the steady state test method mentioned above.

Heat loss in the quasi dynamic test model is expressed by a function that contains a quadratic polynomial, and it depends on the difference between the inlet and outlet mean temperature T_m parabolic trough collector heat-transfer fluid and the ambient air temperature T_a. Furthermore, derivatives containing inlet and outlet mean temperature of heat-transfer fluid of parabolic trough collectors serve as the effective thermal capacities of collectors, in which $dT_m/d\tau$ can be obtained by calculating the difference between the current moment T_m and previous moment T_m and dividing it by the sampling interval of T_{fo} and T_{fi}.

Although the quasi dynamic test method allows the collector thermal performance test to last continuously for several hours together with solar irradiance fluctuation and solar position variation, it still needs to satisfy certain specified allowable deviation of measurement parameters, which are shown in Table 3.5. It is a remarkable fact that the testing system requires to strictly control the inlet and outlet temperature of heat-transfer fluid of parabolic trough collectors and the mass flow of heat-transfer fluid passing through parabolic trough collectors. However, for a large-scale parabolic trough solar thermal collection system, it is difficult to satisfy these test conditions based on its own control equipment.

Based on the inlet temperature of collector and the combination of natural environmental conditions that contain cloudy and clear days, testing sequences recommended by the quasi dynamic test method can be summarized into four types of testing days, and one of them is subject to the condition of partial cloud. The quasi dynamic test method still requires testing inlet temperatures of heat-transfer fluid of at least four evenly separated collectors within the working temperature range of parabolic trough collectors. Furthermore, collector thermal performance quasi dynamic test method requires each testing sequence to last for at least 3 h with an overall testing time of about five testing days.

TABLE 3.5 Requirements for Measurement Parameters Within the Quasi-Dynamic Test Period

Parameter	Value	Parameter	Value
Global solar irradiance	300–1,100 W/m^2	Inlet temperature of heat-transfer fluid within the collector	±1°C
Ambient air temperature	±1.5°C	Outlet temperature of heat-transfer fluid within the collector	1°C more than the inlet temperature of heat-transfer fluid within the collector
Mass flow of heat-transfer fluid	±1% of the set value when being within the testing sequence, ±10% of the set value when being between testing sequences	Mean ambient wind speed	1~4 m/s

However, actual quasi dynamic test time depends on natural environmental conditions of the test site.

Task 4 of the SolarPACES and task 33 of solar heating and refrigeration organization of the IEA jointly constitute the research program of Solar Thermal for Industrial Processes. In this program, according to the quasi dynamic test method, thermal performance test for the parabolic trough solar collector is conducted [28]. However, outlet temperatures of heat-transfer fluid of parabolic trough collectors applied in this program have not exceeded 250°C. Furthermore, there has been no literature to demonstrate that the quasi dynamic test method in European standard EN 12975-2 is completely applicable for the parabolic trough solar collector with an outlet temperature of heat-transfer fluid exceeding 300°C.

3.3.3 Test Methods to Determine the Thermal Performance of the Parabolic Trough Collector

3.3.3.1 *Current Status of Thermal Performance Test Method*

In 1977, America published the ASHRAE 93-2003 standard "Test Methods to Determine Thermal Performance of Solar Collectors," the newest version of which is ASHRAE 93-2010. As indicated in Section Two of the standard method, this standard is applicable for nonconcentrating and concentrating solar collectors; heat-transfer fluid flows into the collector through a single inlet and flows out through a single outlet. Apparently, the parabolic trough solar collector is influenced by the scope of the definition. It defines the thermal efficiency of solar collector as the ratio of collected useful energy and solar energy intercepted by the total area of the collector, and offers the efficiency of concentrating solar collector.

$$\begin{aligned}\eta_g &= (A_a/A_g)F_R\left[(\tau\alpha)_e\rho\gamma - (A_r/A_g)U_L\left(t_{f,i} - t_a\right)/G_{bp}\right] \\ &= \dot{m}c_p\left(t_{f,o} - t_{f,i}\right)/A_g G_{bp}\end{aligned} \tag{3.37}$$

in which for the concentrating collector, A_a refers to the aperture area of collector; A_g refers to the gross area of the collector; A_r refers to the aperture area of receiver of the collector.

For the concentrating collector, Eq. (3.37) generates a linear relationship between the thermal efficiency η_g and parameter $(T_{fi} - T_a)/G_{bp}$. The intercept of this linear equation on y axis is $(A_a/A_g)F_R(\tau\alpha)_e\rho\gamma$, the respective slope is $(A_r/A_g)F_RU_L$. Furthermore, the product $(\tau\alpha)_e\rho\gamma$ varies along with the incidence angle. For many collectors, a linear efficiency curve is sufficient, but for some collectors, it may need a high-order fitting curve.

In order to determine the thermal characteristics of solar collectors, the test shall be conducted under clear weather conditions, while maintaining the incidence of solar radiation near the normal of collector aperture of receiver, namely ensuring the influences of incidence angle on collector thermal efficiency not exceed 2% of the efficiency of collector at aperture of receiver under vertical incidence of solar radiation.

In order to determine the thermal efficiency curve of the collector by applying the two-parameter [$F_R(\tau\alpha)_e\rho\gamma$ and F_RU_L] solar collector thermal performance test model in Eq. (3.37), at least 16 data points shall be measured. These two parameters can be determined on the basis of the regression by applying the least-square method, which is shown in Fig. 3.25. Intercept of the line through regression on the vertical axis and the respective slope of the line are the values of these two parameters. After determining these two parameters, the test model can be used to predict the whole-day solar collector output energy according to different operating temperatures, natural environment data and the IAM by using "hour" as the time calculation unit through accumulation of hourly output energy of the collector, which is of great significance for the solar thermal collection system designer.

As a matter of fact, the comprehensive heat loss coefficient (U_L) that represents the thermal conduction, convection, radiation, and heat exchange losses of receiver of the collector is not a constant, but a function of absorber temperature, ambient air temperature and wind speed. Although U_L obtained through statistical regression analysis is a fixed

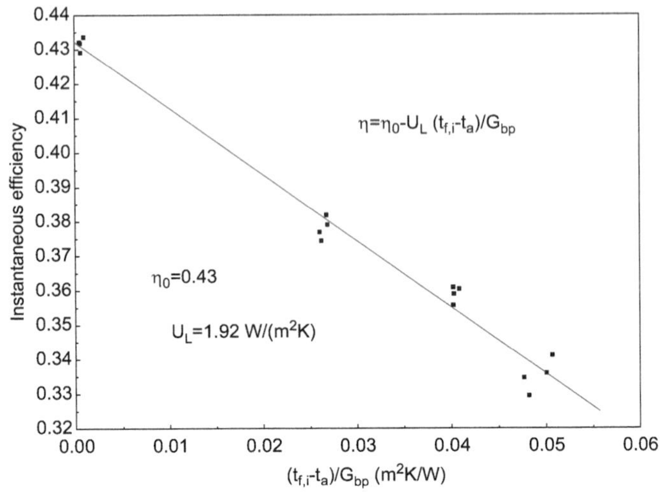

FIGURE 3.25 Collector transient efficiency curve.

value, which is not completely consistent with its strict physical meaning, this method does not have any influence on the prediction of long-term thermal performance of collectors. It is because influences of various factors in the actual process on U_L are insignificant, which have been widely verified. In the ASHRAE 93 standard, a calculation example has offered the testing model based on parameters determined through regression to calculate the hourly output energy of collector and the whole-day output energy of collector after accumulation, the calculation conditions and results of which are shown in Table 3.6 [23].

As described in the section of "significance and application," this test method is intended to provide test data essential to the prediction of the thermal performance of a collector in a specific system application in a specific location. In addition, to the collector test data, such prediction requires validated collector and system performance simulation models that are not provided by this test method. The results of this test method therefore do not by themselves constitute a rating of the collector under test. Furthermore, it is not the intent of this test method to determine collector efficiency for comparison purposes since efficiency should be determined for particular applications." Therefore, such solar collector test method is not offered to conduct quality rating or search for certain quality parameters or indices of a particular component of a collector under specific conditions; the fundamental target is to design a collector application system. As thermal performance of collector is changeable, when comparing two solar collectors, daily output energy and annual cumulative output energy of collector must be considered, instead of

TABLE 3.6 Calculation of Hourly Useful Thermal Collection of Solar Collector [23]

Time/ h	T_a/ °C	G/(W/ m^2)	G_{bp}/ (W/m^2)	$\Delta t/G$/ (m^2 K/W)	θ/ (°)	$K_{\tau\alpha}$	(q_u/A_g)/ (W/m^2)
6	25.56	41	0	0.5141	93.4	0	0
7	26.11	189.24	116.7	0.1086	79.6	0.58	0
8	26.67	394.24	305.93	0.0507	66.2	0.84	78.85
9	29.44	583.48	488.86	0.0296	53.5	0.93	252.31
10	31.11	731.71	633.94	0.0211	42.5	0.96	381.63
11	32.78	826.33	728.56	0.0167	33.4	0.98	469.94
12	33.33	861.02	763.25	0.0155	30	0.98	498.32
13	34.44	826.33	728.56	0.0148	33.4	0.98	476.24
14	35.56	731.71	633.94	0.0187	42.1	0.96	400.55
15	36.11	583.48	488.86	0.0181	53.5	0.93	280.7
16	36.11	394.24	305.93	0.0264	66.2	0.84	119.85
17	35.56	189.24	116.7	0.0593	79.6	0.58	0
18	35	41	0	0.2851	93.4	0	0

q_u, useful energy.

comparing parameters themselves that have been obtained through regression.

Thermal performance test method of solar collector follows a basic idea. Thermal performance test of solar collector aims at designing a collector application system. Therefore, it is necessary to take into consideration main thermal performance characteristics related to the operation of collectors; by measuring various parameters related to thermal performance of collector, according to the respective thermal performance mathematical model, mathematical methods of statistical regression can be used to identify the undetermined coefficient in the model; based on the thermal performance model, other physical conditions can be calculated and predicted, such as the whole-day and annual cumulative useful output energy of collector under solar irradiance, ambient air temperature and system operating temperature of a specific day. It is a remarkable fact that along with the increase of influencing factors for operating conditions of solar collector, it is necessary to use a more complex mathematical model to describe the respective thermal performance on a reasonable basis, and further offer the precise whole-day and annual cumulative output energy of collector under changing input conditions.

There are two basic methods to establish the mathematical model. One is the mechanism analysis method, which is to establish a model by utilizing modeling information or prerequisite provided by modeling assumptions on the basis of the analysis on the internal mechanism of matters; it is normally referred to as the white box, such as the energy balance equation and thermal and mass transfer theory. The other one is the system identification method, which is to establish a model by utilizing modeling assumptions or actual input and output information of matter system offered to system test data while being absolutely ignorant of the internal mechanism of system; it is normally referred to as the black box. According to different principles, model identification methods can be summarized into four categories: (1) the least-square method, including the least-square method, the extended least-square method, auxiliary variable method and the generalized least-square method; (2) gradient correction parameter identification method, such as stochastic approximation method; (3) probability density approximation parameter identification method, such as the maximum likelihood method; (4) new methods that have been recently developed, such as the blur identification method, neural network identification method, wavelet identification method and inheritance identification method.

To sum up, thermal performance test model of solar collector is to establish the mechanism model according to the energy balance theory and heat-transfer principle, and identify the undetermined parameters of the model by applying the least-square method; it belongs to a grey box model. Such test model based on empirical methods is deemed to be able to generate highly precise results with the respective range of applied parameters.

3.3.3.2 Assumed Conditions of Dynamic Test Model

Dynamic test model of parabolic trough solar collector shall satisfy the following assumptions.

1. Based on the requirement of ASHRAE 93 standard on volume flow of heat-transfer fluid being within ±2% of the set value during the test, according to the dynamic test model assumption, volume flow variation of synthetic oil passing through the parabolic trough collector within the test period does not exceed ±2% of the mean value of the period. However, considering that density variation of synthetic oil within the application temperature normally exceeds 10%, such as the density variation of synthetic oil applied under Experimental Conditions of this research, which is shown in Fig. 3.26. Therefore, mass flow of synthetic oil passing through the parabolic trough collector within the test period is assumed to be within ±20% of the mean value without any intensive fluctuation of

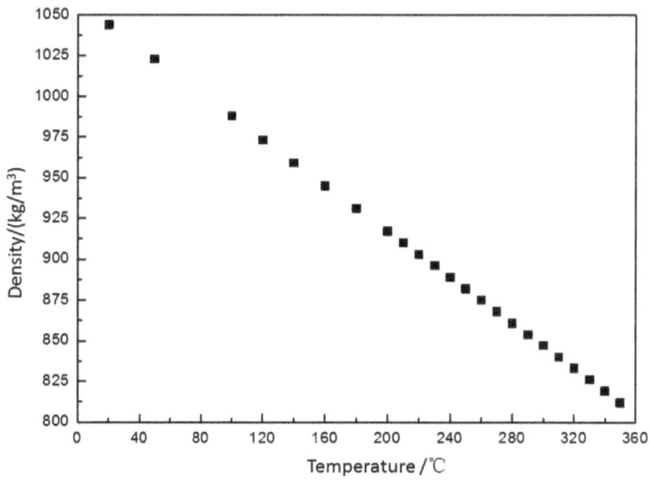

FIGURE 3.26 Density variation of synthetic oil.

abrupt change, such as the stoppage of circulating pump or instantaneous change of valve opening control during normal operation.

2. Based on all experimental data obtained through the test, cleanliness of mirror surface of parabolic trough concentrator and exterior wall surface of glazed shield tube of parabolic trough evacuated tube are consistent with each other, such as cleaning these two surfaces before the test.
3. For the connecting and supporting components at the glass-metal sealing point, thermal stress buffer segment and between two pieces of evacuated tubes, etc., due to the extremely small proportions of their scales against the total length, their influences on the heat-transfer process are not individually considered.
4. Profile defect and installation error of parabolic trough concentrator, and positioning error of parabolic trough evacuated tube, are constants.

3.3.3.3 Model Establishment for Heat-Transfer Process

Based on the Lumped Capacitance Method, energy balance equation for the metal evacuated tube of parabolic trough collector is listed and expressed as follows

$$C_b = \frac{dT_b}{d\tau} = SA_a - A_f U_{bf}(T_b - T_f) - A_{am} U_{ba}(T_b - T_a) \tag{3.38}$$

in which C_b refers to the thermal power of metal evacuated tube; T_b refers to the temperature of metal evacuated tube; T_f refers to the temperature of heat-transfer fluid; T_a refers to the ambient air temperature; A_a refers to the aperture area of parabolic trough collector; A_f refers to the heat exchange area between metal evacuated tube and heat-transfer fluid; A_{am} refers to the heat exchange area between metal evacuated tube and the environment; U_{bf} refers to the heat-transfer coefficient between metal tube and heat-transfer fluid; U_{ba} refers to the comprehensive heat-transfer coefficient between metal evacuated tube and the environment; τrefers to time; S refers to the part absorbed by the exterior wall surface of metal evacuated tube when solar DNI is perpendicular to the aperture of parabolic trough collector. The last one to the right of the equation is to express the heat exchange process between metal evacuated tube and the neighboring environment, which is normally within the glazed shield tube.

Similarly, the energy balance equation can be established for the heat-transfer fluid within the metal tube of parabolic trough collector, which can be expressed as

$$C_f \frac{dT_f}{d\tau} = A_f U_{bf}(T_b - T_f) - \dot{m}c_f(T_{fo} - T_{fi}) \tag{3.39}$$

in which C_f refers to the thermal power of heat-transfer fluid; c_f refers to the specific thermal capacity of heat-transfer fluid; T_{fi} refers to the inlet temperature of the tube; T_{fo} refers to the outlet temperature of the tube; $\dot{m}$ refers to the mass flow.

$1/A_f U_{bf}$ and $1/A_{am} U_{ba}$ are substituted with thermal resistance R_{bf} and R_{ba} separately, then Eq. (3.38) and Eq. (3.39) can be once again expressed as:

$$C_b \frac{dT_b}{d\tau} = SA_a - \frac{T_b - T_f}{R_{bf}} - \frac{T_b - T_a}{R_{ba}} \tag{3.40}$$

And

$$C_f \frac{dT_f}{d\tau} = \frac{T_b - T_f}{R_{bf}} - \dot{m}c_f(T_{fo} - T_{fi}) \tag{3.41}$$

Then Eq. (3.41) is reorganized into

$$\frac{T_b}{R_{bf}} = C_f \frac{dT_f}{d\tau} + \frac{T_f}{R_{bf}} + \dot{m}c_f(T_{fo} - T_{fi}) \tag{3.42}$$

On both sides of the above equation with time τ, the derivative is taken, then

$$\frac{1}{R_{bf}} \frac{dT_b}{d\tau} = C_f \frac{d^2T_f}{d\tau^2} + \frac{1}{R_{bf}} \frac{dT_f}{d\tau} + \dot{m}c_f \left(\frac{dT_{fo}}{d\tau} - \frac{dT_{fi}}{d\tau} \right) \tag{3.43}$$

In order to eliminate T_b from Eq. (3.40), Eq. (3.42) and Eq. (3.43) are substituted into it, then the two energy balance equations can be combined as

$$C_b C_f \frac{d^2 T_f}{d\tau^2} + \frac{C_b R_{ba} + C_f R_{ba} + C_f R_{bf}}{R_{bf} R_{ba}} \frac{dT_f}{d\tau} + \frac{T_f}{R_{bf} R_{ba}} = \frac{A_a S}{R_{bf}} - \frac{R_{ba} + R_{bf}}{R_{bf} R_{ba}} \dot{m} c_f (T_{fo} - T_{fi}) - C_b \dot{m} c_f \left(\frac{dT_{fo}}{d\tau} - \frac{dT_{fi}}{d\tau} \right) + \frac{T_a}{R_{bf} R_{ba}} \tag{3.44}$$

Outlet temperature T_{fo} of heat-transfer fluid within the parabolic trough collector is selected as the lumped temperature of heat-transfer fluid within the metal absorber tube. Thus Eq. (3.44) is changed into

$$\begin{aligned} \frac{1 + (R_{bf} + R_{ba}) \dot{m} c_f}{C_b C_f R_{bf} R_{ba}} T_{fo} = & -\frac{d^2 T_{fo}}{d\tau^2} - \left(\frac{1 + \dot{m} c_f R_{bf}}{C_f R_{bf}} + \frac{R_{bf} + R_{ba}}{C_b R_{bf} R_{ba}} \right) \frac{dT_{fo}}{d\tau} \\ & + \frac{\dot{m} c_f}{C_f} \frac{dT_{fi}}{d\tau} + \frac{(R_{bf} + R_{ba}) \dot{m} c_f}{C_b C_f R_{bf} R_{ba}} T_{fi} \\ & + \frac{1}{C_b C_f R_{bf} R_{ba}} T_a + \frac{A_a}{C_b C_f R_{bf}} S \end{aligned} \tag{3.45}$$

Based on the above, a differential equation can be deduced

$$\frac{d^2 T_{fo}}{d\tau^2} + A \frac{dT_{fo}}{d\tau} + B T_{fo} = C \frac{dT_{fi}}{d\tau} + D T_{fi} + E S + F T_a \tag{3.46}$$

in which

$$A = \frac{1 + \dot{m} c_f R_{bf}}{C_f R_{bf}} + \frac{R_{bf} + R_{ba}}{C_b R_{bf} R_{ba}} \tag{3.47}$$

$$B = \frac{1 + (R_{bf} + R_{ba}) \dot{m} c_f}{C_b C_f R_{bf} R_{ba}} \tag{3.48}$$

$$C = \frac{\dot{m} c_f}{C_f} \tag{3.49}$$

$$D = \frac{(R_{bf} + R_{ba}) \dot{m} c_f}{C_b C_f R_{bf} R_{ba}} \tag{3.50}$$

$$E = \frac{A_a}{C_b C_f R_{bf}} \tag{3.51}$$

$$F = \frac{1}{C_b C_f R_{bf} R_{ba}} \tag{3.52}$$

Apparently, $D = B - F$, then, Eq. (3.46) can be reorganized into:

$$\frac{d^2T_{fo}}{d\tau^2} + A\frac{dT_{fo}}{d\tau} + B(T_{fo} - T_{fi}) = C\frac{dT_{fi}}{d\tau} + ES - F(T_{fi} - T_a) \tag{3.53}$$

Heat losses of parabolic trough collector toward the environment includes not only thermal convection with the surrounding air, but also the radiation heat exchange with the sky; thus the last column that represents heat loss in Eq. (3.53) can be expressed by two terms, one of which takes the form of a quadratic term, then Eq. (3.53) can be modified as

$$\frac{d^2T_{fo}}{d\tau^2} + A\frac{dT_{fo}}{d\tau} + B(T_{fo} - T_{fi}) = C\frac{dT_{fi}}{d\tau} + ES - F(T_{fi} - T_a) - G(T_{fi} - T_a)^2 \tag{3.54}$$

As a matter of fact, for application-level large-scale parabolic trough collectors, inlet temperature T_{fi} and outlet temperature T_{fo} of heat-transfer fluid within the collector that have been measured simultaneously do not correspond to the two parameters of Eq. (3.54) in time; a time lag relationship exists between T_{fi} and T_{fo}, and it is necessary to consider flow time τ_p of heat-transfer fluid from the inlet to the outlet of parabolic trough collector. Therefore, the corresponding actual relationship of them in the dynamic test model can be expressed as:

$$T_{fo}(\tau + \tau_p) = f[T_{fi}(\tau)] \tag{3.55}$$

in which τ_p depends on the length L of parabolic trough collector and mean flow rate v of heat-transfer fluid during the test, which can be expressed as

$$\tau_p = L/v \tag{3.56}$$

However, in order to achieve a concise model expression, in the dynamic test model of this section, $T_{fo}(\tau + \tau_p)$ and $T_{fi}(\tau)$ are no longer specially marked; instead, they will be considered during experimental data treatment, model identification and thermal performance prediction and calculation.

3.3.3.4 Establishment of Optical Model

An optical model is established based on S in Eq. (3.54), which aims at offering the physical relationship between it and solar DNI G_{DN} and the respective mathematical expression. As S is the section absorbed by the exterior wall surface of metal tube when solar DNI is perpendicular to the aperture of parabolic trough collector, it is a parameter that cannot be directly measured; it requires considering influences of parabolic trough concentrator reflection and transmission absorption of evacuated tube. This section of the model involves a parabolic trough collector truncation factor γ, specular reflectance ρ, transmittance τ, absorptance a, cosine factor, end loss correction factor and IAM in optical mechanism analysis.

Based on $K_{\tau a}$ defined in the equation, influence of variation of truncation factor γ along with the incidence angle shall also be integrated in IAM. Thus incidence angle correction factor $K_{\gamma\tau\alpha}$ shall be expressed as Eq. (3.57), in which transmittance-absorptance-absorptance product $(\gamma\tau\alpha)$ is a collective property of parabolic trough collector, and varies along with the incidence angle θ. In addition, when the incidence angle equals zero, vertical incidence of solar beam onto the aperture of parabolic trough collector occurs; then, in this case, the product is $(\gamma\tau\alpha)_n$.

$$K_{\gamma\tau\alpha} = \frac{(\gamma\tau\alpha)}{(\gamma\tau\alpha)_n} \tag{3.57}$$

An incidence angle comprehensive correction coefficient $K(\theta)$ is defined as the product of cosine factor $F_{\cos}$, end loss correction factor F_{end} and incidence angle correction factor $K_{\gamma\tau\alpha}$, which can be expressed as

$$K(\theta) = F_{\cos}F_{\text{end}}K_{\gamma\tau\alpha} \tag{3.58}$$

$K_{\gamma\tau\alpha}$ is calculated by applying the following empirical equation

$$K_{\gamma\tau\alpha} = 1 + a_1\frac{\theta}{\cos(\theta)} + a_2\frac{\theta^2}{\cos(\theta)} \tag{3.59}$$

in which both a_1 and a_2 are constants to be determined through experiments.

Thus S can be expressed as

$$S = G_{\text{DN}}\rho(\gamma\tau\alpha)_{\text{n}}K(\theta) = G_{\text{DN}}\rho(\gamma\tau\alpha)_{\text{n}}F_{\cos}F_{\text{end}}K_{\gamma\tau\alpha} \tag{3.60}$$

Then the calculation equation of cosine factor $F_{\cos}$, calculation equation of end loss correction factor F_{end} and calculation Eq. (3.59) of incidence angle correction factor $K_{\gamma\tau\alpha}$ are substituted into Eq. (3.60), and thus S can be expressed as

$$\begin{aligned} S &= G_{\text{DN}}\rho(\gamma\tau\alpha)_n \cos(\theta)\left[1 + a_1\frac{\theta}{\cos(\theta)} + a_2\frac{\theta^2}{\cos(\theta)}\right]\left[1 - \frac{f}{L}\tan(\theta)\right] \\ &= G_{\text{DN}}\left\{\begin{array}{l} \rho(\gamma\tau\alpha)_n \cos(\theta)\left[1 - \frac{f}{L}\tan(\theta)\right] + \rho(\gamma\tau\alpha)_n a_1\theta\left[1 - \frac{f}{L}\tan(\theta)\right] \\ +\rho(\gamma\tau\alpha)_n a_2\theta^2\left[1 - \frac{f}{L}\tan(\theta)\right] \end{array}\right\} \end{aligned} \tag{3.61}$$

in which $\rho(\gamma\tau\alpha)_n$ refers to the maximum optical efficiency of parabolic trough collector.

In order to simplify the model, three constants are proposed, including

$$E_0 = E_p(\gamma\tau\alpha)_n \tag{3.62}$$

$$E_1 = E_p(\gamma\tau\alpha)_n a_1 \tag{3.63}$$

$$E_2 = E_p(\gamma\tau\alpha)_n a_2 \tag{3.64}$$

Then three functions relevant to incidence angle are defined as:

$$I_0(\theta) = \cos(\theta)\left[1 - \frac{f}{L}\tan(\theta)\right] \tag{3.65}$$

$$I_1(\theta) = \theta\left[1 - \frac{f}{L}\tan(\theta)\right] \tag{3.66}$$

$$I_2(\theta) = \theta^2\left[1 - \frac{f}{L}\tan(\theta)\right] \tag{3.67}$$

Thus ES in Eq. (3.54) can be expressed as

$$ES = [E_0I_0(\theta) + E_1I_1(\theta) + E_2I_2(\theta)]G_{DN} \tag{3.68}$$

In order to reduce the influence of fluctuation of measured solar DNI on the dynamic measurement model, the metal tube is divided into p sections along the flow direction of heat-transfer fluid, with the length of each section being referred to as L_p, which is shown in Fig. 3.27. Value of p depends on the flow time τ_p of heat-transfer fluid passing from the inlet to the outlet of a parabolic trough collector and sampling interval τ_s of experimental data, which can be expressed as

$$p = \tau_p/\tau_s \tag{3.69}$$

From section 1 to section p, solar DNI of heat-transfer fluid within each independent region corresponding to different time points are distinguished from each other. Thus by considering the heat-transfer fluid

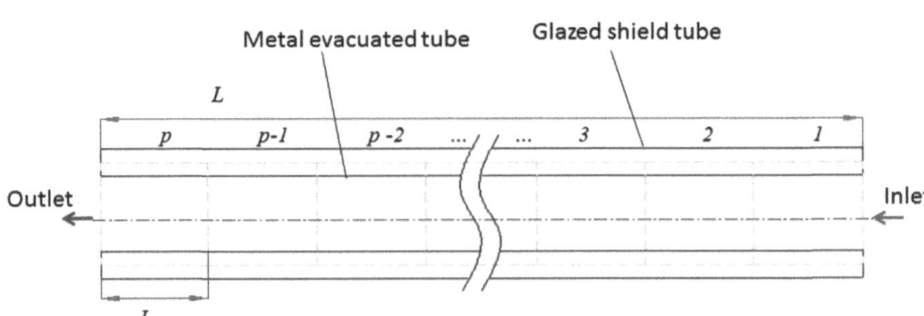

FIGURE 3.27 Section division of metal evacuated tube.

passing through the parabolic trough collector, the solar DNI G_p under averaging effect can be expressed as

$$G_p = \left\{ \begin{array}{l} G_{DN}(\tau) + G_{DN}(\tau + \tau_s) + G_{DN}(\tau + 2\tau_s) + \cdots \\ +G_{DN}[\tau + (p-2)\tau_s] + G_{DN}[\tau + (p-1)\tau_s] \end{array} \right\} \Big/ p \quad (3.70)$$

G_{DN} in Eq. (3.68) is replaced with G_p, and it is substituted into Eq. (3.54), then

$$\frac{d^2T_{fo}}{d\tau^2} + A\frac{dT_{fo}}{d\tau} + B(T_{fo} - T_{fi}) = C\frac{dT_{fi}}{d\tau} + [E_0I_0(\theta) + E_1I_1(\theta) + E_2I_2(\theta)] \\ G_p - F(T_{fi} - T_a) - G(T_{fi} - T_a)^2 \quad (3.71)$$

3.3.3.5 *Dynamic Test Model and Parameter Identification*

Product of G_p and $I_0(\theta)$ is referred to as G_{eni}, and both sides of Eq. (3.71) are divided by this variable. In addition, in terms of experimental data, the second derivative $d^2T_{fo}/d\tau^2$ in Eq. (3.71) may bring uncertainty to the prediction results of dynamic test model, which thus can be removed. So, the ultimate expression of thermal performance dynamic test model of parabolic trough solar collector is

$$\frac{T_{fo} - T_{fi}}{G_{eni}} = e_0 + e_1\frac{\theta}{\cos(\theta)} + e_2\frac{\theta^2}{\cos(\theta)} + a\frac{1}{G_{eni}}\frac{dT_{fo}}{d\tau} + b\frac{1}{G_{eni}}\frac{dT_{fi}}{d\tau} \\ + c\frac{T_{fi} - T_a}{G_{eni}} + d\frac{(T_{fi} - T_a)^2}{G_{eni}} \quad (3.72)$$

in which

$$e_0 = \frac{E_0}{B}$$

$$e_1 = \frac{E_1}{B}$$

$$e_2 = \frac{E_2}{B}$$

$$a = -\frac{A}{B}$$

$$b = \frac{C}{B}$$

$$c = -\frac{F}{B}$$

$$d = -\frac{G}{B}$$

In the equation, e_0, e_1, e_2, a, b, c and d are seven undetermined parameters. It is necessary for them to obtain experimental data for identification by utilizing parabolic trough collector test; G_{eni} is an effectively averaged direct irradiance while considering cosine loss, end section loss of tube and the influences of heat-transfer fluid that passes through the parabolic trough collector. It depends on the measured solar normal direct irradiance G_{DN}, cosine factor F_{cos}, end loss correction factor F_{end}, flow time τ_p of heat-transfer fluid passing from the inlet to the outlet of a parabolic trough collector, and sampling interval τ_s of experimental data.

$dT_{fo}/d\tau$ and $dT_{fi}/d\tau$ are two first-order derivatives, which need to be based on the discretization method in the governing equation of numerical heat transfer by utilizing the differential expression of derivatives deduced through the Taylor expansion method, and handled by applying the mean difference method, then

$$\frac{dT_{fo}}{d\tau}(n) = \frac{T_{fo}(n+1) - T_{fo}(n-1)}{2\Delta\tau}$$

$$\frac{dT_{fi}}{d\tau}(n) = \frac{T_{fi}(n+1) - T_{fi}(n-1)}{2\Delta\tau}$$

in which n refers to the quantity of experimental data during the test ($n > 1$); $\Delta\tau$ refers to the equivalent time interval of two random adjacent numbers of experimental data.

The first three terms to the right of Eq. (3.72) refer to optical characteristics of parabolic trough collector varying along with the incidence angle; the fourth and fifth terms refer to the effective thermal capacity of absorber and heat-transfer fluid of parabolic trough collector; whereas the last two terms refer to heat losses of parabolic trough collector. They are mainly determined by the difference between inlet temperature of heat-transfer fluid within the parabolic trough collector and ambient air temperature, which also include the dual influences of radiation heat exchange loss and convection heat exchange loss. In addition, there is a certain relationship between the dynamic test model and the steady state test model in ASHRAE 93 standard.

Although Eq. (3.72) is not a linear equation, a linear expression can still be obtained through the treatment toward the respective quadratic term, based on which, thermal performance dynamic test model of parabolic trough solar collector applies MLR on the basis of the least-square serial methods as the method to identify the seven undetermined coefficients.

In order to verify the dynamic test model, it shall be applied in the parabolic trough solar collector in this research. Thus by using the experimental condition I test data in Section 3.3.3.6, and applying the mathematical method of MLR, an identified equation for thermal performance dynamic test of parabolic trough collector can be obtained as follows:

$$T_{fo} - T_{fi} = \left[0.182 - 0.00731\frac{\theta}{\cos(\theta)} + 0.000106\frac{\theta^2}{\cos(\theta)}\right]G_{\text{eni}} - 68.379\frac{dT_{fo}}{d\tau} + 33.941\frac{dT_{fi}}{d\tau} - 0.00571(T_{fi} - T_a) - 0.0000217(T_{fi} - T_a)^2 \tag{3.73}$$

Regression results of seven coefficients of e_0, e_1, e_2, a, b, c and d are analyzed. One major index is the coefficient of determination R^2, which is 0.86; it measures the fitting degree of the independent variable of regression toward the dependent variable. Other major indices include least-square estimated value and standard error, which have been listed in Table 3.7.

3.3.3.6 Thermal Performance Prediction of Dynamic Test Model

The dynamic test model assumes the first-order derivative term of outlet temperature of heat-transfer fluid within the parabolic trough collector to be zero, calculates T_{fo} by utilizing the known variables G_{eni}, θ, T_{fi} and T_a, and uses it as the initial value; then it applies the Newton iteration method, and finally predicts a reasonable outlet temperature of heat-transfer fluid within the parabolic trough collector. In order to weaken the influences caused by test conditions fluctuation, time τ_p of heat-transfer fluid passing through the parabolic trough collector is used

TABLE 3.7 Table of Dynamic Test Model Parameter Regression Analysis

Coefficient	Least-Square Estimated Value	Standard Error	t_i	P (>|t|)
e_0	0.182	1.07×10^{-2}	17.042	0
e_1	-7.34×10^{-3}	4.77×10^{-4}	−15.301	0
e_2	1.06×10^{-4}	7.45×10^{-6}	14.198	0
a	−68.379	25.703	−2.660	8.13×10^{-3}
b	33.941	2.190	15.500	0
c	-5.66×10^{-3}	3.84×10^{-4}	−14.856	0
d	-2.17×10^{-5}	1.88×10^{-6}	−11.543	0

t_i refers to the test statistics, which is equivalent to ratio of the least-square estimated value to the standard error of regression coefficient; p refers to the degree of freedom, which is equivalent to the probability of t distribution being larger than the absolute value of t_i under the difference of the quantity of experimental data and that of regression coefficients.

as the reference when implementing smoothing processing toward the predicted outlet temperature data of heat-transfer fluid within the parabolic trough collector.

In order to clearly express the prediction results, difference between the measured and the predicted outlet temperatures of heat-transfer fluid within the parabolic trough collector is offered as the absolute error, and used to divide the measured outlet temperature of heat-transfer fluid within the parabolic trough collector in order to obtain the ratio that can be used as the relative error of prediction results.

Furthermore, in order to demonstrate the working effect of the collector, it is also necessary to calculate the collector efficiency, which refers to the ratio of the output energy from parabolic trough collector to the solar DNI projected into the concentration field during operation. Equation in ASHRAE 93 standard is applied in this book, and solar irradiance G_{bp} that considers the cosine effect is selected as the denominator of efficiency calculation equation of energy projected into the concentration field, which is shown in Eq. (3.74). There is also some literature that has only applied solar direct normal irradiation G_{DN} as the denominator of the efficiency equation without considering the influence of cosine loss.

$$\eta = \frac{\int c_{oil}\dot{m}(T_{fo} - T_{fi})d\tau}{A_a \int G_{bp} d\tau} \tag{3.74}$$

in which c_{oil} refers to the specific thermal capacity of synthetic oil used in this book. The manufacturer has provided its values corresponding to different temperatures, which are shown in Fig. 3.28.

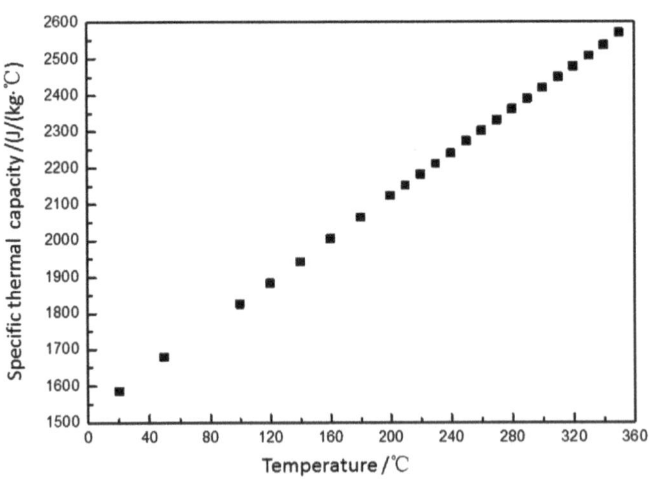

FIGURE 3.28 Variation of specific thermal capacity of synthetic oil.

In order to simplify the calculation, based on the observation, these data have high degrees of linearity. Thus according to these data, a linear equation can be obtained as follows through fitting method:

$$c_{\text{oil}} = 1528.32 + 2.973T_{\text{oil}} \tag{3.75}$$

in which T_{oil} refers to the synthetic oil temperature. In order to specify the reasonability and precision of this method, four typical experimental conditions are discussed separately as follows [30].

3.3.4 Experimental Condition I

Based on the meteorological data of Experimental Condition I and collector inlet fluid data G_{DN}, T_{fi} and T_a, as well as the solar radiation incidence angle θ relevant to solar position and solar irradiance G_{eni} that considers cosine effect and end effect correction, predicted value of fluid outlet temperature of parabolic trough collector can be obtained through iteration calculation by applying Eq. (3.72).

The calculated value and predicted value have been compared in Fig. 3.29. D and M separately refer to outlet temperatures of heat-transfer fluid within the parabolic trough collector obtained through the dynamic prediction method and experimental measurement. Within the test period of Experimental Condition I, namely from 10:07 to 12:46, the predicted value perfectly fits to the experimental value. The obvious difference of both values appears within 2 min after 11:16. The maximum outlet

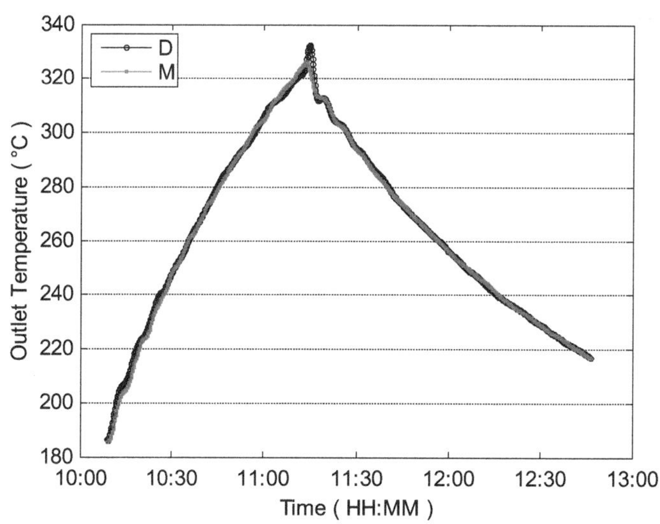

FIGURE 3.29 Predicted value of collector outlet temperature by dynamic model under experimental condition I compared with the experimental measurement.

temperature 326°C of heat-transfer fluid within the parabolic trough collector obtained through experimental measurement appears at 11:16, whereas the predicted maximum outlet temperature 332°C of heat-transfer fluid within the parabolic trough collector by the dynamic test equation appears at 11:17. Such a difference is mainly caused by the necessary time to rotate the aperture of receiver of parabolic trough collector to the poor-light side. Within this tracker adjustment period, the actual incidence of solar direct radiation onto the aperture surface of a parabolic trough collector cannot reduce to zero instantaneously. Thus prediction by applying this dynamic test model may result in lagging and errors.

By drawing the tendency of differences between the measured and the predicted outlet temperatures of heat-transfer fluid within the parabolic trough collector, Fig. 3.30 has clearly indicated the quantity of predicted values that exceeds the measured value by −4°C within the parabolic trough collector adjustment period; however, within the entire test period, most of the differences between the measured value and the predicted value fall in a range of −4 and 2°C, especially for the predicted value in the cooling process, which are within ±1°C of the measured value. As shown in Fig. 3.31, further analysis on the relative error also indicates that the relative error of the predicted value mainly appears within ±1%.

In Fig. 3.32, for the parabolic trough collector output power variation curve D predicted through the dynamic test model and the output power

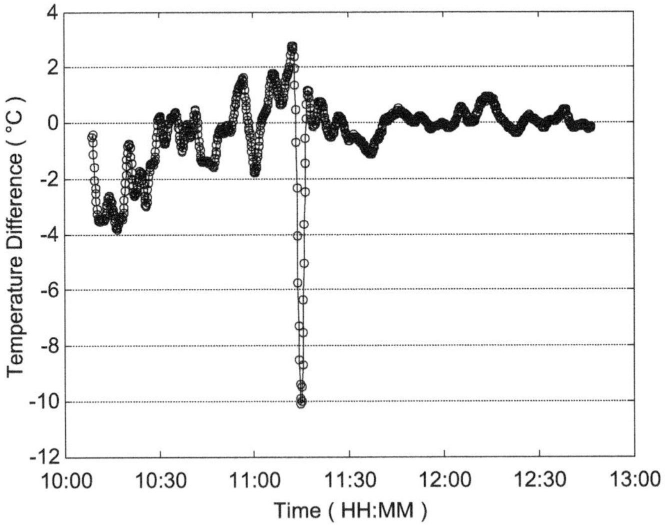

FIGURE 3.30 Absolute error of dynamic model on the predicted value under experimental condition I.

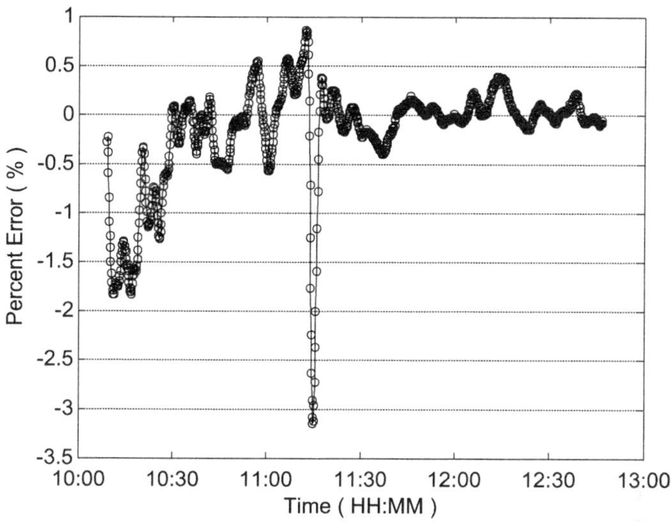

FIGURE 3.31 Relative error of dynamic model on the predicted value under experimental condition I.

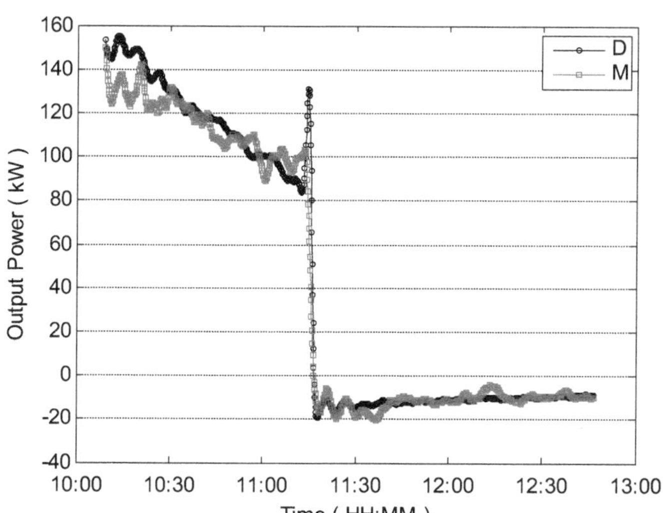

FIGURE 3.32 Predicted value of collector output power by dynamic model under experimental condition I.

variation curve M obtained through the measured value calculation, differences of some values from beginning of the test and parabolic trough collector state adjustment period have exceeded 20 kW, but in general, no matter for the heating process or cooling process of parabolic trough collector, most of the data coincide with each other. This is because an undetermined coefficient of dynamic test model has been obtained through regression of this group of experimental data.

3.3.5 Experimental Condition II

Based on G_{DN}, T_{fi} and T_a of Experimental Condition II, by applying Eq. (3.72) of dynamic test equation for parabolic trough collector, Fig. 3.33 has drawn the dynamic test model prediction curve D and experimental measurement curve M of outlet temperature of heat-transfer fluid within the parabolic trough collector. Within the entire test period, namely from 12:10 to 13:01, parabolic trough collectors have all remained in the tracking status. Thus the measured outlet temperature of heat-transfer fluid within the parabolic trough collector increases from 185 to 307°C, and the outlet temperature of heat-transfer fluid within the parabolic trough collector predicted through the dynamic test model also increases from 185 to 306°C. Except for some slight deviation after 12:39, temperature increase tendency of prediction curve and measurement curve are consistent with each other.

In Fig. 3.34, the specific value corresponding to this deviation has been clearly presented, which is caused by a short-term cloudy. Within 1 min before or after 12:45, the dynamic model has predicted appearance of the maximum absolute error of outlet temperature of heat-transfer fluid within the parabolic trough collector. Some data points indicate that the measured value is 3°C higher than the predicted value. Within the period from 12:53

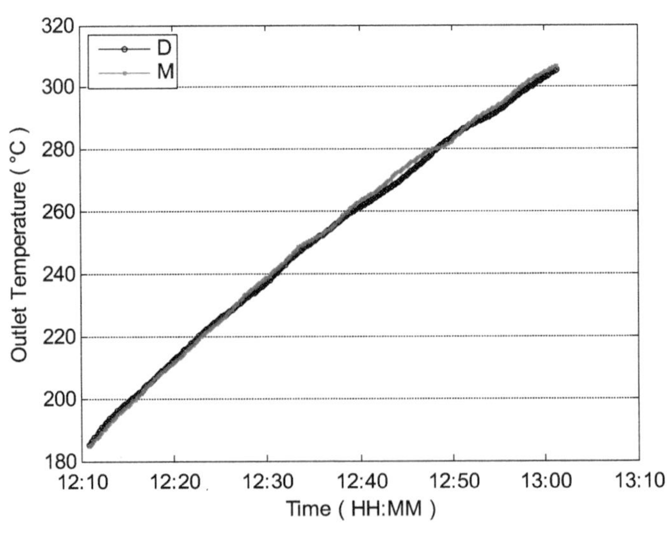

FIGURE 3.33 Predicted value of collector outlet temperature by dynamic model under experimental condition II.

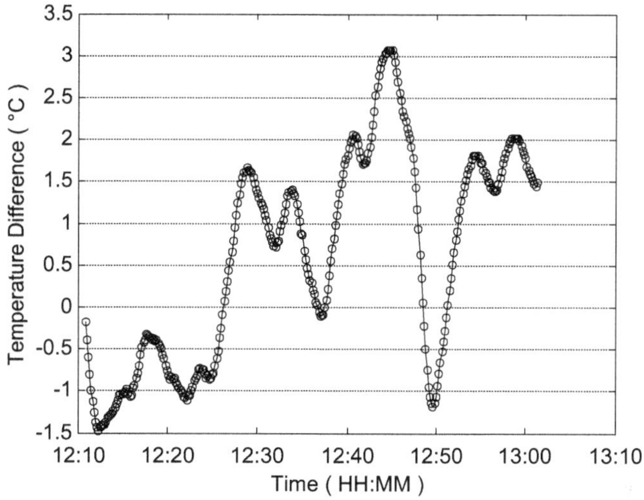

FIGURE 3.34 Absolute error of dynamic model on the predicted value under experimental condition II.

to the end of the test, absolute errors of some predicted values fluctuate between 1.5 and 2°C. However, before 12:40, the difference between the measured and the predicted outlet temperatures of heat-transfer fluid within the parabolic trough collector mainly concentrates within ±1.5°C.

Based on the relative error analysis, as shown in Fig. 3.35, relative error of the predicted values mainly exists within ±0.8%; even the maximum

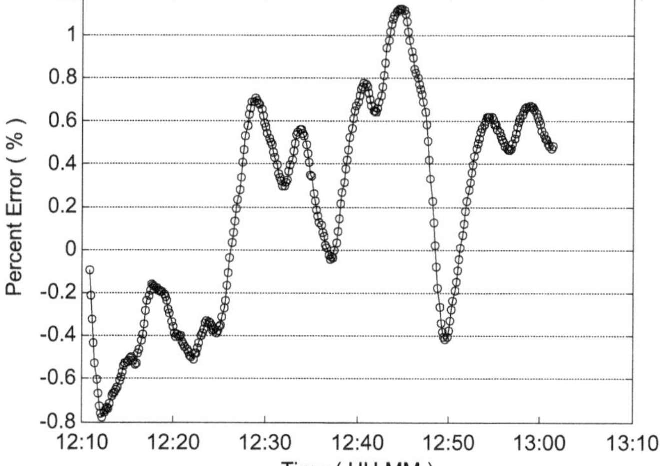

FIGURE 3.35 Relative error of dynamic model on the predicted value under experimental condition II.

relative error within 1 min before or after 12:45 is less than 1.2%, which is also at the nearly same level with the relative error under Experimental Condition I. It indicates that under the natural condition without excessive fluctuations and under the stable operating condition of parabolic trough collectors, even if the inlet temperature of heat-transfer fluid within the parabolic trough collector constantly increases, the dynamic test model for parabolic trough collector is able to show a considerable thermal performance prediction effect.

For parabolic trough collector output power predicted through the dynamic test model and calculated based on the experimental measurement data, as shown in Fig. 3.36, their values mainly vary within a range of 110–130 kW. Although dynamic test model prediction curve D and value calculation curve M through experimental measurement within 1 min before or after 12:45 have been obviously distinguished from each other, it is also the point where the maximum absolute error of predicted values of dynamic test model lies; however, difference between and among these values at the same time point has not exceeded 20 kW.

By focusing on Experimental Condition II, comparison and error analysis of predicted value of thermal performance dynamic test model of parabolic trough collector, experimental measurement data and the respective calculated value have demonstrated that for the test conditions similar to experimental data applied in the coefficient regression of the model, the dynamic test model is able to achieve satisfactory thermal performance prediction results of parabolic trough collector.

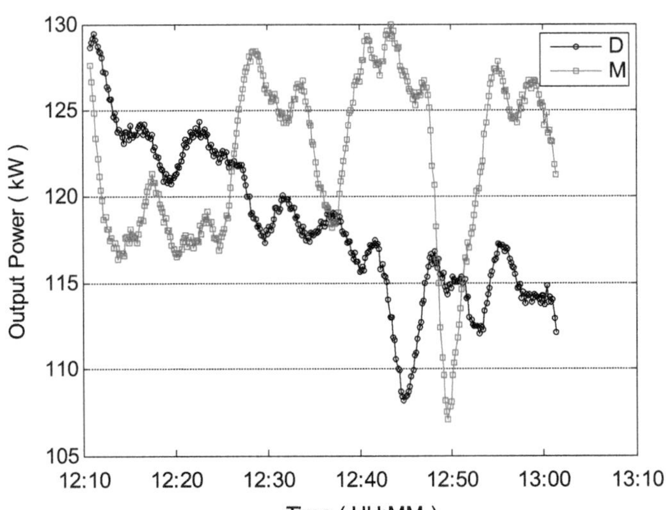

FIGURE 3.36 Prediction of dynamic model on collector output power under experimental condition II.

3.3.6 Experimental Condition III

During the period from 12:32–11:32, as influenced by frequent obnubilation, solar DNI of Experimental Condition III fluctuates greatly between 700 W/m^2 and zero. Due to such fluctuation, no matter for the experiment measured value of inlet temperature of heat-transfer fluid within the parabolic trough collector or the respective calculated value based on G_{DN}, T_{fi} and T_a through the dynamic test model, they have all undergone significant changes, which is shown in Fig. 3.37. Especially for the predicted value of the dynamic test model, such influence is more intensive. For example, within 3 min before or after 10:50, predicted value of the dynamic test model on the inlet temperature of heat-transfer fluid within the parabolic trough collector primarily increases from the minimum value 264°C to the maximum value 287°C, then drops to 266°C. The two changes have both exceeded 20°C. Yet within the same period, the difference between the maximum and the minimum experiment measured values is not more than 10°C. However, it is a remarkable fact that predicted value fluctuation of thermal performance dynamic test model of parabolic trough collector has always been centering the experiment measured value, and both of them enjoy a consistent variation tendency.

Fig. 3.38 has further indicated that although the maximum value of the difference between the measured and the predicted outlet temperatures of heat-transfer fluid within the parabolic trough collector may approach ±15°C, absolute error of the predicted value of dynamic test model mainly exists within ±5°C with an approximate symmetric distribution

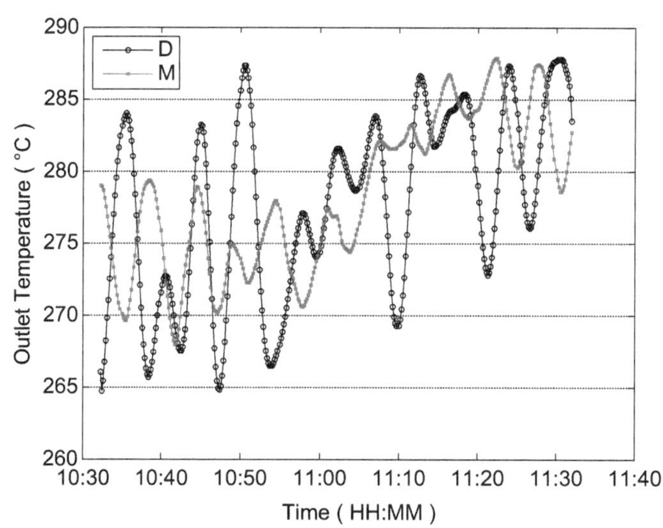

FIGURE 3.37 Predicted value of dynamic model on collector outlet temperature under experimental condition III.

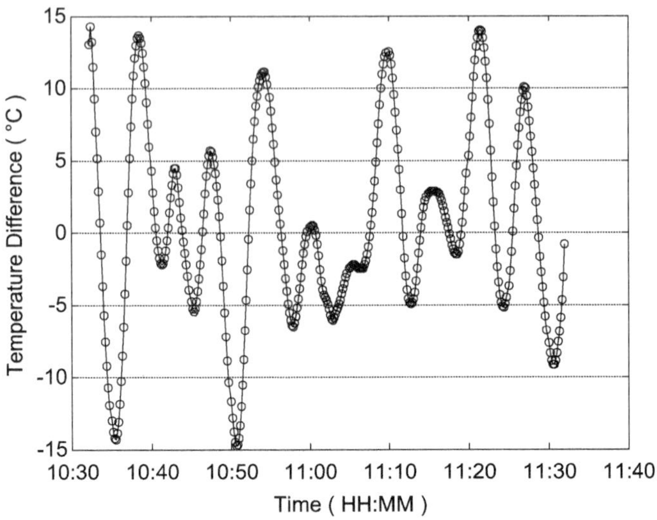

FIGURE 3.38 Absolute error of dynamic model on predicted value under experimental condition III.

around zero. For long-term thermal performance evaluation of parabolic trough solar collector, such distribution is conductive for offsetting some prediction errors.

In Fig. 3.39, according to the relative error analysis, for test conditions containing the significant fluctuation of solar DNI, the maximum

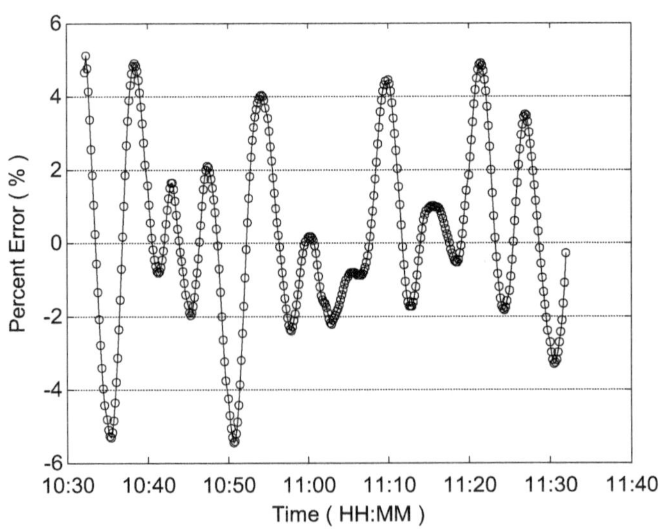

FIGURE 3.39 Relative error of dynamic model on predicted value under experimental condition III.

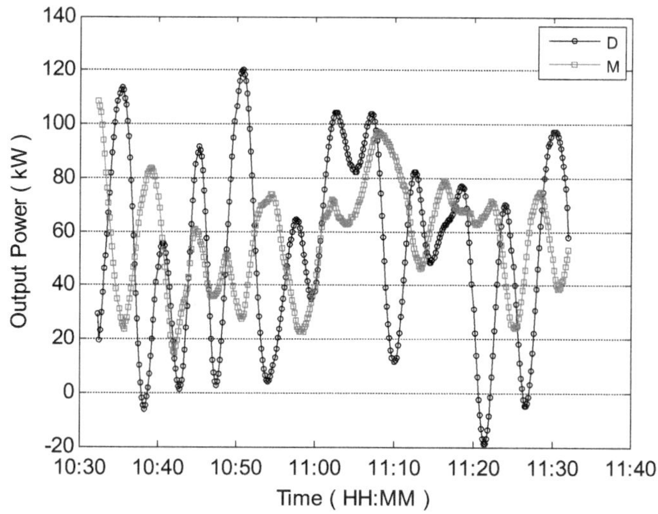

FIGURE 3.40 Prediction of dynamic model on collector output power under experimental condition III.

relative error of outlet temperature of heat-transfer fluid within the parabolic trough collector predicted by the dynamic test model has exceeded ±5%, but the relative error has still mainly concentrated within ±2%; in addition, it has always had approximate symmetric fluctuation around zero.

According to the variation of parabolic trough collector output power within a range of 15–110 kW calculated on the basis of the experiment measured value, as shown in Fig. 3.40, in the predicted value of dynamic test model on parabolic trough collector output power, there are individual data points that are less than zero. This is obviously incorrect. It has also indicated that in case of significant fluctuation of solar DNI, dynamic test model is not able to ensure the precision of every transient value of parabolic trough collector output power.

Based on test conditions under Experimental Condition III, by comparing the predicted value of thermal performance dynamic test model of parabolic trough collector and the experiment measured value as well as the respective calculated value, absolute error, and relative error, they all indicate that in case of being influenced by adverse condition of frequent obnubilation, the dynamic test model is not able to ensure the high precision of every transient predicted value. However, its predicted data are able to fluctuate around the measured value, so that the long-term thermal performance prediction effect of parabolic trough collector can be ensured.

3.3.7 Experimental Condition IV

Based on the measured data under Experimental Condition IV, for two cooling processes of parabolic trough collector, outlet temperature of heat-transfer fluid of parabolic trough collector predicted by the dynamic test model separately decreases from 338 to 166°C and from 231 to 168°C, the corresponding experiment measured values separately decrease from 325 to 168°C and from 222 to 170°C, which are shown in Fig. 3.41. Although there is a significant deviation of the predicted value and experiment measured value around 10:23 and 14:12, namely during tracking status variation of parabolic trough collector, both of them coincide with each other as a whole.

In Fig. 3.42, within the period of 10:09–10:23, namely the primary thermal charging stage of parabolic trough collector, the absolute error of outlet temperature of heat-transfer fluid within the parabolic trough collector predicted by the dynamic test model exceeds the experiment measured value by $-5\sim-10$°C; it even exceeds by -15°C at certain data points. However, for the subsequent primary cooling stage of parabolic trough collector, absolute error of the predicted value of dynamic test model mainly exists within ± 1°C. When once again adjusting the parabolic trough collector to be in the tracking status, the reheating stage of parabolic trough collector appears between 13:24 and 14:12. Absolute error of the predicted value of dynamic test model concentrates within ± 10°C, and presents an approximate symmetric distribution around their zero value. In the last cooling stage of parabolic trough collector, predicted value of dynamic test model is 3°C less than the experiment measured value.

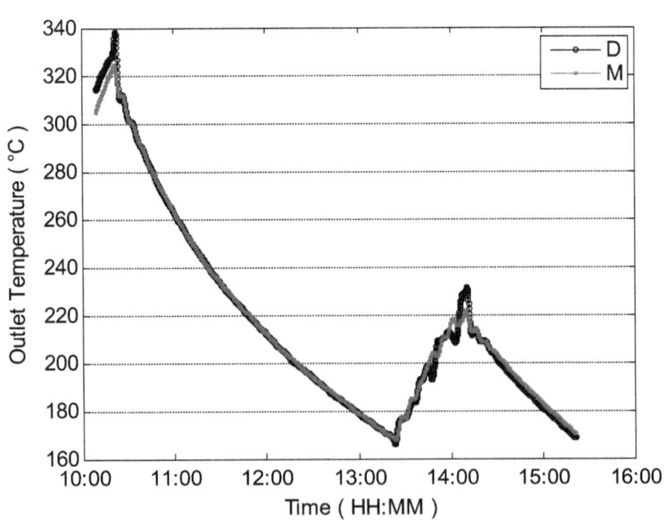

FIGURE 3.41 Predicted value of dynamic model on collector outlet temperature under experimental condition IV.

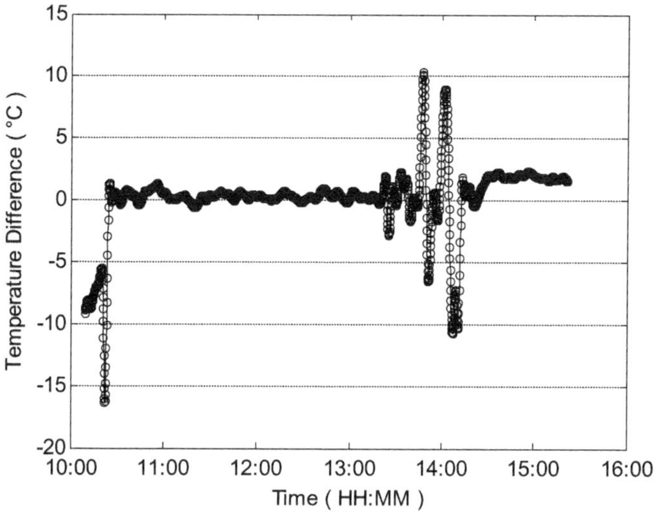

FIGURE 3.42 Absolute error of dynamic model on predicted value under experimental condition IV.

In Fig. 3.43, according to the relative error analysis, relative error of outlet temperature of heat-transfer fluid within the parabolic trough collector predicted by the dynamic test model does not exceed ±5%, in which the two data points when the relative error approaches ±5% mainly appear in the status adjustment of parabolic trough collector and

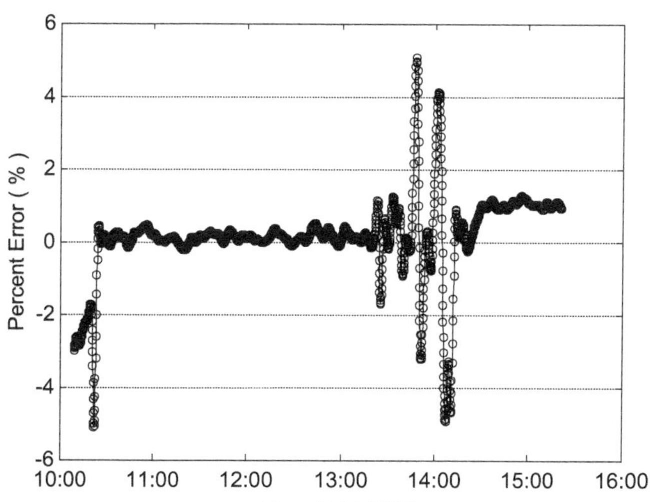

FIGURE 3.43 Relative error of dynamic model on predicted value under experimental condition IV.

the reheating stage under influences of intensive cloud. However, for two cooling processes of parabolic trough collector, relative error of the predicted value of dynamic test model mainly exists within ±1%. As shown in Fig. 3.44, for the primary heating process of parabolic trough collector, predicted value of the dynamic test model on the output power of parabolic trough collector mainly varies between 145 and 183 kW; whereas the calculated value of experiment test data falls in a range of 100–111 kW, and this is the testing stage with the largest deviation of these two values. During 10:23–13:24, in the primary cooling process of parabolic trough collector, these two values fall in a range of −2 ~ −19 kW with a consistent variation tendency. The dynamic test model has manifested excellent prediction effects.

Based on test conditions under Experimental Condition IV, by comparing and analyzing the predicted value of dynamic test model and the experiment measured value as well as the respective calculated value, the absolute error of ±3°C and relative error of ±1% have indicated that thermal performance dynamic test model of parabolic trough solar collector is able to ensure the prediction effect in the cooling process of parabolic trough collector.

In order to further demonstrate the effect of thermal performance dynamic test model of parabolic trough collector in engineering applications, based on the four typical experimental conditions, and according to Eq. (3.74), output energy and mean thermal efficiency of parabolic trough collector corresponding to each test period can be separately calculated, results of which have been listed in Table 3.8.

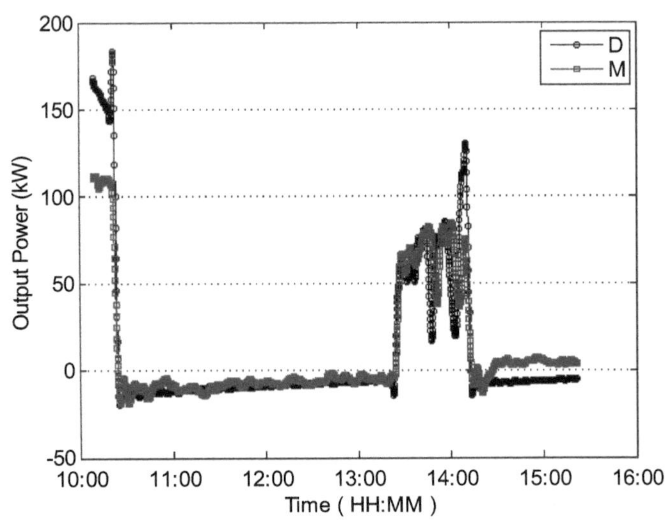

FIGURE 3.44 Prediction of dynamic model on collector output power under experimental condition IV.

TABLE 3.8 Predicted Value of Dynamic Test Model and Experiment Measured Value and Respective Calculated Value

Experimental Condition	Time	Model-Predicted Output Energy/MJ	Output Energy Calculated Based on Measured Value/MJ	Solar Energy/MJ	Model-Predicted Thermal Efficiency/%	Thermal Efficiency Calculated Based on Measured Value/%	Error/%
1	10:09–12:47	418.56	394.92	1296.9	32.27	30.45	1.82
2	12:11–13:01	359.84	371.56	1006.4	35.75	36.92	−1.16
3	10:32–11:32	194.94	206.58	676.18	28.83	30.55	−1.72
4	10:09–5:21	192.03	198.88	783.57	24.51	25.38	−0.87

By integrating the above analysis, based on the prediction effect of dynamic test model on parabolic trough collector, experimental requirements and process of performance dynamic test of parabolic trough collector can be summarized: Thermal performance dynamic test of parabolic trough solar collector shall be conducted under a clear days; the whole testing process is required to enjoy a solar DNI over 700 W/m^2; variation of volume flow of heat-transfer fluid passing through the parabolic trough collector shall be within ±2%. According to the scope of application of parabolic trough collector, a specific variation scope of outlet temperature of heat-transfer fluid within the specific parabolic trough collector can be determined. The testing process consists of two stages, namely outlet temperature ascent process of heat-transfer fluid within the parabolic trough collector under normal tracking concentrating conditions, and outlet temperature descent process of heat-transfer fluid within the parabolic trough collector under conditions when adjusting the aperture of receiver of parabolic trough collector to the poor-radiation side. In addition, the test shall be conducted around the time of the day when the maximum incident angle appears; for example, for the most common parabolic trough solar collector arranged horizontally on the north-south axis, the tracking test for parabolic trough collector shall be conducted within 2 h before or after high noon every day.

3.4 BASIC DATA REQUIRED BY POWER PLANT DESIGN

1. Solar radiation data and meteorological data of CSP plant, neighboring areas and local meteorological stations.
2. Soil and Water Conservation Program Report
3. Water Resources Argumentation Report
4. Geological Survey Report
5. Geological Hazard Risk Evaluation Manual
6. 1:50,000 topographic map within 10 km of extended site scope and 1:2000 topographic map within site scope
7. Process Gas and Water-Supply Conditions
8. Correspondence of Uncovered, Identified Major Mineral Resources within Land Use Scope of Site Selection during Project Construction
9. Grid Access System Report
10. Environmental Impact Assessment Report
11. Price data of local construction materials, equipment and labor cost

3.5 MAJOR PARAMETERS AND PRINCIPLES OF DESIGN

1. Installed capacity of power plant, estimated annual power generation.
2. Concentrating mode: Tower-type, parabolic trough-type or others, concentrator area.
3. Thermal-absorbing and heat-transfer medium and the respective maximum working temperature: Water/steam, synthetic oil and the respective maximum working temperature, molten salt, and its maximum working temperature
4. Requirements on energy storage device: Quantity of thermal storage tanks, thermal storage capacity, temperature, category of materials, category and quantity of circulating pumps
5. Evaporator: Heat exchange pattern, volume, applicable medium, circulating pump
6. Auxiliary boiler: Whether there is any auxiliary heating system, type of auxiliary fuel (fuel or natural gas), annual consumption and source of auxiliary fuel.
7. Shape and dimension of receiver: Cavity diameter and cylinder size of tower-type; length and diameter of the evacuated tube of parabolic trough-type.
8. Rated steam inlet parameters of steam turbine (main steam pressure and temperature), rated power, minimum stable load, unit thermal efficiency.
9. Heat-transfer fluid circulating equipment: Circulating pump, expansion tank etc.
10. Control: Concentration field control mode; entire field control mode.
11. Concentrator cleaning facilities and method: Manual cleaning, machine cleaning, dry cleaning, washing.
12. Steam turbine condensed mode, dry-cooled or wet-cooled, annual water consumption.
13. Efficiency: Annual mean efficiency of concentration field, annual maximum efficiency of concentration field, annual mean efficiency of receiver, annual mean efficiency of power plant.
14. Power plant initiation time under typical solar irradiation and meteorological conditions.
15. CSP plant access system voltage grade and boost mode.
16. Thermal power plant grid access metering gateway point designed at the division point of property.
17. Permanent engineering land use area.

18. Maximum annual water consumption during project construction and total annual water consumption during project operation.
19. Power plant construction period: In northwestern areas of China, the construction period of a 50 MW power plant normally does not exceed 30 months; whereas that of a 100 MW power plant normally does not exceed 42 months.
20. Power plant operation period: Normally to be 25 years

3.6 DESCRIPTION OF GENERAL PARAMETERS OF THE POWER PLANT

Below are the items described in the overall technical parameters of an CSP plant; after completion of design, explanation to the following items shall be provided.

1. Concentrator: Area and dimension of aperture of a single concentrator. Tracking modes include azimuth/altitude angle, working wind speed corresponding to open-loop/close-loop design precision, protective wind speed.
2. Concentration field: Concentrating forms include parabolic trough and tower, etc. Maximum power projected on the receiver from the concentration field, concentration field aperture area.
3. Receiver: Structural forms mainly include cavity-type, cylinder-type, and evacuated tube-type. Receiver water inlet temperature, receiver outlet pressure, receiver outlet temperature, receiver fluid flow rate.
4. Steam turbine generator: Inlet steam pressure of steam turbine; inlet steam temperature of steam turbine; maximum temperature of condensate water; generator outlet voltage.
5. Thermal storage: Thermal capacity, temperature, volume, necessary time for full-load power generation of steam turbine.
6. Backup emergency power supply: power, AC uninterruptible power supply and 220 V DC power supply
7. Power access: Voltage, transformer power
8. Grid access: Voltage, current, and time period

3.7 CALCULATION OF ANNUAL POWER GENERATION

In a regular project, it is normally the owner who firstly proposes to construct a power plant with an expected capacity, and the design unit shall calculate annual power generation of the power plant and concentration field area based on the capacity requirement proposed by the owner. Annual power generation of the power plant is related to solar to electricity conversion efficiency at the design point, concentration field area, and solar irradiation resources. The specific calculation process has been explained in details as follows.

3.7.1 Calculation by Applying the Design Point Method

In annual power generation calculation, except for such invariable factors as solar irradiation resources and weather, others are all variables.

The key point for annual power generation calculation is to determine the collector field power, which includes the concentration field area and receiver power. The concentration process, thermal receiver, thermal storage, and heat exchange are coupled to power generation, which shall be calculated by using multiple factors simultaneously on the basis of system energy balance. At a typical time point, such as the design point, the concentration field area depends on the rated input of steam turbine, rated input of thermal storage, and operating mode of power plant corresponding to the design point. It is normally required that at the design point, the concentration field provides energy to the receiver, output power of which shall be not less than the sum of rated input of steam turbine and rated input of thermal storage. At this moment, the power plant is basically able to ensure the full-load power generation during the day and several hours of power generation during the night. In this case, the rated input power of thermal storage shall be the ratio of the capacity of thermal storage to its thermal-charging operation hours.

Fig. 3.45 shows the annual power generation calculation process. Firstly, irradiation and meteorological conditions shall be determined; then, a concentration field area shall be assumed. When considering the output of concentration field as the input of receiver, the output power of receiver shall be equivalent to the sum of required input power of steam turbine and thermal storage. In case of the condition not being satisfied, it is necessary to reassume the concentration field area until the request is satisfied.

After determining the concentration field area, the system efficiency can be calculated based on the concentrating efficiency, receiver efficiency, generating efficiency, etc. Then, annual power generation of the system can be calculated.

Thus the basic idea for calculating the annual power generation is that under specified geographic location, meteorological conditions, solar irradiation conditions, steam turbine capacity and thermal storage capacity, the concentration field area and the annual mean generating efficiency of CSP generation system can be determined.

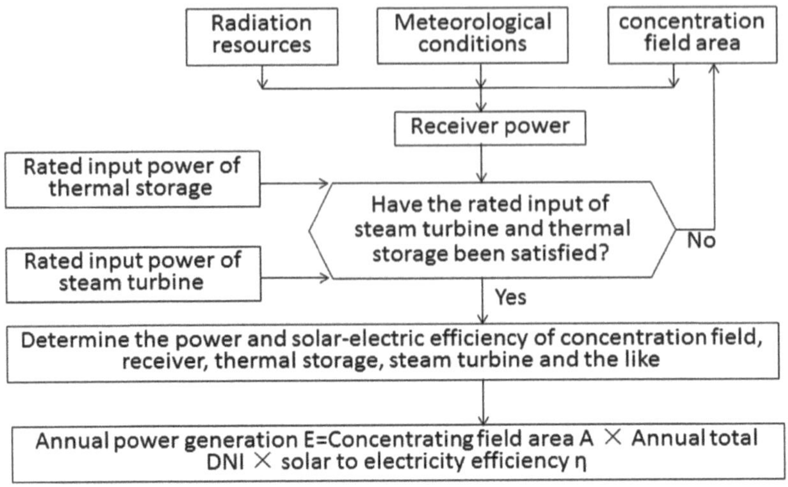

FIGURE 3.45 Annual power generation calculation process.

Thermal storage capacity normally depends on the night peak modulation power supply capacity (product of power and generation hours). In case of only focusing on night power generation, concentration field area shall satisfy the rated thermal request of steam turbine corresponding to night generation hours. However, such operational condition has been rarely seen. In case of no thermal storage, output of concentration field at the design point is the rated input of steam turbine.

Step I: The design point irradiation and meteorological conditions, and the steam turbine rated power P_{TURBINE} are defined.

Step II: The concentration field area is assumed to be A.

Step III: Power plant output power at the design point can be calculated according to the process described in Table 3.1. In case that P satisfies:

$$|P - P_{\text{TURBINCE}}| < \varepsilon \tag{3.76}$$

Then the calculation process ends, and concentration field area A is obtained. Otherwise, a new concentration field area shall be assumed for another calculation, until Eq. (3.76) is satisfied. ε refers to the allowed margin of error of iteration, which shall be determined before calculation, such as $\varepsilon = 50$ kW.

System photoelectric efficiency η at the design point is calculated as follows: assuming DNI $= 1000$ w/m^2 at the design point

$$\eta = \frac{P_{\text{TURBINE}}}{1000\ \text{A}} \tag{3.77}$$

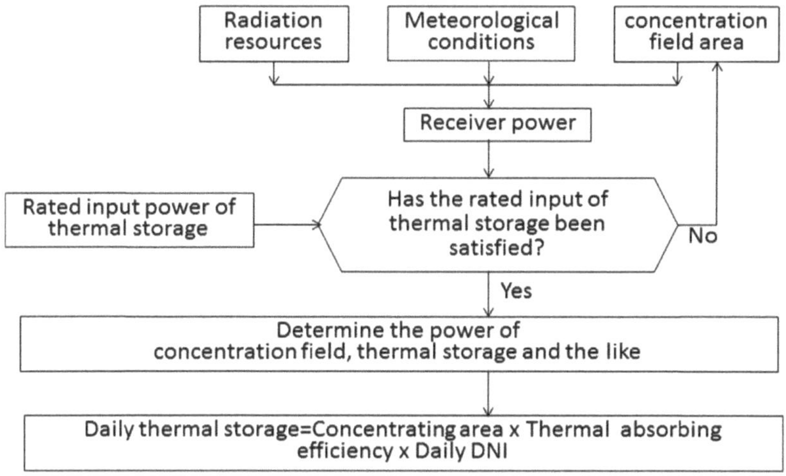

FIGURE 3.46 Determination of thermal storage and concentrating area.

In case of substituting the annual mean efficiency with the photoelectric efficiency of the system, the annual power generation E can be calculated as follows:

$$E = \text{DNI} \times A \times \eta \tag{3.78}$$

In case of the system being equipped with a thermal storage, after determining the corresponding concentration field area of steam turbine, the corresponding concentration field area of thermal storage can be calculated, which is shown in Fig. 3.46. Thermal storage is calculated in a different way with thermal radiation in the denomination of "day." For a system equipped with a thermal storage, the total area of the concentration field equals to the sum of the concentration field area calculated according to Fig. 3.45 corresponding to the steam turbine and the required concentration field area corresponding to thermal storage.

3.7.2 Calculation Example for Annual Power Generation

It is known that in a certain location, the annual total DNI = 1850 kWh/m^2, annual mean solar irradiance is 750 W/m^2, and annual mean ambient air temperature is 10°C. Steam turbine parameters of the 50 MW power tower plant equipped with 4 h thermal storage are shown in Table 3.9.

Provided that thermal energy is accumulated through 4 h of full-load operation of steam turbine, the heliostat concentration field area and annual power generation of power plant are required to be calculated.

TABLE 3.9 Parameters of Steam Turbine

S/N	Content	Unit	Value
1	Power Range	MW	30–50
2	Sliding Pressure Working Scope (load)	%	30–110
3	Steam Parameter Range	MPa	3–9
4	Rated Power	MW	50
	Main Steam Pressure	MPa	9.2
	Main Steam Temperature	°C	360–383
	Rated Air Inflow	t/h	226

Analysis and solution: Midday of the spring equinox is taken as the design point, solar irradiance takes the value of annual mean solar irradiance of 750 W/m^2, design point ambient air temperature takes the value of annual mean ambient air temperature of 10°C. Output power of collector field at the design point is required to exceed the sum of required input power of generator unit and thermal storage power.

Electricity generation capacity corresponding to the project is 50 MW, and the rated input thermal power is 150 MW.

The rated input storage thermal power is calculated as follows:

Daily required thermal storage = 4 h × 150 MW = 600 MWh. Assuming the thermal storage working duration is 6 h during the daytime, then the storage thermal power equals 600 MWh/6 h = 100 MW.

150 MW generate power during the daytime directly, another 100 MW using for the night. Required output thermal power from the receiver at the design point can be obtained as

$$150 + 100 = 250(\text{MW}) \tag{3.79}$$

Step I: The required heliostat concentration field area is assumed to be 100,000 m^2, and the receiver is cylinder. Based on the concentration field design software, the efficiency of concentration field at the design point can be calculated, which is 68%, and the intercept factor of receiver is 100%.

In this case, output power $P_{\text{concentrator}}$ of mirror field is:

$$P_{\text{concentrator}} = 68\% \times 100\% \times 100{,}000 \times 0.75 = 51\ (\text{MW})$$

According to the method specified in 3.2, the efficiency of receiver can be calculated, which is 90%. In this case, the output P_{receiver} of receiver is:

$$\text{P}_{\text{receiver}} = 51 \times 90\% = 45.9\ (\text{MW})$$

Comparing with Eq. (3.79): 250 MW/45.9 MW ≈ 5.44.

Step II: Considering the concentration field area is basically proportional to the output, and the decrease of concentration field efficiency after increase of the scale of concentration field, the concentration field area is reassumed to increase by 5 times and reach up to 500,000 m^2. Based on the concentration field design software, the output efficiency of concentration field at the design point can be calculated, which is 63%, and the intercept factor of receiver is 95%.

In this case, the output power $P_{concentrator}$ of concentration field is:

$$P_{concentrator} = 63\% \times 95\% \times 500{,}000 \times 0.75 = 224.4\ (\mathrm{MW})$$

According to the method specified in 3.2, the efficiency of receiver can be calculated, which is 85%.

In this case, the output $P_{receiver}$ of receiver is:

$$P_{receiver} = 224.4 \times 85\% = 191\ (\mathrm{MW})$$

The result is less than 250 MW of Eq. (3.79).

Step III: The concentration field area is reassumed to increase by 35% and reach up to 675,000 m^2. Based on the concentration field design software, the output efficiency η_{hel} of concentration field at the design point can be calculated, which is 62%, and the intercept factor η_{int} of receiver is 94%.

In this case, the output power $P_{concentrator}$ of concentration field is:

$$P_{concentrator} = 62\% \times 94\% \times 675{,}000 \times 0.75 = 295\ (\mathrm{MW})$$

According to the method specified in 3.2, the efficiency $\eta_{receiver}$ of receiver can be calculated, which is 84%.

In this case, the output $P_{receiver}$ of receiver is:

$$P_{receiver} = 295 \times 84\% = 247.8\ (\mathrm{MW})$$

The result is nearly 250 MW in Eq. (3.79), which satisfies the calculation requirement.

From the above, the concentration field area of the power plant is 675,000 m^2. After substituting the number into Table 3.1, and calculating term by term, results can be obtained in Table 3.10.

$$\eta_T = \eta_{hel} \times \eta_{int} \times \eta_{receiver} \times \eta_{turbine} = 62\% \times 94\% \times 84\% \times 30\% = 14.7\%$$

From Fig. 3.45, $E = 1850\ \mathrm{kWh/m^2} \times 14.7\% \times 675{,}000\ \mathrm{m^2} =$ 187,310,000 kWh; namely the annual power generation approximates 187 million kWh.

Full-load generating hours of the power plant = 187,310,000 kWh/ 5×10^4 kWh = 3746 h.

TABLE 3.10 Energy Balance at Example Design Point

S/N	Item	Inputted Power	Lost Power	Residual
1	Solar Irradiance Input to Concentration Field at the design point	675000 × 0.75 = 506 MW	0	
2	Concentration Field Loss: Shade, cosine, reflectance, atmospheric transmission loss, receiver intercept factor		211 MW	
3	Inputted Power of Receiver	295 MW		
4	Receiver Loss: Reflection, convection, radiation, conduction		47 MW	
5	Output Power of Receiver			248 MW
6	Input to Thermal storage	248 × 98% = 243 MW		
7	Output of thermal storage and HX			243 × 97% = 236 MW
8	Steam Turbine Input Power	150 MW		
9	Steam Circulation Loss		100 MW	
10	Output Power			50 MW
11	Loss: Auxiliary system consumption and loss (rated operating conditions)		11.1 Steam Turbine 100 kW 11.2 Concentration Field and Communication 20 kW 11.3 DCS 20 kW 11.4 Circulation Pump 2000 kW 11.5 Others 100 kW	2.2 MW
12	Net Output of Power Plant			47.8 MW

3.7.3 Thermal Power Plant Capacity Optimization

Thermal power plant system calculation in the previous section has been based on the fact that the capacity of a steam turbine has been specified in advance. Furthermore, an alternative calculation method is that as the generating capacity is the core index to determine the price, it is possible to firstly assume an initial investment and expected price, then the corresponding annual generating capacity can be inversely deduced based on these two factors. As the annual generating capacity is relevant to the system efficiency, after determining the generating capacity, it is also possible to calculate the annual generating efficiency based on local meteorological conditions.

In case of the power price of the power plant being considered as a modifiable quantity, technical and economic optimization for the power plant can also be conducted according to the method shown in Fig. 3.47, and a reasonable power price can be obtained.

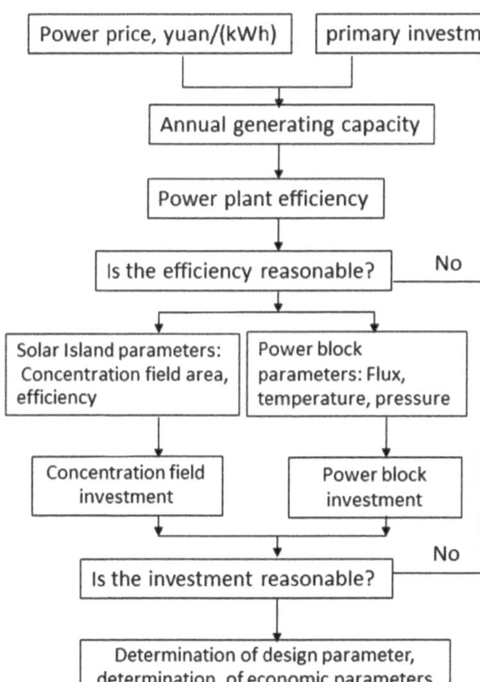

FIGURE 3.47 Power plant technical and economic optimization method.

3.7.4 Power Generation Calculation Methods Based on Hourly Simulation

The key point of this method is that the hourly solar irradiance and meteorological conditions data are necessary. Through the simulation on the basis of system energy balance, relationships between the collector field output and rated input of a steam turbine corresponding to different concentration field areas can be calculated, so that the annual maximum generating capacity corresponding to the steam turbine capacity can be obtained, and the "optimum concentration field area," thermal storage, auxiliary boiler capacity, etc. can be determined respectively. The "optimum concentration field area" refers to the power plant determined according to the constructed capacity of a certain area. Such concentration field area configuration intends to maximize the annual generating capacity of the power plant.

By applying this method, comparing with the design point method, more abundant information can be obtained, such as energy distribution among different units of the power plant system under various kinds of meteorological conditions; logical connection among various units, which serves as the foundation of power plant DCS preparation; energy flow and control information flow among various units during power plant initiation, standby, and stoppage, which are extremely important for process design. As the design point method corresponds to the typical hours in a year and typical meteorological conditions, in fact, it is merely to calculate a "typical point."

Simulation software that has been broadly applied right now is TRNSYS, which is capable of analyzing system principles, system composition, component model, operating mode, operating status, control logic, etc. In order to facilitate the reader on understanding this process, principle system of Beijing Badaling Experimental CSP plant at IEE-CAS has been established in this section and is shown in Fig. 3.48. The corresponding full system simulation TRNSYS model is shown in Fig. 3.49.

Simulation model of the system mainly consists of the meteorological module (Type15-2), heliostat concentration field module (FeffMatx, Type394), receiver module (CenRec, Type395), high- and low-temperature thermal storage module (Type 5b, Tank-Type 14) and steam turbine module of Rankine Cycle (Stage, Type 318), condenser, deaerator, various kinds of pump and generator modules. This section mainly introduces the approximate method of system simulation. The specific modeling process for the mathematical models of the above modules and basic mutual coupling logics will not be described here.

By applying the HOC (the heliostat optical code of IEE-CAS) software, as shown in Fig. 3.50, heliostat concentration field of Badaling power plant

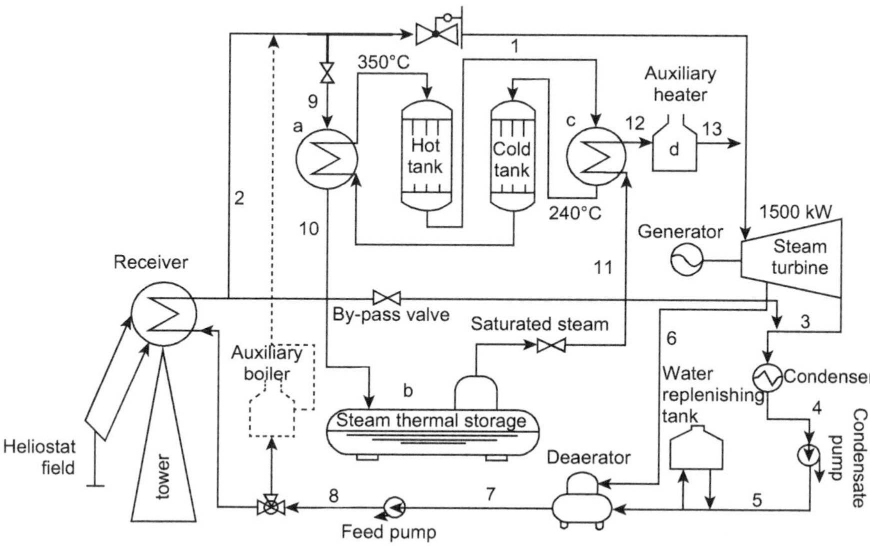

FIGURE 3.48 Principle thermal system of Badaling power plant (numbers in the figure refer to serial number of flow passage).

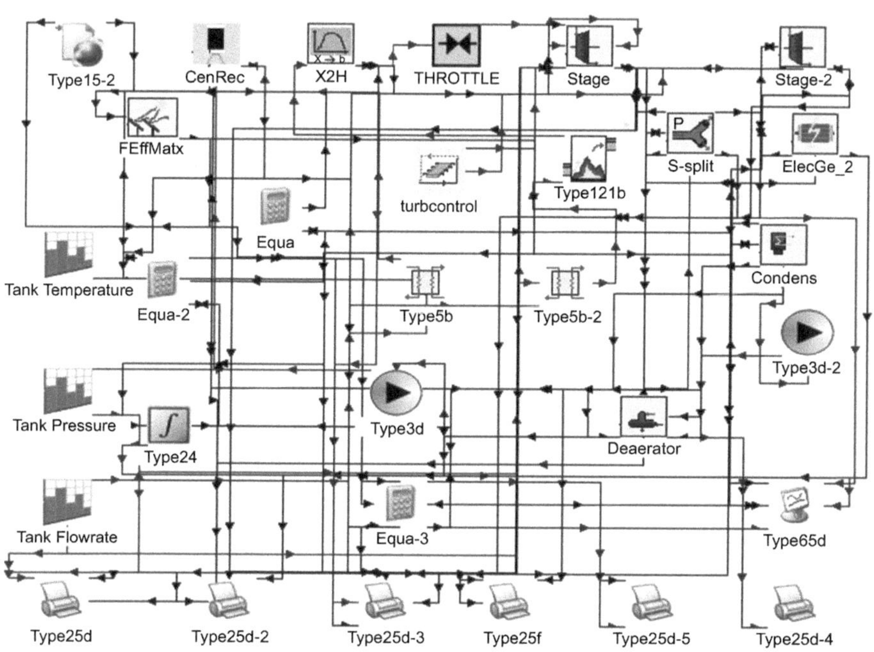

FIGURE 3.49 Full system simulation TRNSYS model of Badaling power plant.

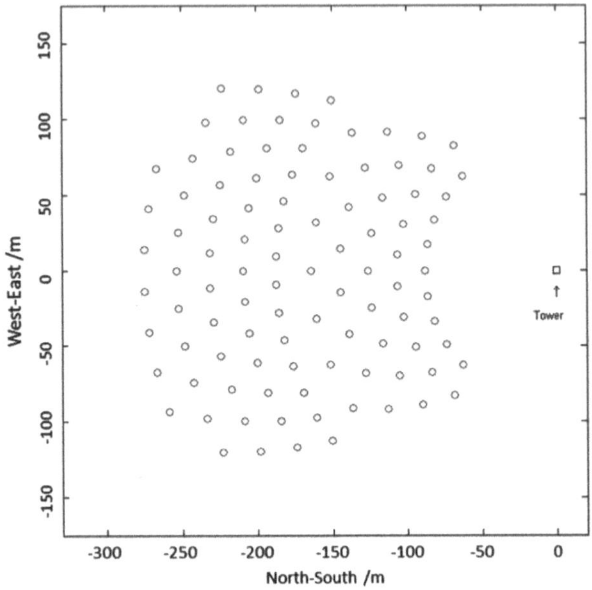

FIGURE 3.50 One layout of heliostat concentration field of Badaling power plant (sector radial-staggered layout of concentration field).

TABLE 3.11 Major Parameters of Heliostat Concentration Field of Badaling Power Plant

Location	**40.4°N, 115.9°E**	**Layout of Heliostat Concentration Field Optical Efficiency**	**Radial-staggered pattern**
Heliostat Surface Profile	Ideal sphere	Ring 1 and Tower Spacing Coefficient	1
Annual Mean Optical Efficiency of Concentration Field	66.6%	Rated of Concentration Field at the design point	81.72%
Length and Width of Mirror Surface of One-sided Heliostat	10m × 10 m	Heliostat Number	100 pieces

has been established. Various kinds of design parameters and efficiency calculation results of the concentration field are shown in Table 3.11.

Figs. 3.51 and 3.52 show the solar thermal collection part of power plant simulated on the basis of Fig. 3.49 [31]. The relationship between thermal power concentrated by heliostat concentration field onto the surface of aperture of receiver (Curve 1 in the figure) and DNI (Curve 2 in the figure)

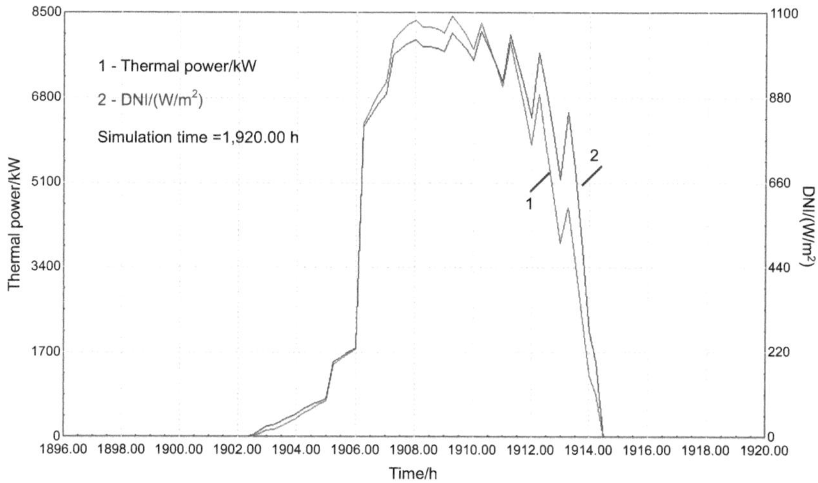

FIGURE 3.51 Relationship between the concentration power projected by heliostat concentration field onto the receiver and direct normal irradiance on the spring equinox.

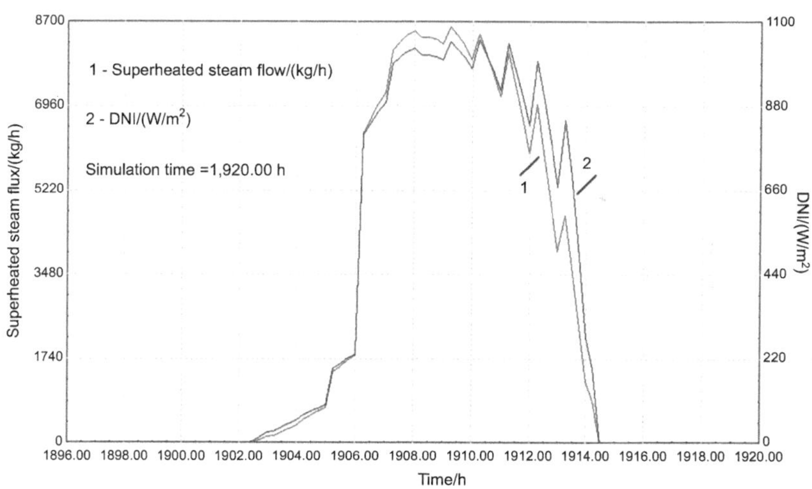

FIGURE 3.52 Relationship between the outlet superheated steam flow of receiver of Badaling power plant and direct normal irradiance on the spring equinox.

can be calculated, which is shown in Fig. 3.51. Under the premise of the inlet and outlet temperatures being set, the relationship between the outlet superheated steam flux of receiver (Curve 1 in the figure) and DNI (Curve 2 in the figure) is shown in Fig. 3.52.

The relationship between the whole-day generating capacity (Curve 1 in the figure) and solar DNI (Curve 2 in the figure) is shown in Fig. 3.53.

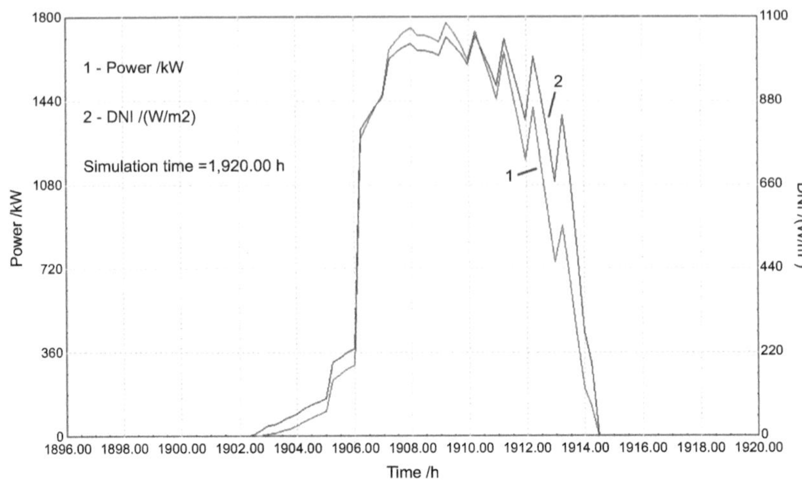

FIGURE 3.53 Relationship between generating capacity of Badaling power plant and direct normal irradiance on the spring equinox.

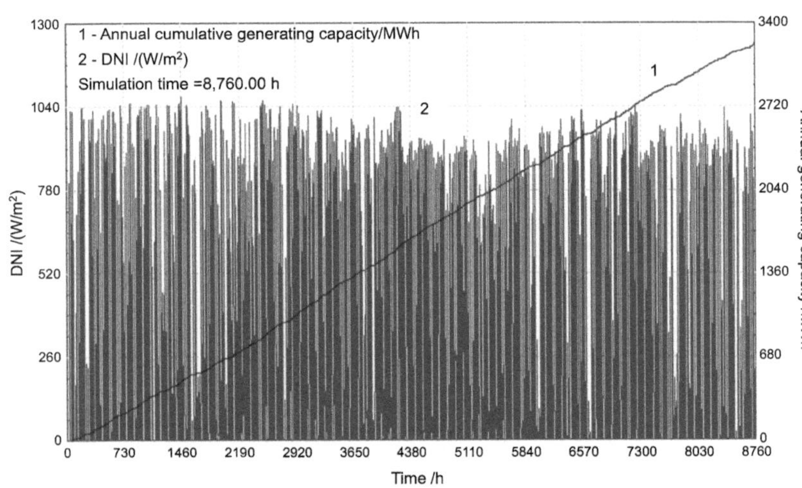

FIGURE 3.54 Relationship between the annual generating capacity and daily solar direct normal irradiance of Badaling power plant.

As shown in Fig. 3.54, by utilizing the hourly meteorological data, annual generating capacity of Badaling CSP plant has been analyzed and researched. The figure has indicated the annual cumulative generating capacity, as shown in Curve 1 in the figure. Curve 2 refers to the annual hourly DNI.

Relationships between outlet steam thermal variation, outlet steam pressure variation, temperature variation of steam thermal storage in the

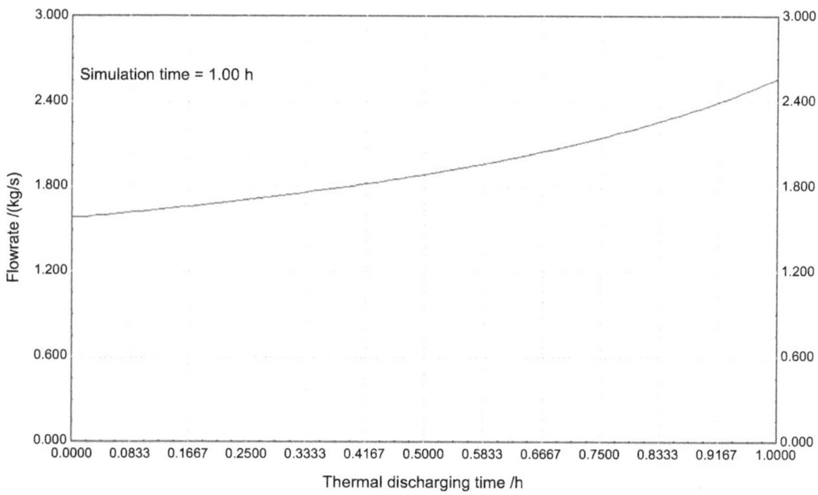

FIGURE 3.55 Variation of outlet steam flow of steam thermal storage along with thermal discharging time.

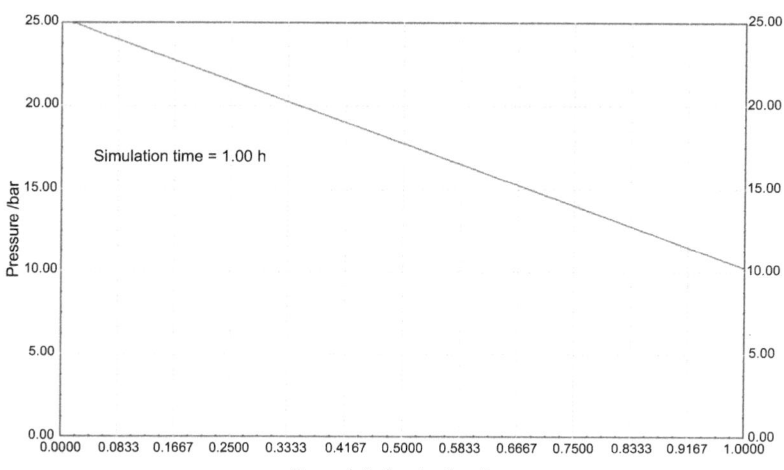

FIGURE 3.56 Variation of outlet steam pressure of steam thermal storage along with thermal discharging time.

thermal storage system, and thermal discharging time have been separately shown in Figs. 3.55–3.57 [31]. The designed thermal charging and discharging time is 1 h; therefore, in Figs. 3.55–3.57, the designed time on the horizontal axis is 1 h.

At the design point of the Badaling power tower plant (namely 12:00 on the midday of the spring equinox, DNI = 1000 W/m^2), major thermodynamic parameters calculated based on the above values of

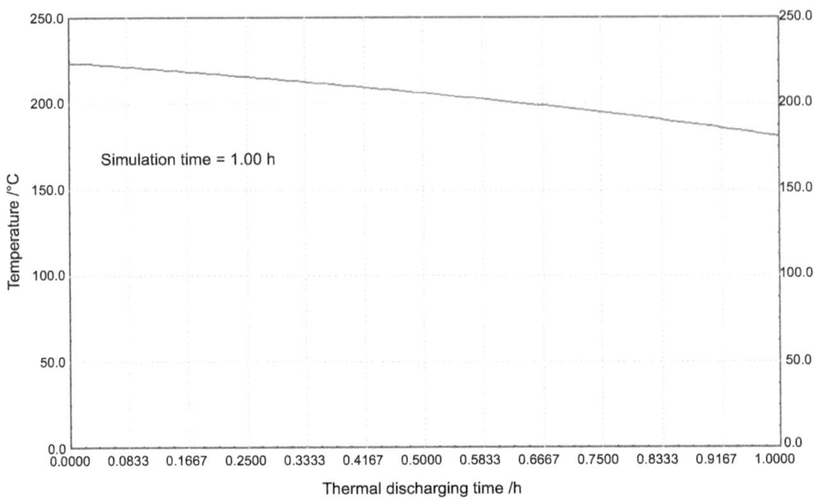

FIGURE 3.57 Variation of outlet steam temperature of steam thermal storage along with thermal discharging time.

various spots of Badaling power plant indicated in Fig. 3.48 are as follows: The selected steam turbine's relative internal efficiency is assumed to be $\eta_{ri} = 0.8$, mechanical efficiency $\eta_m = 0.98$, and generator efficiency $\eta_g = 0.98$. Through calculation, various parameters can be obtained, including steam turbine inlet superheated steam pressure $P'_{in} = 2.354$ MPa, steam turbine inlet superheated steam temperature$T'_{in} = 390°C$, and steam turbine final outlet exhaust pressure$P'_{out} = 0.0073$ MPa. As shown in Fig. 3.48, steam turbine intermediate extraction pressure $P_b = 0.3$ MPa. According to the definition of the relative internal efficiency η_{ci} of steam turbine and the water/steam diagram, thermodynamic parameters corresponding to various points indicated in Fig. 3.48 can be obtained as follows:

Point 1
$p_1 = p'_{in} = 2.354\ \text{MPa},\ T_1 = T'_{in} = 390°\text{C},\ h_1 = 3220.1\ \text{kJ/kg},$
$s_1 = 7.014\ \text{kJ/(kg·K)};$
Point 3
$p_3 = p'_{out} = 0.0073\ \text{MPa},\ T_3 = T'_{out} = 39.784°\text{C},\ h_3 = 2390.684\ \text{kJ/kg},$
$s_3 = 7.677\ \text{kJ/(kg·K)};$
Point 6
$p_6 = 0.3\ \text{MPa},\ T_6 = 183.2°\text{C},\ h_6 = 2831.27\ \text{kJ/kg},\ s_6 = 7.239\ \text{kJ/(kg·K)}$

Feed water in Fig. 3.48 is assumed to pass through the condensate pump, the thermodynamic parameters of point 5 at deaerator inlet are as follows:

$$p_5 = 0.12\ \text{MPa},\quad T_5 = 41°\text{C},\quad h_5 = 171.82\ \text{kJ/kg},\quad s_5 = 0.586\ \text{kJ/(kg·K)}$$

Feed water design parameters of point 7 at deaerator outlet are as follows:

$$p_7 = 0.12\ \text{MPa},\quad T_7 = 104.0^\circ\text{C},\quad h_7 = 435.99\ \text{kJ/kg},\quad s_7 = 1.352\ \text{kJ/(kg}\cdot\text{K)}$$

Steam turbine final outlet exhaust at point 3 in Fig. 3.48 enters into the condenser and experiences the 3-4 constant-pressure cooling process in the condenser, in which steam is cooled and completely condensed into saturation water through phase change while discharging thermal; water from the environment is used as the refrigerant. From the above, thermodynamic parameters of point 4 in Fig. 3.48 are as follows:

$$p_4 = 0.0073\ \text{MPa},\quad T_4 = 39.78^\circ\text{C},\quad h_4 = 166.64\ \text{kJ/kg}$$

So far, the entire thermodynamic process has been successfully simulated. From the above, by applying this thermodynamics-based method, calculation of the annual generating capacity can be accomplished. Furthermore, other parameter groups that are comparatively comprehensive can be obtained as well. Whereas by applying the design point method, only system energy parameters corresponding to one time point can be calculated, and the mathematical model is steady state model.

3.7.5 Influences of Location on Calculation of Parabolic Trough Collector Efficiency

Two definitions of transient efficiency of parabolic trough collector:

$$\eta_{\text{DNI}} = \frac{\dot{Q}}{DNI \times A}$$

$$\eta_{Gbp} = \frac{\dot{Q}}{G_{\text{bp}}A}$$

Efficiencies calculated based on these two definitions vary significantly. Considering cosine influences, η_{Gbp} is used in order to precisely reflect the actual thermal energy.

A calculation example on the efficiency of parabolic trough collector has been provided in this section by selecting three locations:

Badaling, Beijing (40°22′N, 115°56′E);

Sanya, Hainan (18°15′N, 109°30′E);

Hohhot, Inner Mongolia (41°N, 111°45′E)

Annual mean efficiencies of parabolic trough collector in above areas can be calculated, from which, it can be concluded that different geographic locations have resulted in different incidence angles; influences of latitude on the efficiency can also be seen. The local solar hour is assumed to be 8:00–16:00 and the annual mean DNI of three locations is the same fixed value (such as 800 W/m^2). Parabolic trough collectors have been arranged on the north-south axis, with a length of 100 m. Collector's output power $\dot{Q}$ is constant, and the efficiency of collector in case of solar normal incidence is 60%.

According to the assumption, $\eta_{Gbp} = 60\%$, and based on

$$G_{bp} = DNI \times \cos(\theta)\left[1 - \frac{f}{L}\tan(\theta)\right]$$

it can be further deduced, that $\eta_{\mathrm{DNI}} = \frac{\dot{Q}}{\mathrm{DNI} \times A} = \frac{\eta_{Gbp} G_{bp}}{\mathrm{DNI}} = \eta_{Gbp}\cos(\theta)$ $\left[1 - \frac{f}{L}\tan(\theta)\right]$

The incidence angle is assumed to remain unchanged within each minute.

1. Cosine value of daily maximum incidence angle, which is shown in Fig. 3.58.
2. Cosine value of daily mean incidence angle, which is shown in Fig. 3.59
3. Daily cosine value variation on a typical day, which is shown in Fig. 3.60, in which 172, 266, and 356 separately refer to the summer solstice, autumnal equinox and winter solstice.
4. Annual mean corrected value of beam incidence angle
   a. Badaling

$$\cos(\theta)\left[1 - \frac{f}{L}\tan(\theta)\right] = 0.8013$$

Then

$$\eta_{\mathrm{DNI}} = \eta_{Gbp}\cos(\theta)\left[1 - \frac{f}{L}\tan(\theta)\right] = 0.6 \times 0.8013 = 0.4808$$

   b. Sanya

$$\cos(\theta)\left[1 - \frac{f}{L}\tan(\theta)\right] = 0.9209$$

Then

$$\eta_{\mathrm{DNI}} = \eta_{Gbp}\cos(\theta)\left[1 - \frac{f}{L}\tan(\theta)\right] = 0.6 \times 0.9209 = 0.5525$$

   c. Hohhot

$$\cos(\theta)\left[1 - \frac{f}{L}\tan(\theta)\right] = 0.7971$$

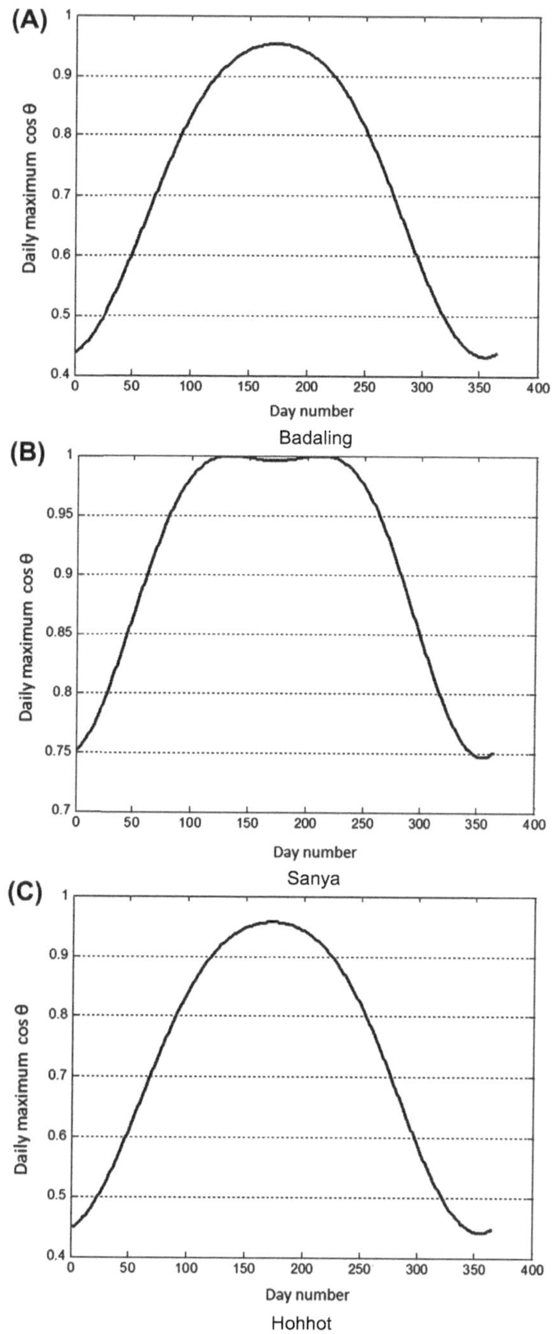

FIGURE 3.58 Cosine value of daily maximum incidence angle.

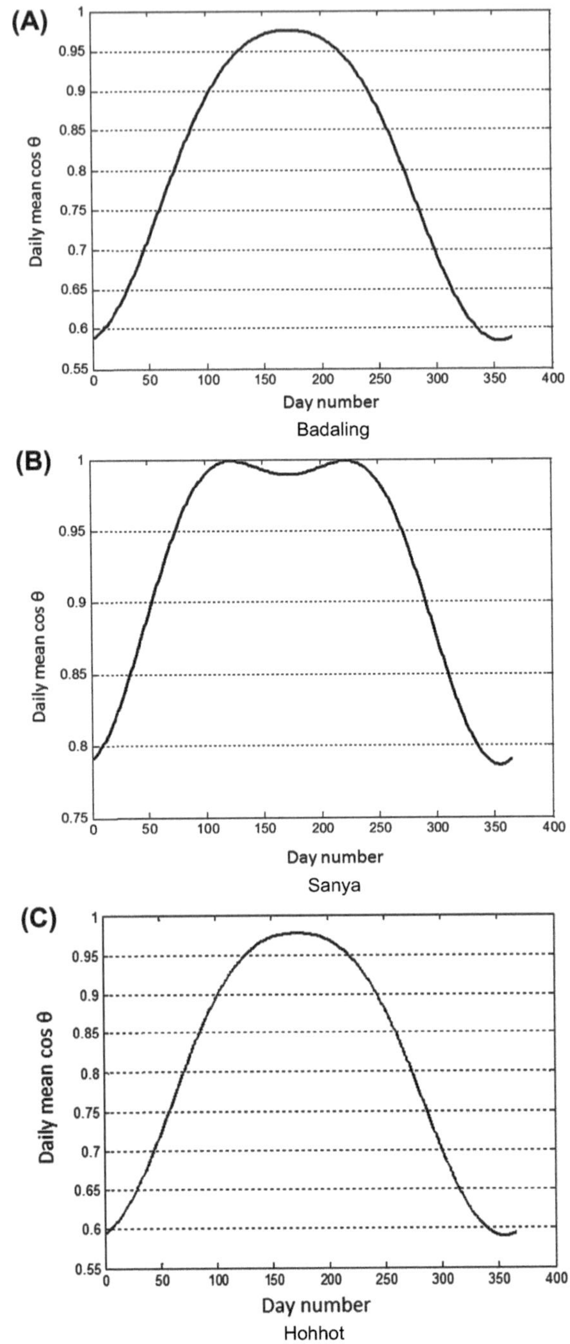

FIGURE 3.59 Cosine value of daily mean incidence angle.

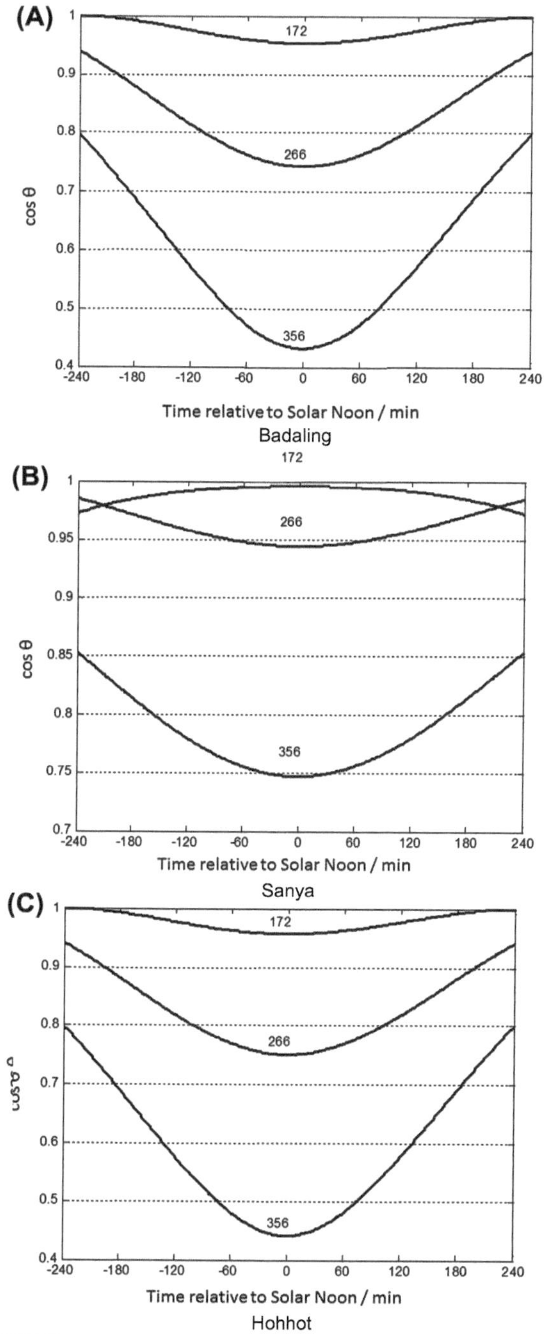

FIGURE 3.60 Daily cosine value variation on a typical day.

Then

$$\eta_{\mathrm{DNI}} = \eta_{Gbp}\cos(\theta)\left[1 - \frac{f}{L}\tan(\theta)\right] = 0.6 \times 0.7971 = 0.4782$$

Badaling is close to Hohhot in latitude, thus the respective calculation results are almost the same.

3.8 DETERMINATION OF THERMAL STORAGE RESERVE

Thermal storage mainly aims at guaranteeing the stable and continuous operation of power plant, so as to satisfy the request of power grid and achieve the maximum economy.

3.8.1 Principles of Selecting Thermal Storage Capacity

Determination of thermal storage mainly depends on the distribution of feed-in tariff against time and the requirement of power grid on peak modulation. Duration of thermal storage shall be merely determined by the full-load generating hours in sunless intervals and economy of power price. As great investments are involved, it must be calculated on a prudent basis.

Step I: Preliminary determination of thermal storage period according to the feed-in tariff and time difference of sunset.

Fig. 3.61 has indicated that time interval with high power price after sunset is more or less 6 h. Thus thermal storage period can be preliminarily set as 6 h.

However, due to the application of thermal storage system, initial investment may be increased, and the generating cost of power plant may grow. Thus it is necessary to conduct Step II.

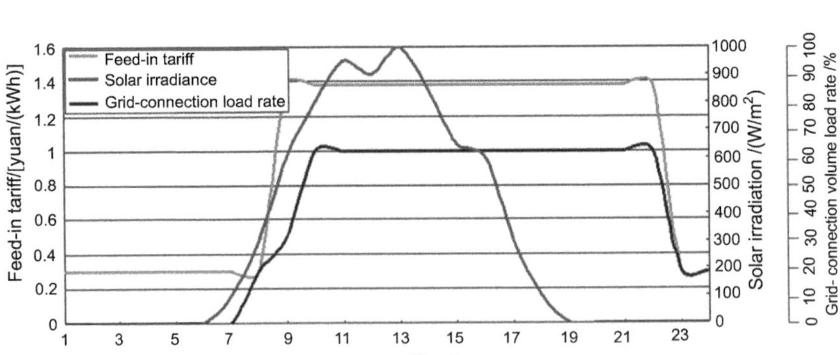

FIGURE 3.61 Determination of thermal storage period.

Step II: Calculation of influences of different thermal storage periods on generating cost.

The value of the thermal storage period direct influences on the variation in the initial investment cost of the solar power plant, which further influences the variation of generating cost. Fig. 3.62 displays the variation of levelized cost of electricity (LCOE) under different thermal storage periods in a 50 MW parabolic trough CSP plant located in Ordos, China, in which along with the increase of thermal storage period, LCOE of the power plant drops. When the thermal storage period of the power plant in this case is 10 h, the respective power price can reach the minimum level. Value of this minimum level mainly depends on the power plant capacity, primary investment on the thermal storage unit and local solar irradiation resources.

To sum up, based on the above analysis, setting thermal storage period of the power plant as 6 h is determined to be economically efficient, which is able to help reduce the power price. So, is it necessary to further increase the thermal storage period? According to Fig. 3.62, LCOE corresponding to the thermal storage period of 6 h is 1.32 yuan/(kWh); in this case, grid purchase price of the power plant indicated in Fig. 3.61 is 1.38 yuan/(kWh); after exceeding this time interval, namely starting from 22:00, the grid purchase price decreases to 0.30 yuan/(kWh). Thus it makes no sense to further increase the thermal storage period. According to Fig. 3.61, LCOE still can be reduced, but the feed-in tariff has exceeded 0.30 yuan/(kWh).

In case that it is necessary to optimize power price in terms of thermal storage period, such as the power price bidding project, for the case shown in Fig. 3.62, it is obviously conductive by setting the thermal storage period as 10 h of full-load power generation of a steam turbine.

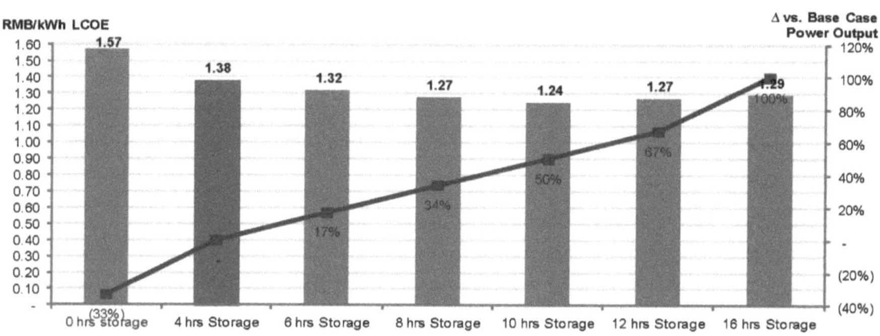

FIGURE 3.62 Relationship between thermal storage period and LCOE. *Data Source: "China Solar Thermal Power Industry Research Report 2013," China National Solar Thermal Energy Alliance.*

3.8.2 Principles of Selecting Thermal Storage Capacity

Thermal charging and discharging power of thermal storage shall be equivalent to the output power of collector field and input power of a steam turbine. Output power of collector field at the design point is equivalent to the sum of rated input thermal power of a steam turbine and the transient thermal storage power. However, input power of thermal storage still has been designed according to the maximum thermal charging power, and the thermal discharging power has been designed according to the maximum load requirement of a steam turbine.

When calculating thermal storage capacity solar multiple acts as an important concept, which comes from concentration field. In case of considering the concentration field while supplying power to the steam turbine and thermal storage, area of concentration field at the design point must be considered to be large enough. Relationship between thermal storage and receiver has been given in Sections 3.7.1 and 3.7.4. While in Section 3.7.4, the more precise value of thermal storage can be obtained by applying TRNSYS, and also, the thermal storage unit can be designed and the respective working status can be understood. In addition, during design of thermal storage, self-consumed energy of thermal storage unit must be considered as well, including fossil fuel backup and synthetic oil or molten-salt anti-coagulation measures.

3.9 MAIN POINTS FOR POWER PLANT SITE PLAN

The overall planning can be conducted based on the location of urban areas, meteorological conditions, access route connection, HV outlet, water source, road, land coverage of concentration field and other external conditions as well as features of the plant area by integrating with specific concentrating power generation process, in which the following points shall be specified: power plant capacity, concentration field layout, steam generating area, thermal storage area, conventional generating area, location and transportation of the power plant, HV outlet orientation and passageway, water source of power plant, fuel transportation, drainage of power plant, power plant general layout and general elevation planning factors.

1. Power plant capacity. Current scale and reserved scale for site expansion shall be determined, especially for the concentration field; when the site is arranged, long-distance transmission losses of high temperature fluid and wind direction shall be considered. For a power tower plant, necessary site conditions for receiver tower installation and receiver maintenance service shall be considered.

2. Concentration field layout. Concentration field has the largest land coverage in an CSP plant. Normally, area of a heliostat concentration field is five times the sum aperture area of heliostats. Land coverage of a parabolic trough concentrator is about three times the total area of aperture of receiver of a parabolic trough concentrator.

In general cases, except for the efficiency of concentration field, fluid transportation-related problems shall also be considered during the layout of concentration field, especially the layout of parabolic trough concentration field.

An example of concentration field layout in a 50 MW parabolic trough power plant is shown below. In terms of parabolic trough thermal collection, solar radiations are concentrated in the primary loop and converted into thermal through the collector consisting of a parabolic trough concentrator with the concentration ratio of 75 times and a evacuated tube receiver; the respective heat-transfer medium is synthetic oil with the working temperature of 400°C. The secondary loop is for oil-water heat exchange, which produces superheated steam in order to drive the highly efficient steam turbine to work and generate power. A solar collector field consists of 160 parallel "thermal collection loops"; each "thermal collection loop" consists of four rows of parabolic trough "collector units" with a length of 150 m (Fig. 3.63) through series combination, which is shown in Fig. 3.64. Length of thermal collection loop is determined by the solar irradiation and meteorological conditions of the working site of collectors. Normally the temperature increase of outlet/inlet loops of synthetic oil is 105°C.

FIGURE 3.63 A parabolic trough collector unit with a length of 150 M [32].

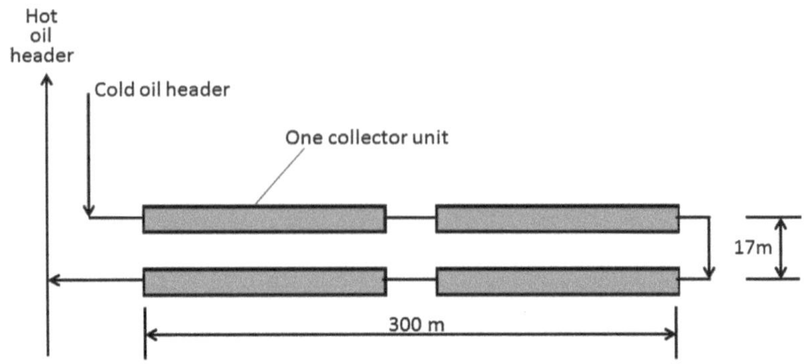

FIGURE 3.64 Parabolic trough collector loop.

Center-to-center distance between collector units in the thermal collection loop is normally three times of the width of aperture of parabolic trough concentrator. For example, in case of the width of aperture of concentrator being 5.7 m, the center-to-center distance shall be about 17–20 m.

Layout of power tower concentration field will be explained in details in Section 4.2. Normally, there are two ways, namely the northward concentration field and encircled concentration field. Fig. 3.65 displays an example of the northward concentration field.

3. Steam generating area. A steam generating area includes such major equipment as feed water preheater, evaporator, superheater, and reheater. In case of the heat-transfer medium

FIGURE 3.65 Concentration field of PS10 power tower plant in Spain [32].

FIGURE 3.66 Molten-salt/steam generator of achimide power plant, Italy [2].

being fluid, like synthetic oil, molten salt, or air, it is possible to transmit the heat within the area from the primary loop to the feed water facilities in the secondary loop through a heat exchanger in order to produce qualified superheated steam, and drive the generator unit of a steam turbine to work (refer to Fig. 3.66).

4. Thermal storage area. Currently, thermal storage medium of the large-scale power plant is normally molten salt. A thermal storage system includes such major equipment as cold molten-salt storage tank, hot molten-salt storage tank, cold molten-salt circulating pump, hot molten-salt circulating pump, and oil-salt heat exchanger. Due to the high temperature of thermal storage area, space and access for emergency measures taken in case of any leakage shall be reserved, which shall be placed at the downwind area of the power plant. In the case that the distance of fluid entering into the thermal storage tank is too long, it may result in great heat losses; based on this point, thermal storage area shall not be located far away from the thermal collection area. For a parabolic trough power plant, it should be specially noted that headers connecting various outlets of thermal collection loops shall not be located too far away from the thermal storage area. Receiver of the power tower plant has been close to the thermal storage, thus the transmission distance is not a huge problem.
5. Conventional generating area. As shown in Fig. 3.67, it is an example of a power plant in Spain, according to which, partial generating units have been mounted in the middle of the concentration field.

FIGURE 3.67 Generating unit layout of Andasol-I power plant, Spain [32].

6. Location and transportation of the power plant. Identification of location of traffic hub and the distance between the traffic hub and the high-grade highways corresponding to the proposed power plant is of great significance to power plant construction. Due to the access of large-scale overheight and overweight equipment, like the thermal storage tank, steam turbine, and boiler, roads are required to have strong carrying capacities; due to the transportation of a large quantity of glass reflectors, roads are required to have high evenness. Thus grade of road, as well as dimension of roads and bridges along the line, shall be given special attention to; otherwise, it may lead to great expenses during infrastructure construction.

   Upwind area of power plant shall be free of large-scale air pollution sources; otherwise, mirrors will be contaminated by dusts.
7. HV outlet orientation and passageway. Diagrams and texts for receiving station of power plant outlet, conditions of cable passageway, conditions of power outlet, etc. shall be provided.
8. Water source of power plant. Name and distance of the water supplier shall be specified, as well as the respective daily water supply capability, water price description, actual daily water supply and annual total water supply.
9. Fuel transportation. Transportation method for engineering liquid, gas or solid fuels, distance between the fuel supply point and the plant area, whether there is any road of the appropriate grade in Gobi areas for transportation; road with an uneven surface may cause danger to fuel transportation.

10. Drainage of power plant. Drainage of a solar power plant includes production water supply and domestic water supply. As the amount of water for washing the surfaces of concentrators is insignificant, site drainage can be neglected. Normally, for power plants located in Gobi areas and deserts, rainwater can be drained off in natural ways. After being treated, power condenser water can be used for irrigation and for washing mirror surfaces.
11. Power plant general lay-out planning factors. A thermal power plant normally consists of four sections, including the collector field, thermal storage area, generating area and office area.

    A collector field contains both solar concentrator and thermal receiver; dimension of the concentration field shall be indicated on the general layout. For a parabolic trough system, combination pattern of thermal collection units shall be indicated, such as the quantity of units in order to constitute a fluid serial loop, and the quantity of loops in order to constitute a power plant concentration field through parallel connection. As the heliostat, concentration field layout drawing shall be prepared. The concentration field can use a northward sector layout or circumferential circular/oval layout. Position and height of receiver tower shall be marked on the concentration field drawing accordingly.

    Core area of a power plant normally consists of the main powerhouse, control room, thermal storage area, power tower, etc.

    The entire plant area layout shall be based on the process requirements and architectural function division while fully utilizing the limited land within the plant area.

    A plant area consists of two sections, the concentrator field and the generating area. Due to large landscape in the concentration field that cannot be combined into the floorage, it shall be considered accordingly when calculating the floor area ratio of the power plant; otherwise, it may result in the conflict between it and the floor-area ratio of construction land required by the state in regulations and norms. It shall be calculated based on the forestation of the power plant and land coverage of the generating area.
12. Power plant general elevation planning factors. The concentration field of a tower power plant does not have strict requirement on the site evenness; even the slope of mountains can be utilized wisely to facilitate concentration field layout, which is very conductive for taking advantage of the mountain resources in China. For the concentration field of a parabolic trough power plant, due to the flow passage consisting of the evacuated tubes and requirements

on tracking, slope of installation site of a parabolic trough concentrator is required to be not more than 1%. Normally, in the concentration field of power plant located in the northern area of China, it is not necessary to design a drainage ditch.

A power tower plant normally uses the horizon of the receiver tower as the zero elevation, and center of the tower as center of the heliostat concentration field. For a northward concentration field, in case of being a highland to the south of the concentration field, it will facilitate the reduction of building height of the receiver tower.

It is also possible for a power tower plant to mount all equipment within the tower in order to save land. As shown in Fig. 3.68, interior of the receiver tower contains all parts except for the heliostat concentration field of the power plant (thermal receiver, thermal storage, steam generation, power generation). Such layout pattern has great reference value for the construction of a power plant while saving land. Of course, due to the increase of load, building costs of the tower will be increased; yet for areas with comparatively higher prices for land use, such layout pattern can be applied.

Concentration field of a parabolic trough power plant has a strict requirement on land evenness; the standard horizon of the entire plant area can be selected to be consistent with the horizon elevation of the concentration field. In case of concentration field wind wall being designed in the power plant, special attention shall be paid to the design

FIGURE 3.68 Solar tower and heliostats of the power tower plant of the Korean Institute of Energy Research (Located in DAEGU, Korea, October 2011).

of the shape, height of the wall, and the distance between the wall and the concentration field; otherwise, the negative pressure area created by the wind striding over the wall may aggravate the dust accumulation in the concentration field.

3.10 NOTICES FOR CONCENTRATION FIELD LAYOUT

A concentration field serves as the unit for collecting, reflecting, and concentrating solar energy. During design of a heliostat field, attention shall be paid to the harms to its neighboring buildings and personnel. It is better not to construct any tall buildings on other directions around the tower besides the north direction in the concentration field.

Road layout in the concentration field and the space between concentrators shall be designed by considering the service request of equipment. For large-area heliostats, requirements for 20 t-level crane car land and road widths shall be fully considered; whereas for small-area heliostats, less large-scale transportation equipment is required for erection and hoisting. However, due to large quantity of mirrors, multiple linkage mechanisms and numerous fault points, the space necessary for frequent service shall also be fully considered.

Concentrator in the concentration field has extremely high requirement on precision; a slight shaking may have an impact on the precision. Thus when analyzing the geological conditions of the concentration field, construction site shall be free of significant geologic hazards, such as earthquake and underground cave collapse. It shall be fully considered during the foundation design for the concentrator and receiver tower.

Diameter and volume of the receiver tower shall be as small as possible. It is better to apply the scheme of steel structure tower that is permeable to light, which could minimize the shade caused the tower during functioning of the northward concentration field.

Communication and power cables in the concentration field shall be rat-proof. By using rat-proof heavy-armored cables, cable costs will be greatly increased. For areas with severe rat problems, it is suggested to apply wireless communication to achieve the connection between the concentration field and the host computer.

For a parabolic trough system, the concentration field has been integrated with the collector field; thus soil pollution caused by evacuated tube oil leakage and the respective safety preventive measures shall be fully considered. In order to facilitate the treatment for oil leakage and prevent the spilled materials from being spread, it is normally not suggested to perform any concrete or asphalt hardening on the ground below the parabolic trough concentrators, so as to facilitate the timely replacement of contaminated soil.

Due to the short distance to the ground, and small wind resistance of concentrator, FRESNEL concentrator can be arranged without considering the introduction of wind resistance facilities while the receiver tubes over the reflectors should be considered with wind resistance.

CHAPTER

4

Design of the Concentration System

4.1 GENERAL SYSTEM DESCRIPTION

4.1.1 Constitution of the Concentration System

The solar-concentrating system consists of the solar concentrator as well as other units with various functions: solar field control, communication, power supply, optical precision measurement and calibration, security monitoring, solar irradiance–ambient air temperature–wind speed measurement, concentrator cleaning, etc.

The qualified power plant shall also be equipped with cloud-cover and wind-speed forecast units, which normally provide an additional 5 min before strong wind and cloud-cover events so that security and safety measures and can be taken to mitigate the risks of personal and property damage.

4.1.2 Principles and Modes for Concentrator Adjustment and Control

The concentrator serves as a light source for the receiver, the aimpoint adjustment of which shall be based on the receiver's working status. Electrical interlocking, signaling, and communication facilities shall be set between the concentrator and receiver. In the event of overheating on the heat absorber surface in a solar tower power plant, the action of the heliostat field must be considered so as to guarantee the safety of the receiver, the solar tower, etc. Due to the comparatively large area of the heat absorber plus the changing conditions of the meteorological environment and changing solar position, it is difficult to locate the overheating spot on an entire piece of heat-absorbing surface using

Design of Solar Thermal Power Plants
https://doi.org/10.1016/B978-0-12-815613-1.00004-3

thermocouples; temperatures of various heat-absorbing surfaces can be determined, however, through the infrared camera monitoring method. Fig. 4.1 shows a thermal image of a receiver at work in a solar tower power plant. According to the color distribution, the temperature distribution of the heat-absorbing surface inside the cavity receiver can be detected. In the event that the absorber's surface temperature exceeds the upper limit, the relevant heliostats should be moved so that temperature increases of the corresponding part can be restrained.

4.1.3 Modes of Concentration Field Control

Due to the large land coverage of the solar-concentrating field, the concentrator should be locally controlled during regular work. A host computer can only monitor the relevant status and give "start/stop" commands; if possible, centralized control should also be applied. The control system of the concentrating solar heat collection system is connected to the distributed control system (DCS) of the power plant through a fieldbus. The solar collector field functions under control of the concentrating field supervisory controller (CFSC), as a whole or by sector. The CFSC is a computer system located in the central control room that is capable of communicating both with the control unit of each concentrator and with the DCS. The CFSC collects and supervises the status and position information of various concentrators and receives instructions from the DCS, as well as giving general control instructions to the concentrating field. During the day and when the power plant is well prepared for it, the CFSC gives starting instructions; at night and in the event of strong winds, the CFSC gives stopping instructions including shutoff of the concentrators.

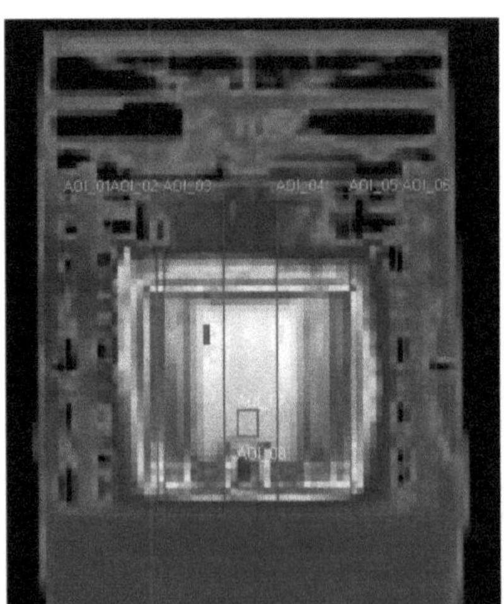

FIGURE 4.1 Thermogram of the receiver at Badaling power tower plant [20].

A meteorological station near the power plant can offer climatic information relevant to the solar heat-collecting island. Direct solar radiation data are used to determine the performance of the solar heat-collecting section. The collection of wind speed data is essential because under strong winds conditions, the relevant equipment of the solar heat-collecting section must be in standby or shutdown status to prevent damage. Based on the experiences of the Yanqing Dahan solar tower power plant, the sampling period for wind speed data should be 10 s. When the wind speed exceeds the set value for two consecutive periods, the CFSC will give the instruction to stop the concentrating field and report the instruction to the DCS.

The CFSC transmits real-time data with the power plant's DCS, which coordinates and controls the overall operations of the power-generating area, heat-transfer pump-valve system, and solar heat-collecting section. The DCS operator station supervises certain operational parameters related to the solar collector and gives control instructions such as stoppage, standby, hedging, and shutdown through the CFSC. In the event of power-off of the communication control equipment or a CFSC malfunction, the solar-concentrating field's local controller must be able to automatically give turnoff instructions to all concentrators in order to prevent burnup of the receiver and solar tower.

4.1.4 Location and Environmental Requirements of the Concentrating Field Supervisory Controller Control Room

The CFSC control room shall not be located in a dusty area with great vibration or much dust, but rather in a position within the solar-concentrating field that is conductive to signal transmission; meanwhile, fire hazards, leakage, and toxicity of the heat-transfer fluid and heat-storage system also must be considered. A parabolic trough power plant should be located on the windward side of the concentrating field. The heliostat field control room of a solar tower power plant shall be located far away from the solar heat-absorbing tower in order to avoid falling objects from the tower as well as high-temperature leakage.

4.1.5 Calculation Methods for Concentration Field Output

Concentrating field energy output acts as receiver input; thus concentrating field output power field shall be determined in accordance with receiver input power requests while considering the capacity of the heat-storage system.

The input power of the solar-concentrating field and receiver may can be determined two ways: (1) according to the annual peak value of the input–output relationship and (2) according to the annual mean input–output relationship.

In the first mode, it is feasible to determine the area of the heliostat field before receiver design so that the corresponding rated input power is equivalent to the peak value output of the heliostat field; alternatively, it is feasible to determine receiver input power before heliostat field design so that the peak value of heliostat field output is equivalent to the rated receiver input.

In the second mode, receiver input power and solar-concentrating field output power have an equivalent annual average. The problem with this method is that when concentrating-field output exceeds the annual mean value, partial concentrators must be shut down so that receiver input does not become supersaturated. Therefore the annual mean calculation method is usually not applied. For parabolic trough and Fresnel collector systems, receiver power depends on the Nusselt number of the fluid inside the heat-absorbing tube. It is required that the heat-transfer system is designed to adapt to the concentrator's maximum output power.

While performing power matching, energy inside the receiver must be distributed on a reasonable basis, especially for a "water/superheated steam" type receiver with a phase-change process; the concentrating field shall feature high optical precision and good flexibility, and be capable of projecting solar radiation onto different positions inside the receiver at different times; otherwise, dry-heating can easily occur on the superheated section of the heat absorber, especially during receiver startup. Because complex processes are involved, the control design techniques corresponding to this method are especially difficult.

4.1.6 Influences of Dust Accumulation on Mirror

Dust accumulation on the mirror may result in a sharp decline in mirror reflectivity. Attenuation of reflectivity is related to both time and location. Fig. 4.2 shows the record of dust accumulation's influence on mirror reflectivity during the period from August 28, 2011, to May 24, 2012. During the test period, no manual cleaning were engaged; mirrors were cleaned only by rainwater.

According to Fig. 4.2, the degree of dust accumulation on the mirror is related to the angle in which the mirror is placed; the reflectivity attenuations of mirrors placed perpendicularly to the ground and those with their reflective surface facing downward are comparatively slow; due to the cleaning effect of rainwater, the reflectivity attenuation of a mirror placed perpendicularly to the ground is slower than it is for one with its reflective surface facing downward. According to Fig. 4.2, after being washed by rainwater, all mirror reflectivities increased significantly; the reflectivity of mirrors placed facing the

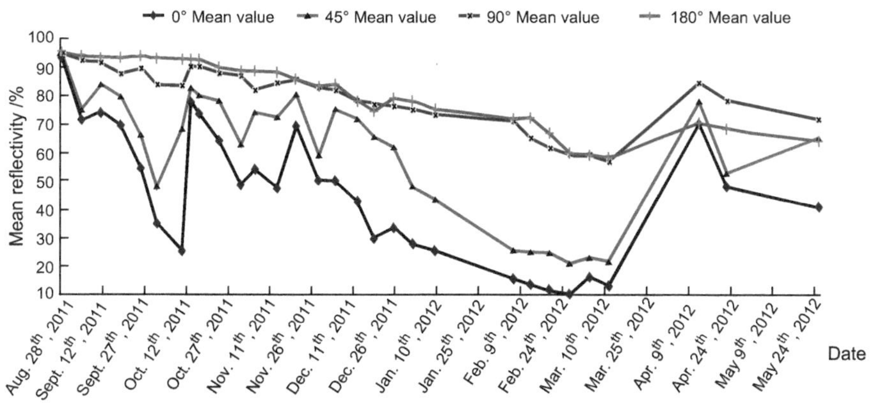

FIGURE 4.2 Influence of mirror dust accumulation on reflectivity. *Provided by the Institute of Electrical Engineering, Chinese Academy of Science.*

sky (namely at 0°) increased from 10% before to 85% after the rain on April 9, 2012.

The difference between the mean value in the case of 45° and that in the case of 0° is insignificant. In areas with severe dust accumulation, the difference approximates 10%.

4.2 PRINCIPLES FOR CONCENTRATION FIELD LAYOUT

4.2.1 Basic Knowledge of the Heliostat

The heliostat normally consists of five major parts: the mirror, support frame, pedestal, drive mechanism, and tracking control system. The heliostat normally has two orthogonal rotation axes capable of continuously tracking the movement of the Sun; one rotation axis serves as the fixed axis and is fixed to the ground foundation, whereas the other rotation axis serves as the slave axis and revolves around the fixed axis along with the mirror surface of the heliostat. Lipps and Vant-Hull listed some typical two-axis tracking modes, such as azimuth-elevation tracking (Fig. 4.3), pitch-roll tracking with a horizontally placed fixed axis, polar tracking, and spinning-elevation tracking with the fixed axis pointing to the target point (Fig. 4.4) [33].

In azimuth-elevation tracking, the azimuth axis serves as a fixed vertical axis, and the elevation axis serves as a slave axis that is always horizontal.

In pitch-roll tracking, the pitch axis serves as a horizontally placed fixed axis, and the roll axis serves as a slave axis for the left–right tilt of the mirror surface; the plane of the roll axis and the vertical direction are always perpendicular to the pitch axis.

FIGURE 4.3 Azimuth-elevation tracking heliostat [20].

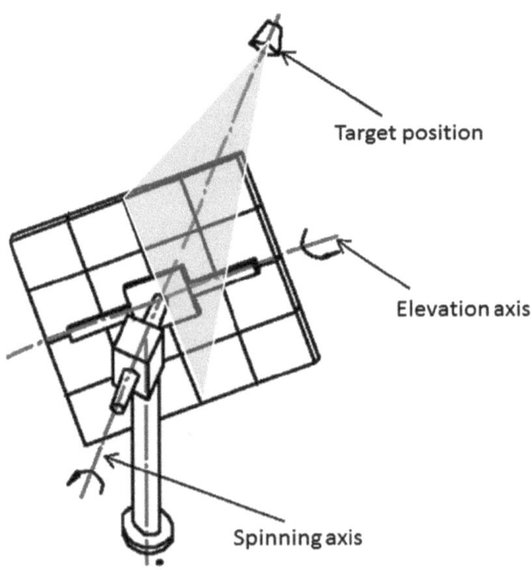

FIGURE 4.4 Operational principle of the spinning-elevation tracking heliostat [33].

In polar two-axis tracking, the fixed axis is parallel to the Earth's axis. When regularly tracking the movement of the Sun, the tracker revolves around the fixed axis clockwise at a constant speed.

In spinning-elevation two-axis tracking, the spinning axis serves as the fixed axis pointing to the target location, with the elevation axis perpendicular to the spin axis (Fig. 4.4). The elevation axis is fixed along the sagittal direction of the mirror surface on the support frame of the heliostat and revolves around the spinning axis together with the frame and the mirror surface.

The azimuth-elevation tracking mode is the most common two-axis tracking mode and is applied by most solar trackers, parabolic solar dish concentrators, and heliostats. Analysis by Schramek and Mills concluded that compared with azimuth-elevation tracking, pitch-roll tracking is capable of making the heliostat field layout more compact. Spin-elevation two-axis tracking can be integrated with a toroidal mirror surface. The different curvature radii of a toroidal heliostat (Figs. 4.4 and 4.5) in the tangential and sagittal directions can be used to correct off-axis astigmatism and thus improve the heliostat's concentration effect.

The most common shapes for heliostat mirror surfaces are planar, spherical, and paraboloidal. In order to reduce spherical or paraboloidal surface astigmatism effects on off-axis incident sunlight, the heliostat's mirror surface can be designed into a nonrotational, symmetric high-order curved toroidal surface to improve light-concentration performance. A heliostat's general mirror surface can be a mirror facet or a compound mirror surface consisting of several mirror facets. A heliostat consisting of only one mirror facet is normally small; a heliostat with a large mirror surface area shall be formed using several mirror facets through the frame to form an approximately spherical or paraboloidal surface.

FIGURE 4.5 Spinning-elevation tracking heliostat [33].

4.2.2 Concentration Astigmatism of a Spherical Heliostat

The most common heliostat type, the spherical mirror (Figs. 4.4 and 4.5), has been applied in the Solar One, Solar Two, Dahan, PS10, PS20, and Gemasolar solar tower power plants. Fig. 4.6 shows the concentrated solar image of a spherical heliostat at the Dahan power plant. To address the off-axis astigmatism problem of the spherical heliostat, E. A. Igel and R. L. Hughes have carried out in-depth theoretical and experimental research that for the first time has laid a foundation for the subsequent development of a series of heliostats with astigmatism-correction curved surfaces. For a spherical mirror with a spherical radius of $R = 2f$, in the case of off-axis incidence of the parallel beam with an incident angle of θ, the tangential and sagittal focal distances of the mirror surface are $f_t = f\cos\theta$ and $f_s = f/\cos\theta$ respectively. It is thus clear that only in the case of incidence of a parallel beam along the axis of a spherical mirror ($\theta = 0°$), $f_t = f_s = f$; namely the tangential and sagittal focal distances are equal to the axial focal distance f of the spherical mirror.

For the case of a spherical mirror with diameter D on the target plane with distance L to the mirror surface center, the focal spot's height in the tangential direction and width in the sagittal direction are respectively shown as follows.

$$h_1 = \frac{D}{f}(L - f\cos\theta) \text{ and } h_2 = \frac{D}{f}(f - L\cos\theta)$$

FIGURE 4.6 Concentrated solar image of a spherical heliostat [20].

Refer to Fig. 4.7.

In the case of considering the divergent angle effect of the solar beam, the dimension of the defocused spot of the off-axis incident solar beam by the spherical mirror is:

$$H_1 = h_1 + \beta L = D\left(\frac{L}{f} - \cos\theta\right) + \beta L \tag{4.1}$$

$$H_2 = h_2 + \beta L = D\left(1 - \frac{L}{f} - \cos\theta\right) + \beta L \tag{4.2}$$

in which β refers to the beam angle of the solar cone, which is equivalent to 9.3 mrad.

From Eqs. (4.1) and (4.2), for the case that $L = f$—namely the distance of the light target to the mirror is equivalent to the focal distance of the mirror—the focal spot is circular:

$$H_1 = H_2 = D(1 - \cos\theta) + \beta L = 2D\sin^2\left(\frac{\theta}{2}\right) + \beta L \tag{4.3}$$

Thus, there is a requirement to adapt the spherical heliostat slant-concentrating distance L so that its value is as close as possible to one-half the distance of curvature radius R in the solar tower concentrating system. Of course, according to Eq. (4.3), when $\theta = 0°$—namely when the Sun, target, and heliostat are on the same line—the heliostat concentrated spot's diameter reaches the minimum value, βL; with increases in the incident angle θ of the solar beam, the diameter of the focal spot increases.

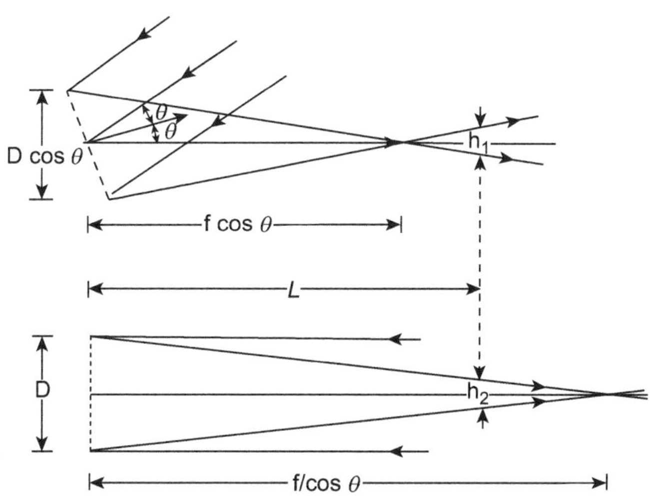

FIGURE 4.7 Focal spot's height h_1 in tangential Direction and width h_2 in sagittal direction on a focal plane with an axial distance of L.

4.2.3 Brief Introduction to the Toroidal Heliostat

A toroid, also known as a toroidal surface, has two symmetric sections perpendicular to each other, namely the tangential and sagittal sections; the arcs of these two sections have different curvature radii and therefore are nonrotational symmetric curved surfaces.

For a fixed incident angle θ, the value of the tangential curvature radius of the toroidal surface is:

$$R_t = 2f_t = 2L/\cos\theta_0$$

and the sagittal curvature radius is:

$$R_s = 2f_s = 2L\cos\theta_0$$

Then according to Eqs. (4.1) and (4.2), the beam spot on a focal plane with toroidal surface slant distance L has the minimum area, no astigmatism, and a diameter of βL, from which the minimum value of the spot size of the spherical mirror appears when $\theta = 0°$, and the toroidal surface can shift the minimum value of the spot diameter from $\theta = 0°$ to $\theta = \theta_0$.

In reality, the surface shape of the heliostat is fixed, whereas the incident angle θ of the solar beam varies over time. Therefore, in order to determine the surface shape of the toroidal surface, an appropriate θ_0 shall be selected, which is referred to as the designed incident angle of the toroidal surface. Once the slant distance L and designed incident angle θ_0 have been determined, the surface shape of the toroidal surface can be determined accordingly.

R. Zaibel et al. proposed the astigmatic-corrected target-aligned heliostat with a fixed rotation axis pointing to the target position in 1995, which combines a toroidal surface with a two-axis tracking mode with a fixed rotation axis pointing to the target position so that during whole-day sun tracking and solar concentrating by the heliostat, the incident plane of the solar beam at the mirror surface center (including normal and incident solar beam at the mirror surface center) always coincides with the tangential plane of the mirror surface (including the principal optic axis of the mirror surface and the normal direction of the mirror surface center). In this case, further optimization design of the heliostat mirror surface shape is feasible, and the corresponding method for optimization is also simple, thus making practical application of the toroidal heliostat feasible. Figs. 4.5 and 4.8 display a two-axis tracking heliostat with a fixed rotation axis pointing to the target position. The fixed axis points to the center of the receiver aperture; the slave rotation-axis is fixed to the heliostat frame and is perpendicular to the fixed axis. The design method for the surface shape of an astigmatism-corrected toroidal heliostat uses the variation range of incident angle θ of the solar beam over time on the heliostat mirror surface to select the incident angle range $[\theta_{min}, \theta_{max}]$, introduce the three design parameters $s = f_s/L$, $t = f_t/L$, and $h = H_s/H_t$ (H_s and H_t are the width and height of the mirror surface respectively), and derive the optimization expression that provides the appropriate h, R_t, and R_s, namely:

$$h = H_s/H_t = (\cos\theta_{min} + \cos\theta_{max})/2, \quad R_t = 2f_t = 2L/h, \quad R_s = 2f_s = 2hL$$

In this case, on a focal plane with the slant distance of L, the focal plane light spot corresponding to $\theta = \theta_{min}$ and $\theta = \theta_{max}$ is circular with the equivalent radius. At this time, the designed incident angle of the toroidal surface is:

$$\cos\theta_0 = (\cos\theta_{min} + \cos\theta_{max})/2 \tag{4.4}$$

On a more regular basis, the toroidal surface is a special cylindrical surface. As shown in Fig. 4.9, the toroidal surface is created by the conic section (generatrix) from the y–z plane revolving around the axis that is parallel to the y-axis; the y-z plane serves as the tangential plane of the toroidal surface, and the cross section of the toroidal surface with the y–z plane is the tangential curve. Tangential curve expression of the toroidal surface is:

$$z = g(y) = \frac{Cy^2}{1 + \sqrt{1 - (1+K)C^2y^2}} \tag{4.5}$$

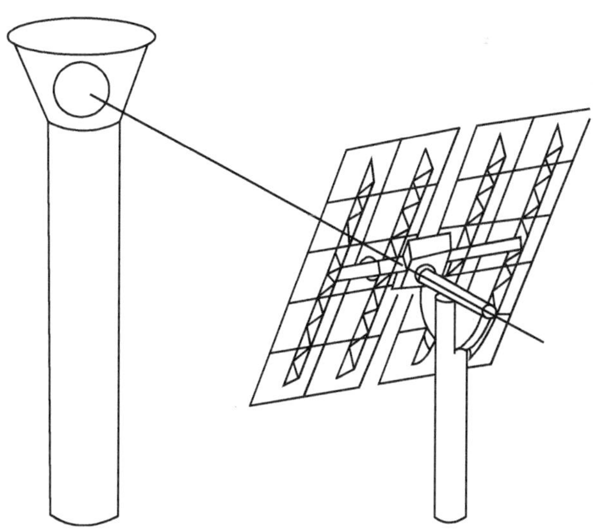

FIGURE 4.8 Astigmatic-corrected target-aligned heliostat conceived by R. Zaibel et al. in 1995.

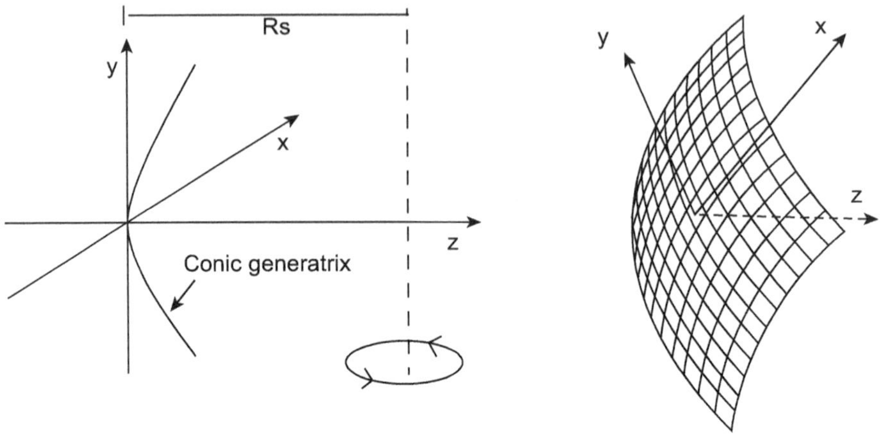

FIGURE 4.9 Toroidal surface created by the conic section on the y–z plane revolving around the axis parallel to the y-axis.

The toroidal surface can be expressed as follows:

$$z = f(x, y) = R_s - \sqrt{[g(y) - R_s]^2 - x^2} \tag{4.6}$$

Here, $C = 1/R_t$ is the tangential curvature corresponding to the vertex; R_t refers to the tangential curvature radius; R_s refers to the sagittal (horizontal) curvature radius corresponding to the vertex; and K is a constant for the conic curve of the tangential section transversal.

When Zemax optical design software is used to design the surface shape of a toroidal heliostat, tangential section expression of the toroidal surface meridian transversal is adopted as follows:

$$z = g(y) = \frac{Cy^2}{1 + (1 + K)C^2y^2} \tag{4.7}$$

Eqs. (4.7) and (4.5) are slightly distinguished from each other in terms of expression. In this case, for the same conic section, the value of K obtained from Eqs. (4.7) and (4.5) may be different. As for the selected incident angle range $[\theta_{min}, \theta_{max}]$, the light spot areas corresponding to θ_{min} and θ_{max} can be calculated through the ray tracing method; different values of parameters K, C, and R_s should be processed to determine light spot areas corresponding to θ_{min} and θ_{max} that are equivalent.

4.2.4 Optical Losses of the Parabolic Trough Concentrator

The working principle of a parabolic trough concentrator is basically the same as that of the heliostat, which is to utilize solar energy to the largest extent by sun tracking and concentrating direct solar radiation. However, the working geometries of heliostats and parabolic trough concentrators are different from each other. A heliostat reflects and concentrates solar radiation onto a fixed position that does not change over time, whereas in a parabolic trough system, the heat-absorbing tube moves with the concentrator to track direct solar radiation, and thus the target heat-absorbing tube is moving. The parabolic trough concentrator applies a single axis to track solar radiation, whereas the heliostat applies two axes to track solar radiation.

Incidence angles are generated by solar positions at different time points on the aperture of the parabolic trough solar collector, resulting in cosine effect, end-loss effect, and so on. All of these may influence whether the parabolic trough solar collector receives sufficient energy.

For most parabolic trough solar power plants that use collectors on a north–south axis that track Sun movement from east to west and morning to dusk, the incident sunbeam is never normally from directly above. Therefore, three major influences of oblique incidence of the sunbeam on the collector aperture must be considered, including cosine effect, end-loss effect, and incident angle impacting factor.

Direct normal solar irradiance *(DNI)* is irrelevant to the tracking position of a parabolic trough solar collector. The part of *DNI* perpendicular to the aperture area of the collector is the available normal irradiance (*ANI*) for the solar field collector and can be calculated as the product of *DNI* and the cosine value of the incident angle, namely $ANI = DNI\cos\theta$. In the case where the incident angle is 0, solar radiation is parallel to the normal of the collector aperture, namely $ANI = DNI$. In the case where the incident angle does not equal 0, cosine effect modification shall be conducted. Thus it can be seen that although *DNI* resources in winter are still quite high, available solar resources are greatly reduced by the influences of cosine effect, and therefore solar resources in winter are normally much less than those in summer. In the case of establishing a parabolic trough system in a high-latitude area, special attention must be paid to the influences of *ANI* on annual power generation.

In addition to considering the cosine effect caused by solar incident angle, a certain geometrical relationship exists between the incident angle and the reflective concentrator and absorber in the case of nonvertical incidence of sunbeam onto the reflective concentrator. This may result in a situation where solar radiation that is reflected and concentrated by the parabolic trough mirror surface is missed by an absorber end piece when the absorber end is close to the Sun during the parabolic trough operation, as shown in Fig. 4.10. L_{end} refers to the length of the parabolic trough

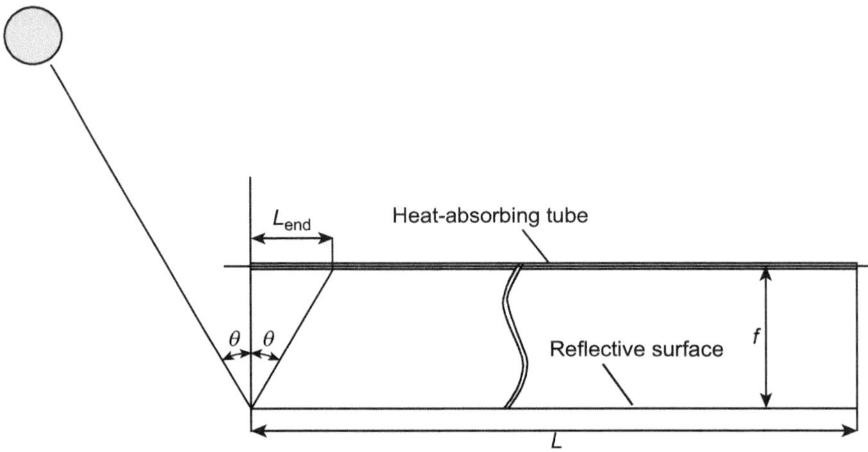

FIGURE 4.10 Nonvertical solar beam incidence on a parabolic trough solar collector.

collector under end-loss effect, and the relevant end-loss effect modifying factor F_{end} can be calculated through the following equation:

$$F_{end} = 1 - \frac{f}{L}\tan\theta \tag{4.8}$$

in which f refers to the focal distance of the parabolic trough solar collector and L refers to the length of one section of a parabolic trough collector. Apparently, the parabolic trough collector end-loss effect is related to parabolic focal distance and its structural parameters. Furthermore, even with the same focal distance, two rows of collectors may undertake different influences of end-loss effects due to the different lengths of their collector sections.

In the case of nonvertical incidence of solar beam onto the concentrator, except for various kinds of geometrical effects, the optical characteristics of the reflective and heat absorber surfaces also vary with incident angle. The parameter that measures this variation is called the incident angle modifier (*IAM*) and refers to the ratio of nonvertical incident solar beam to normal incident solar beam onto the vacuum tube. A parabolic trough concentrator with a west–east axis always remains in vertical incidence status in the altitude angle direction, in which direct normal solar radiation at midday is parallel to the normal of the aperture plane of the parabolic trough concentrator. In this case, $IAM = 1$; otherwise, $IAM < 1$. IAM is calculated considering the variation of optical properties along with incidence angle, especially for the transmittance of the vacuum tube glass wall and the absorptance of the heat-absorbing tube, which can be calculated through theoretical analysis. However, considering errors on the mirror surface of the concentrator and during tracking makes the solar beam concentration calculation for a concentrator with errors extremely difficult; also, precision is low, so experimental values may be comparatively more precise. Due to the different temperature strains of the concentrator support frame under different temperatures, a good equation must consider the influence of temperature on optical concentration precision or give separate calculation equations for winter and summer in estimating concentrator output energy during winter and summer:

4.2.5 General Principle for Heliostat Field Layout

According to Figs. 4.11–4.14, there are a few typical heliostat fields for solar thermal tower power plants.

1. For the heliostat field in a solar tower power plant, for the case where the receiver's shape (cylinder or cavity), location, and aperture have been determined, the heliostats shall be arranged following the basic principle of achieving optimal annual efficiency of the heliostat field. In this case, tower height, the distances between heliostats and the tower, and solar shading and blocking between various rows of heliostats shall be considered in determining heliostat locations.

   In cases where the receiver's aperture and location have not been determined, four factors—the receiver's geographic location, position, and aperture, as well as the solar concentrating field—shall be comprehensively considered while striving for solar-concentrating field design optimization. From the 1970s through the 1990s, a variety of heliostat field design software was developed,

FIGURE 4.11 North sector field of heliostats at the Beijing Dahan power plant (1 MW, Beijing, China).

FIGURE 4.12 Surrounding heliostat field at the gemasolar power plant (20 MW, Spain). *Gemasolar plant owned by Torresol Energy, Spain. Picture provided by SENER, Spain, 2018.*

FIGURE 4.13 Heliostat field at work at the gemasolar power plant (20 MW, Spain). *Gemasolar plant owned by Torresol Energy, Spain. Picture provided by SENER, Spain, 2018.*

FIGURE 4.14 North heliostat field at the daegu power plant (200 kW, Korea).

such as HELIOS, ASPOC, HFLCAL, RCELL, DELSOL, MIRVAL, and Solergy. [20] Early solar-concentrating field design software focused on the modeling and analysis of already-arranged heliostat fields and specified receiver locations, while user interfaces featured comparatively low degrees of visualization, inconvenient usage, and great difficulty upgrading the software. These past deficiencies have led to the successful development of new heliostat field design software in recent years. In 2001, PSA of the CIEMAT organization in Spain developed the WinDELSOL 1.0 software, which runs in a Windows environment on the basis of DELSOL3 with an improved degree of visualization in the software interface. Within the same year, Siala et al. from the Center For Solar Energy Studies, Libya, proposed the concept of no-blocking radial staggered layout of the heliostat field and wrote the heliostat field design software MUEEN to analyze heliostat field performance; however, it was not equipped with a heliostat field optimization function. In 2003, the National Renewable Energy Laboratory in the United States developed SolTrace software, which applied Monte Carlo ray tracing to calculate interception efficiency and energy flux density; it simulated and compared complex optical systems and analyze their optical performance; however, it also was not equipped with a heliostat field optimization design function. In 2005, SENER Corp. of Mexico developed the powerful SENSOL software, which performed simulation, analysis, and optimization design for various types of solar-concentrating power generation systems. During the period 2005–2011, with support from the national "11th Five-year Plan" 863 Program, the Changchun Institute of Optics, Fine Mechanics and Physics, Chinese Academy of Sciences, and the Institute of Electrical Engineering, Chinese Academy of Sciences, jointly developed the heliostat field design software HFLD (Heliostat Field Layout Design). The Institute of Electrical Engineering, Chinese Academy of Sciences designed the 10,000 m^2 heliostat field of the Beijing Badaling solar tower power plant using this software. The main feature of HFLD was analysis of the solar field performance for known mirror coordinates and receiver locations. It automatically arranged the heliostat field according to parameters set by the user and conducted performance analysis and optimization design of the heliostat field. This version was not equipped with a joint optimization function with the receiver. Separately, the Institute of Electrical Engineering, Chinese Academy of Sciences also developed a code for a set of solar tower optics in 2013, named heliostats optical code (HOC), that can both optimize the layout of the heliostat field and analyze a given heliostat field's optical performance.

2. The level-land requirement of a solar tower's heliostat field is lower than that of a parabolic trough as long as the adjacent rows of heliostats satisfy no-shading and no-blocking conditions. In the case where mountains are located south of the heliostat field, construction costs for the heat-absorbing tower can be reducing the building height of the tower. In the case where mountains are located to the north, the blocking of front-and-back neighboring heliostats in the heliostat field will be comparatively small.
3. The total mirror area of a solar tower heliostat field shall be determined in accordance with the requirements of the design point while satisfying the capacities of a steam turbine and thermal energy storage system and the requests of time-dependent grid-connected power generation.
4. The azimuth of the rotating axis of a parabolic trough concentrator shall be determined by considering local longitude and latitude and seasonally stepped grid-connected electricity prices. Generally, the efficiency of a west–east axis parabolic trough concentrator during summer is higher than for one with a north–south axis, and the annual average efficiency of a north–south axis concentrator is higher than for one with a west–east axis.
5. For the case of considering the growing of green plants within the heliostat field during the local plant growing season (usually April–September annually in northern areas), the sunshine percentage on the land where plants grow should not be less than 70%, which can be designed using solar-concentrating field optimization software.
6. When considering the growth of ground plants, the water used for cleaning mirror surfaces shall be free of alkaline and grease constituents. For a parabolic trough power plant using heat-transfer oil as its medium, oil-unloading and firefighting channels shall be designed within the solar field.
7. The solar-concentrating field layout should consider controller grounding, lightning protection, and the intersection of the communication cables with the high-temperature fluid pipeline. The pipeline should be designed with consideration given to cleaning vehicle access.
8. Maintenance space for concentrators should be reserved during design. Major maintenance includes maintenance of the gearbox and supporting frame deformation. Maintenance for mirror damage and replacement requires a small space, and thus it is not necessary to preserve additional space.
9. Due to the long length of the high-temperature, high-pressure pipeline of a parabolic trough concentrator, leakage monitoring and alarming facilities as well as solar field firefighting facilities

should be designed within the parabolic trough field. Especially in the case of a heat-collection system that uses oil as its heat-transfer medium, a fire-prevention trench shall be designed between the solar field and main powerhouse, where weeds and other inflammable goods shall be cleaned regularly.

10. For peripheral parabolic solar trough connectors within the solar field, due to the need for wind resistance, a large wind-resistance load shall be designed, including within the transmission design. The concentrators in the field center suffer weak wind load so the design wind load for these concentrators can be reduced to save structure cost.
11. The length of each line of concentrators in a parabolic trough concentrator field can be determined according to the temperature of the working fluid and the local ambient temperature using technical and economic comparisons. A long concentrator requires a greater site leveling cost and has a higher requirement for axial installation precision of the concentrator.
12. Rodent pest prevention within the solar-concentrating field is also a topic to focus on during cable-laying design. In the event of a rodent pest problem, cable trenches must not be used and the piping shaft must be tightly sealed. The wiring of the concentrator's control cabinet shall be protected from rodents. Qualified projects can either consider applying wireless communication in place of wire transmission or substituting the cable layout by using photovoltaic cells to drive the concentrator.

4.3 DESIGN OF THE SOLAR TOWER POWER PLANT CONCENTRATING FIELD

The heliostat is important equipment in the solar thermal power generation system during the initial stage of energy conversion. In a solar tower power system, hundreds or thousands of heliostats are typically used to continuously track solar radiation through respective independent control systems and concentrate solar energy onto the receiver mounted on top of the tower to utilize energy in the form of heat. The design of the heliostat field serves as a major step in the design of a solar thermal tower power generation system and can reduce the overall scale of electricity-generating costs while demonstrating a certain theoretical foundation for the application of solar thermal power generation techniques.

4.3.1 Basic Operation Modes of the Heliostat Field and Basic Design Parameters

4.3.1.1 Operating Modes of the Heliostat Field

The function of a heliostat field is to provide a receiver with as much energy as possible while ensuring system safety and service life. Local overtemperature conditions and frequent thermal shocks to a receiver's heat absorber have adverse effects on heat absorber servicing and safety.

By following the above principles, the control modes of the heliostat field can be divided into manual and automatic modes (refer to Table 4.1 for details). The manual mode includes a standby mode and an operation control mode, and the automatic mode includes a strong wind mode and a breeze mode. The breeze mode includes a normal operating mode and a beam characterization system (BCS) calibration mode. The normal operating mode can be divided into daytime and night modes based on time of day. The daytime mode can be further divided into a "ready point" mode and a normal tracking mode. The special modes include a cleaning mode and an overhaul mode.

4.3.1.2 Design Parameters

The heliostat field control system controls the heliostat field by setting the working modes of each heliostat, either one by one or in sections and groups.

1. Working conditions of heliostat field design: in general, the normal working ambient air temperature range is −10 to 45°C. For wind speeds not more than 13 m/s, normal heliostat functioning can be ensured; for wind speeds not more than 20 m/s, the heliostat should not be damaged under any working status; in the protection state, the heliostat should not be damaged for cases in which the wind speed is not more than 36 m/s.
2. Viability under special extreme conditions
   a. Extreme temperature tolerance results in no damage to heliostat components from −40 to 60°C.
   b. Seismic tolerance intensity is 8°, especially for heliostat foundation stability.
   c. Weather resistance: various components of the heliostat can withstand rain, snow, sandstorm invasion, and outdoor temperature variation without damage during service life; various metal components shall be free of any corrosion.
   d. Hail resistance: the reflective surface of the heliostat shall withstand hail impact with hail diameter of 20 mm along the direction parallel to the normal of the mirror surface without any damage.

TABLE 4.1 Mode Categories of the Heliostat Field

Local Mode	Users Can Operate Heliostats Using Buttons on the Local Control Cabinets			
Remote Mode	Remote manual mode	User can control rotation of heliostats through control keys on the host computer		
	Remote automatic mode	Initial mode	Rotate to initial position	
		Cleaning mode	Rotate heliostat to facilitate cleaning	
		Strong wind mode	Rotate horizontally if wind speed exceeds set value	
		Emergency action mode	For power-off and communication blackout conditions, rotate heliostat horizontally	
		Fault mode	In case of heliostat malfunction, the host computer sends an alarm	
		Tracking mode	Receiver Mode	Target point is the receiver
			Ready Point Mode	Target point is the ready state point
			Tracking errors Inspection Mode	Target point is the target board

4.3.2 Methods for Heliostat Field Optimization Design

The investment cost of a heliostat field usually accounts for 40%–50% of the investment cost for an entire solar thermal tower power system; thus the optimization design of the heliostat field can offer favorable conditions for reducing investment and power-generating costs and further promote commercialization and scale-up of solar thermal tower power-generation techniques.

4.3.2.1 Current Status of Heliostat Concentration Field Design Software

To stay consistent with the development pace of solar thermal tower power generation techniques, some countries have been developing software programs since the 1980s, such as UHC, HFLCAL, RCELL, DELSOL, MIRVAL, FIAT LUX, and SolTrace, and these programs have been used in the analysis and design of heliostat fields and entire solar thermal tower power generation systems [34]. The mathematical models of MIRVAL, FIAT LUX, and SolTrace can be used only for detailed heliostat field energy collection calculations and not for heliostat field optimization. The mathematical models of HFLCAL and DELSOL, however, can be used for the entire solar thermal tower power generation system, including heliostat field optimization design; they can be directly used for estimating the annual average optical performance of a large-scale heliostat field; but for small-scale heliostat fields, their calculation accuracies are comparatively low [34]. The main features of the above performance-analysis programs are shown in Table 4.2 [34].

From 2003 to 2005, SENER Corp. of Spain successfully developed SENSOL software [35], which was written in Fortran and used for thermal economic analysis of solar thermal tower power generation systems. The coordinate position of the heliostat can be determined based on the level of economy. This software was applied in the system design of the Solar Tres program in Spain.

The main heliostat field design codes in china include HFLD and HOC. Several main heliostat field design codes are compared in Table 4.2.

4.3.2.2 Basic Idea of Concentrating Field Design

The basic idea is to apply a radial-staggered layout pattern, conduct heliostat field layout optimization for a solar thermal tower power generation system using conventional dual-axis tracking under the premise of avoiding mechanical collision between adjacent heliostats while collecting maximum energy or achieving optimal economy, as well as optimizing results to obtain a Pareto curve under dual-objective coordinate axes that consist of various optimal heliostat field layout schemes. The optimized heliostat field features not only a low-unit energy cost and good economy, but also uniform and reasonable energy distributions.

1. Required space for heliostat free rotation [13]: a heliostat is by nature a mirror (reflective mirror). Conventionally, a rectangular heliostat rotating around a fixed vertical axis, which has been the most widely used as shown in Fig. 4.15, continually tracks variations in solar position in order to reflect solar radiation onto the fixed target of the receiver. Because the diameter of the vertically placed barrel-shaped cylinder created by its free rotation around the azimuth axis is equal to the length of the heliostat's diagonal, during heliostat field design, feature parameter D_m is defined as the length of the diagonal plus a safety clearance of 0.3 m. During the placement of heliostats, the spacing between two heliostats must not be less than this value.

TABLE 4.2 Main Features of Five Codes for Concentrated Solar Flux Calculation [34]

Name of Code	UHC-RCELL	DELSOL	HFLCAL	MIRVAL	FIAT LUX	SolTrace
Research team	Houston University	SANDIA	GAST project	SANDIA	CIEMAT.PSA	NREL
Development start	1974	1978	1986	1978	1999	1999
Programming Language	Fortran/C++	Fortran/basic	Fortran	Fortran	MATLAB	Delphi5
Flux calculation method	Hermite polynomial expansion/convolution	Hermite polynomial expansion/convolution	Simplified convolution of each heliostat's flux	Monte Carlo ray tracing	Normally distributed random value of slope error	Monte Carlo ray tracing
Receiver type	Flat, cavity or external cylinder	Flat, cavity or external cylinder	Flat, cylindrical or conical	Flat, cavity or external cylinder	Flat	Almost any receiver
Annual performances	Yes	Yes	Yes	Yes	No	No
Optimized components	Heliostats layout and boundary, tower height, receiver geometry	Heliostat boundary, layout, tower height, receiver size, storage capacity	Heliostat layout, tower height, receiver area and orientation	Heliostat layout	Not available	Not available
Optimization criteria and constraints	Energy or cost criteria with allowable flux/land constraint	Cost criteria with optional flux/land constraints	Energy or cost criteria	Energy criteria	Not available	Not available

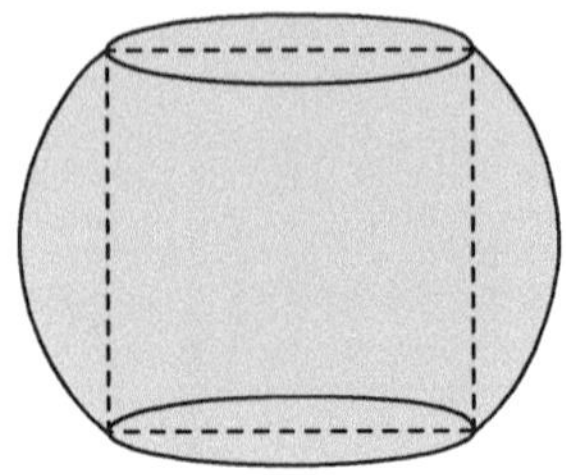

FIGURE 4.15 Required space for collisionless rotation of heliostat.

2. Layout pattern of the heliostat: the heliostat field is arranged by applying radial-staggered pattern, the advantage of which is the avoidance of significant optical losses caused by reflected solar beams from heliostats that are blocked by adjacent heliostats located straight ahead of them. In a radial-staggered design, the heat-receiving tower is located at the origin of the coordinates while the heliostats are placed on rings with varying distances from the receiving tower. In this method, an essential ring is defined as a ring with a heliostat on the axis straight ahead of the tower, and a staggered ring is defined as a ring without any heliostat on the axis straight ahead of the tower, as shown in Fig. 4.16. The first ring is defined as the essential ring, the radius of which is normally related

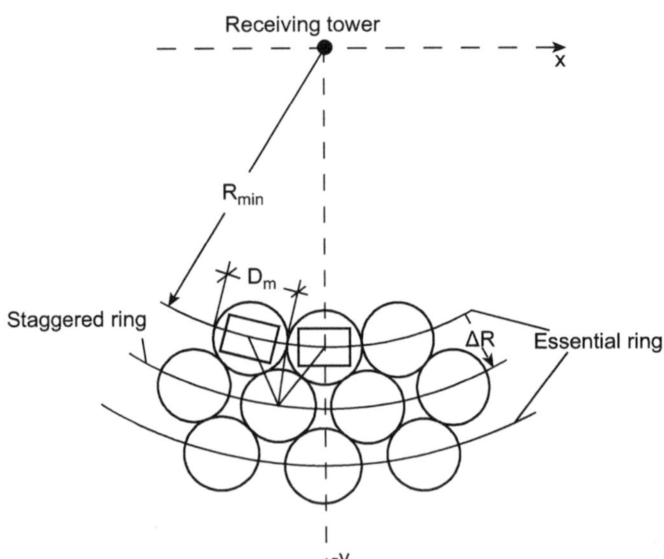

FIGURE 4.16 Schematic diagram of radial-staggered layout.

to the height of the receiving tower; the radii of other rings is determined by the value of radial spacing between adjacent rings. Two heliostats on the same ring or adjacent rings shall be kept from each other by a certain azimuthal angle, the value of which is related to the azimuthal spacing coefficient.

3. Calculation of Heliostat Radial Spacing
   a. Calculation method for minimum radial spacing: minimum radial spacing of heliostats shall ensure no mechanical collision between adjacent heliostats and that all heliostats can rotate freely without interference. Thus the radial spacing of heliostats shall be determined while considering maximum barrel diameter (Fig. 4.17):

$$\Delta R_{\min} = R_{m+1,\min} - R_m = D_m \cos 30^\circ \cos \beta_L \tag{4.9}$$

   b. Maximum radial spacing: when heliostat spacing exceeds maximum radial spacing, there is no blocking loss between heliostats:

$$z_m = R_m \tan \beta_L + H_h \tag{4.10}$$

$$d = \sqrt{R_m^2 + (H_c - z_m)^2} \tag{4.11}$$

$$\gamma = \arcsin\left(\frac{D_m}{2d}\right) + \arcsin\left(\frac{R_m}{d}\right) - \beta_L \tag{4.12}$$

$$\Delta R_{\max} = R_{m+1,\max} - R_m = D_m/\cos \gamma \tag{4.13}$$

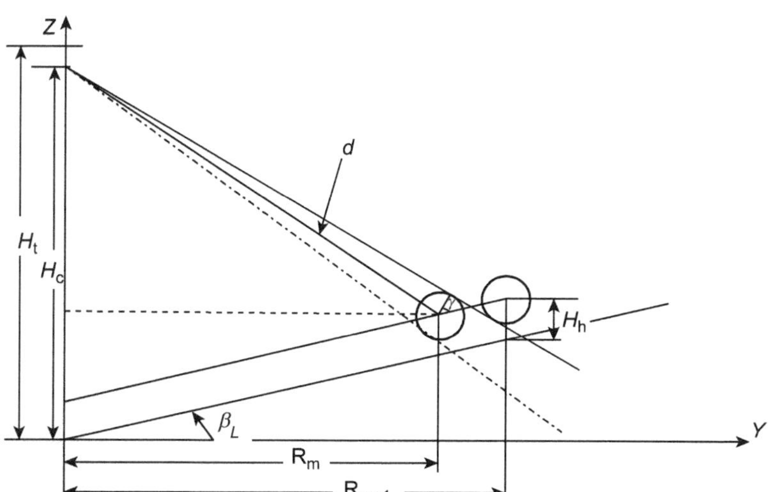

FIGURE 4.17 Schematic diagram of calculation of radial spacing of heliostats without any blocking losses.

The value of ΔR_{max} may be great so that the area of the heliostat field is great.

During program design, as a decision variable, the radius of the first ring where the heliostat has been placed shall be optimized before selection, whereas the radii of other rings shall be calculated based on the radial spacing between each ring and the ring previous to it:

$$R_{m+1} = R_m + \Delta R_{min} + R_{min-max}(\Delta R_{max} - \Delta R_{min}), \quad 0 < R_{min-max} < 1 \tag{4.14}$$

4. Calculation of circumferential spacing: in design, the first heliostat on the essential ring is placed on a negative section of the y-axis. The angle between it and the negative section of y-axis is assumed to be θ; then the positions of other heliostats can be calculated based on the azimuthal angle between them and adjacent heliostats (especially those located in the left front).

a. Calculation method for minimum azimuthal spacing: as the heliostat must reflect the solar beam onto the receiving tower, there shall be a certain azimuthal angle between adjacent front-and-back heliostats that avoids significant shading and blocking losses.

For a staggered ring:

$$\begin{aligned}\theta(m,n)_{min} &= \left|\text{Angle}(m,n)_{min} - \text{Angle}(m-1,n)\right| \\ &= \arcsin[D_m/(2R_{m-1})]\end{aligned} \tag{4.15}$$

For the second and other heliostats on the first ring:

$$\begin{aligned}\theta(m,n)_{min} &= \left|\text{Angle}(m,n)_{min} - \text{Angle}(m+1,n-1)\right| \\ &= \arcsin[D_m/(2R_m)]\end{aligned} \tag{4.16}$$

For the second and other heliostats on the essential ring:

$$\begin{aligned}\theta(m,n)_{min} &= \left|\text{Angle}(m,n)_{min} - \text{Angle}(m-1,n-1)\right| \\ &= \arcsin[D_m/(2R_{m-1})]\end{aligned} \tag{4.17}$$

As shown in Fig. 4.18, the solid black circle refers to well-placed heliostats, the hollow solid line circle refers to the position of heliostats on adjacent rings with the minimum azimuthal angle, and the dotted-line circle refers to the position of heliostats with the maximum azimuthal angle.

b. Calculation of maximum azimuthal spacing: the designed maximum azimuthal spacing in this book ensures no blocking in

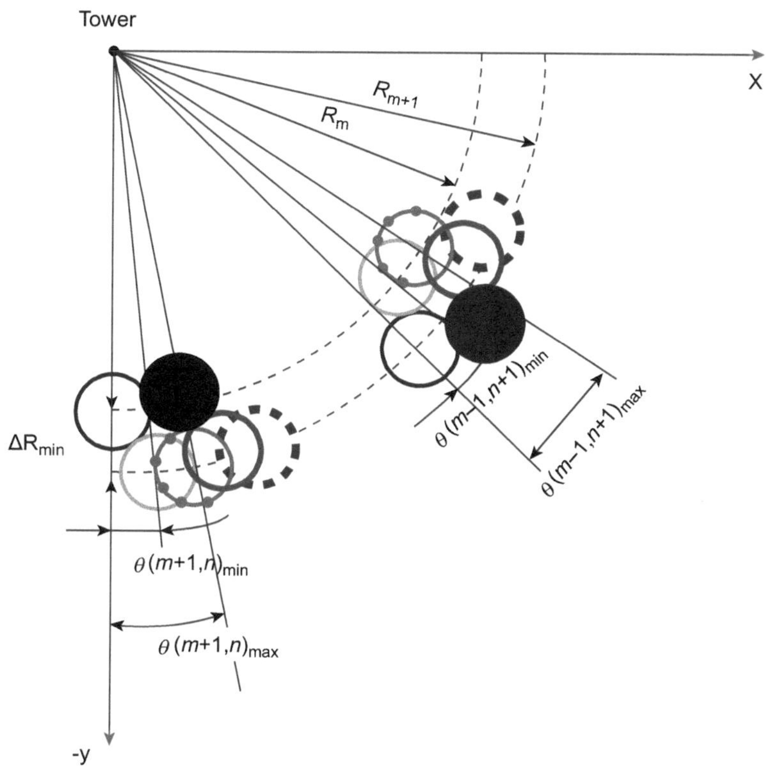

FIGURE 4.18 Schematic diagram of circumferential spacing calculation.

the direction of reflective beams between adjacent front-and-back heliostats. Calculation of the azimuthal spacing between heliostats is shown in Fig. 4.18.

For heliostats on the staggered ring:

$$\begin{aligned}\theta(m,n)_{\max} &= \left|\text{Angle}(m,n)_{\max} - \text{Angle}(m-1,n)\right| \\ &= \arcsin[D_m/(2R_{m-1})] + \arcsin[D_m/(2R_m)]\end{aligned} \tag{4.18}$$

For the second and other heliostats on the first ring:

$$\begin{aligned}\theta(m,n)_{\max} &= \left|\text{Angle}(m,n)_{\max} - \text{Angle}(m+1,n-1)\right| \\ &= \arcsin[D_m/(2R_m)] + \arcsin[D_m/(2R_{m+1})]\end{aligned}$$

For the second and other heliostats on the essential ring:

$$\begin{aligned}\theta(m,n)_{\max} &= \left|\text{Angle}(m,n)_{\max} - \text{Angle}(m-1,n-1)\right| \\ &= \arcsin[D_m/(2R_{m-1})] + \arcsin[D_m/(2R_m)]\end{aligned} \tag{4.19}$$

Thus during optimization design, the azimuthal positions of the heliostat can be calculated through the following equation:

$$\text{Angle}(m,n) = \text{Angle}(m,n)_{\min} + A_{\min-\max}\left[\theta(m,n)_{\max} - \theta(m,n)_{\min}\right],\quad 0 < A_{\min-\max} < 1 \tag{4.20}$$

Symbols in Eqs. (4.9–4.20) and Fig. 4.17 have been listed in the following table:

Angle (m, n)	Angle of Heliostat From y-Axis	Rad
$A_{\min-\max}$	Angular spacing coefficient	
d	Distance	m
D_m	Characteristic dimension of heliostat	m
H_c	Height of receiver center	m
H_h	Height from center of heliostat to ground	m
H_m	Height of heliostat	m
H_t	Height of the receiving tower	m
m, n	Index number "m" of the ring, azimuthal position number "n"	
R	Radius of a ring	m
$R_{\min-\max}$	Radial spacing coefficient	
Z_m	Vertical height of center of heliostat	m
β_L	Slope angle of heliostat field against receiving tower, here, $\beta_L = 0$	rad
r	Angle	rad
θ	Angle	rad

In terms of heliostat layout except for the calculation of spacing, a heliostat field with a fixed area has an optical efficiency related to its dimensions, the quantity of heliostats, and the height of the heat-receiving tower. In order to improve the Pareto curve, the receiving tower's height shall be increased at an appropriate scale. The higher the receiving tower, the smaller the blocking loss for adjacent front-and-back heliostats in the course of solar beam reflection, and thus the smaller the distances between mirrors and the tower, so that more energy can be collected (refer to Fig. 4.19).

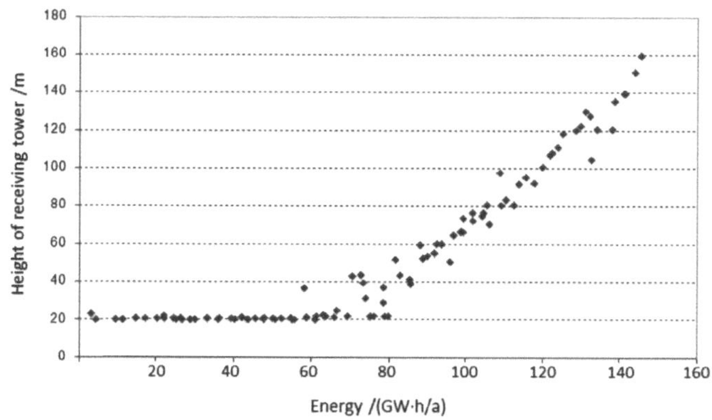

FIGURE 4.19 Relationship between receiving tower height and collected energy.

4.3.3 Design of the Solar Tower Receiver

1. Output of the heliostat field during the day is an unsteady quantity that varies constantly due to variations in solar position. The solar field's output at the design point shall satisfy receiver requirements under full-load working conditions.
2. The receiver's aperture dimensions shall satisfy the requirement that it intercept over 90% of the energy concentrated by the heliostat field at the design point. The receiver's heat-absorbing surface design shall satisfy the boundary conditions of the maximum concentrated flux density of the heliostat field, normally 1–2 MW/m^2.
3. Design of the heliostat field's concentration ratio shall allow the energy flux density on the heat-absorbing surface during normal operation to be 300 kW/m^2 (water/steam working fluid), 600 kW/m^2 (molten salt), or 750 kW/m^2 (ceramics/air).
4. Sun tracking of the heliostat field can be calibrated in either an open-loop or a closed-loop manner. The operational orientation of a heliostat can be coordinated with the temperature control of the receiver's heat-absorbing surface. If the temperature of the heat-absorbing surface becomes too high, the main control host computer shall immediately instruct the heliostats corresponding to the overtemperature region to gradually move away one by one until the temperature of the dangerous point on the absorber surface has been reduced to within a reasonable range. It is necessary to conduct real-time temperature monitoring of the heat-absorbing surface.

FIGURE 4.20 Temperature-monitoring picture of receiver at the Dahan power plant [20].

5. In addition to measurement by thermocouple array, the heat-absorbing surface's temperature can be measured by noncontact methods, such as an infrared camera. When an infrared camera is used, attention must be paid to calibrating it with the receiver in advance. This is because thermal emittance of a solid surface is relevant to the thermophysical properties of the surface, and these properties are relevant to temperature. Normally, thermal emittance increases along with temperature increases on the surfaces of objects. Fig. 4.20 shows the temperature-monitoring picture of a running receiver in the Dahan power plant.
6. If the heat-transfer fluid inside the receiver is cut off due to an event such as power-off of the entire field, a pump fault, or pipeline leakage, the heliostat field must be stopped. Such a stoppage instruction is collected by the flowmeter on the incoming fluid main pipeline or the tower's drum level gauge and sent to the solar field host computer; afterward, the host computer gives instructions to the heliostat to return to their original positions until the fluid loop fault is eliminated. To avoid false alarms, two flowmeters should be placed in series in the main loop, which is better done at the receiver inlet. Also, the suggested quantity for the drum level gauge is "two."
7. For an air receiver, the working status of the heliostat field can be detected using receiver airflow rate or heat absorber surface temperature. Due to the high temperature of heat coming from the air receiver, temperature measurement becomes extremely difficult, especially in the event of an accident or a failure. It is normally

suggested that airflow measurements should be used for estimating surface heat flux conditions and the control of heliostat actions.

a. A sealed glass window fault or sudden stoppage of the air compressor will cause the pressurized air receiver's air resistance to drop suddenly. In such an event, the heliostats must be moved away immediately to prevent the receiver from being burnt up. The heliostats and flow-resistance measurement system inside the receiver shall be designed with a linkage device.

b. In the event of fan failure of a nonpressurized air receiver, the receiver temperature rises sharply or the flow rate becomes zero. In this case, the heliostat field must be immediately stopped. The flowmeter should be placed behind the receiver to measure receiver airflow or can be placed at the outlet of an induced draft fan; due to multistage heat exchange and storage, air temperature at this point will already be comparatively low.

8. In the event of freezing blockage inside the transmission pipeline of a molten-salt receiver, the molten salt will not flow into the receiver and the surface temperature of the heat absorber will rise. Based on the surface temperature of the absorber, heliostat actions can be controlled based on absorber surface temperature and through the flow of molten salt. As a molten-salt flowmeter is easy to break down, flowmeter failure diagnosis can be made by combining the heat-absorbing surface temperature and molten-salt flow rate of the receiver. When either one occurs, the heliostats can be moved away to a safe position.

9. Before receiver startup, pipelines shall be preheated. Before heliostats project energy into the receiver, the receiver shall first be filled with heat-transfer fluid to avoid dry burning of the receiver. The fluid temperature shall be close to the heat absorber temperature that corresponds to the same time. If the receiver contains a superheated section, focus must be placed on monitoring the temperature the superheated section and adjusting it according to input energy directions from the heliostat field. The difference between the temperature of the superheated section of the receiver and that of the water-cooled wall of the receiver shall not exceed 100°C [20].

a. For a molten-salt receiver, the heat absorber's surface temperature shall exceed the freezing point of molten salt by over 50°C. Heliostats shall be controlled in groups and their concentrated beams gradually moved to corresponding positions on the heat-absorbing surface one by one so that the temperature of the heat absorber can rise accordingly. All heliostats must be moved to normal working positions within 60 min. A molten-salt system should be circulated 24 hours a day to achieve an antifreezing effect. Such circulation shall be conducted through temperature control when no power is generated. Electric auxiliary heating of the transmission pipeline is not reliable, as salts are easily frozen in partial areas. When applying this method, heat tracing of all narrow parts should be fully completed, mainly including the valve, flange, sensor interface, elbow, and pumping hole. Good insulation measures shall be taken on the aperture surface of a molten-salt receiver.

For an external cylinder receiver (Fig. 4.13), the respective antifreezing energy consumption of molten salt is comparatively great. However, due to geometrical features, a larger concentration ratio can be obtained for an external receiver than using a cavity receiver.

b. For water/steam receivers, only when temperature of the receiver's heat absorber exceeds zero can water be added to make it start working. Heat absorber preheating can be conducted by integrating solar concentration using a small number of mirrors and absorber surface temperature measurements.

c. For air receivers, before heliostats are put into operation, the fan or compressor should first be turned on and the heliostat aimpoints adjusted according to the surface temperature of the heat absorber or the air temperature at the receiver outlet.

4.4 CONTROL DESIGN OF THE HELIOSTAT FIELD OF A SOLAR TOWER POWER PLANT

4.4.1 Technical Conditions for the Heliostat Field Control System

1. Control requirements of the heliostat field shall be determined according to solar irradiation conditions, wind speed, ambient temperature, and characteristics of the heat-transfer working fluid while meeting the following requirements:
   a. Division of the heliostat field. The solar-concentrating field consists of many heliostats. The receiver shall be free of major thermal shock during starting and stopping stages, which in a heliostat field can be divided into several sections. During startup, solar energy is gradually concentrated into the receiver by sections at different time intervals; the corresponding operation time point and time interval of each section of heliostats are given by instructions from the solar field master controller. During operation control of a heliostat, energy input to the receiver shall be estimated, and the position of the projected solar beam inside the receiver shall be determined.

b. Concentrated solar irradiation control. The heliostat field controller determines the quantity of heliostats to be input into the receiver according to external meteorological conditions, heat-transfer loop emergency alarm, and the like, which can also be determined on the basis of instructions from the power plant's main controller.

c. Constitution of the heliostat field control system. The controller consists of hardware and software to control the actions of heliostats as well as the BCS (beam characterization system) for inspecting the tracking precision of heliostats. Input signals of the controller include wind speed, temperature, heat absorber temperature, drum pressure, receiver inlet flow, and fluid temperature at the receiver outlet.

2. Control structure of the heliostat field. The heliostat has two rotation axes. The commonly used transmission equipment includes the gear transmission, linear actuator, and hydraulic transmission. Each heliostat is equipped with a local controller for controlling actions of the rotation axis; the heliostat field controller controls rotation of the heliostat through the local controller:

a. The heliostat local controller either calculates the heliostat's rotational position corresponding to each time point by using the astronomical formula of the Sun or receives heliostat tracking position instructions distributed from the host controller, which has more powerful computational capabilities; heliostats themselves can be only equipped with an emergency response function.

b. The heliostat field controller is connected to the solar concentration field wind speed sensor and the receiver safety alarm device in order to provide heliostat orientation control in emergencies. The heliostat field controller can also be connected to the power plant's main controller.

3. Heliostat grounding can be divided into two sections, lightning protection grounding and control electrical appliance grounding, both of which shall be executed in accordance with respective national standards. Heliostat field grounding shall be considered together with grounding of the entire power plant; the grounding electrodes of all power plant electrical equipment shall be connected in an equipotential manner. Grounding materials shall be selected while fully considering the chemical constituents and nature of soil to ensure that the designed service life of the power plant can be achieved. The grounding resistance shall be designed while considering winter soil freezing and summer meltdown.

4. The logical relationship between the heliostat field controller and power plant control. The power plant's control host receives heliostat output energy information and the like from the heliostat field controller as well as offering a variety of information to the heliostat controller, such as meteorological conditions and the working status of the receiver, thermal storage unit, and steam turbine. The heliostat field controller determines the working status of each heliostat based on such information. All instructions for any actions of the heliostat will be sent by the heliostat field controller, which means that every heliostat only receives action instructions from the solar field controller.

The control of the heliostat field can also be not connected to the host computer of the power plant; instead, it receives direct signals from the temperature sensor and meteorological conditions from the receiver to control the actions of the heliostat.

4.4.2 Correction of Heliostat Tracking Errors

Heliostat tracking precision is a key index for the solar thermal tower power generation system. The heliostat's current position can be calculated in accordance with the astronomical formula of the Sun while achieving very high calculation precision. Nevertheless, in the case of manufacturing, installing, and operating a heliostat, it is inevitable that various kinds of errors will be encountered. For example, the azimuth rotation axis of the heliostat shall be perpendicular to the horizontal plane, and the elevation rotation axis shall be parallel to the horizontal plane; yet during the manufacture and installation process, absolute perpendicularity and parallelism cannot be achieved. Furthermore, the higher the requirement for precision, the higher the respective cost. Because of various influencing factors, heliostat tracking precision is normally low. Although it does not deviate greatly from the target center, it is unable to satisfy the request for power generation. Therefore, alternative deviation calibration methods must be used to improve tracking precision. If deviation corrections cannot be performed in time, deviation of the concentrated solar spot from the target point may result and lead to burnup of the support structure of the heat-absorbing tower.

Equipment used in heliostat tracking deviation correction includes a single-heliostat control system, a CCD (charge-coupled device) image-acquisition camera, an image processing and analyzing system, a full heliostat field control programmable logic controller, and a host monitoring system for the entire heliostat field (refer to Fig. 4.21).

It applies fully closed-loop inspection and correction of tracking deviation, recording of historical deviation correction curve, interpolation calculation, successive approximation, and the like, which features good deviation correction effects and satisfactory adaptability.

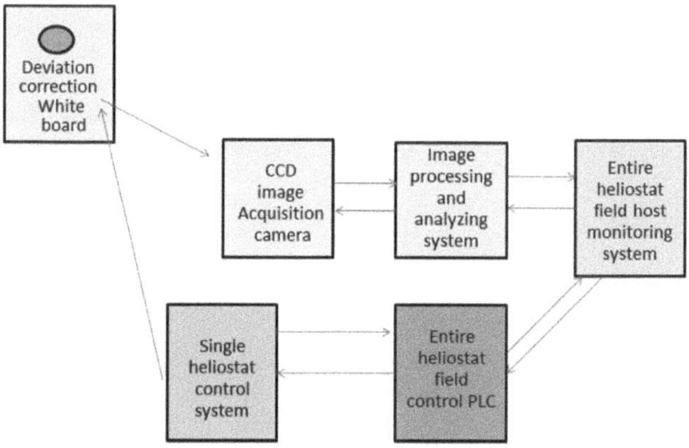

FIGURE 4.21 Constitution of the heliostat tracking deviation correction system.

The current tracking angle of the heliostat consists of the initial angle, rotation angle, and tracking deviation angle of the heliostat.

By inspecting heliostat tracking errors multiple times daily for several days during the year, heliostat tracking deviation at a typical time can be obtained through analysis and processing of one year or multiple years of heliostat tracking deviation angle data and curve fitting, each day's corresponding tracking deviation curve can be obtained. In this way, the corresponding daily tracking deviation curve for each heliostat can be obtained. By utilizing this tracking deviation curve, the current tracking angle of each heliostat shall be adjusted so that the solar spot of each heliostat can be more precisely projected to the target position.

Based on the feature of insignificant variation of the heliostat tracking deviation angle during a short period (such as half an hour) and the feature of insignificant variation of the tracking deviation angle at the same time point for consecutive days (such as 15 days), a day can be divided into several time intervals of equal duration. During each time interval, a tracking-deviation angle is obtained through deviation correction inspection.

In order to facilitate reader understanding, the heliostat deviation correction method applied at the Badaling Dahan solar tower power plant is used for demonstration in the example that follows.

After the heliostat and deviation correction equipment in the Dahan solar tower power plant are initiated, the measured azimuth deviation angle and elevation deviation angle are input into the database while recording the completion date and time of deviation correction for interpolation calculations (refer to Fig. 4.22).

1. Manual tracking deviation correction. The operator shall input the serial number of the heliostat that requires deviation correction

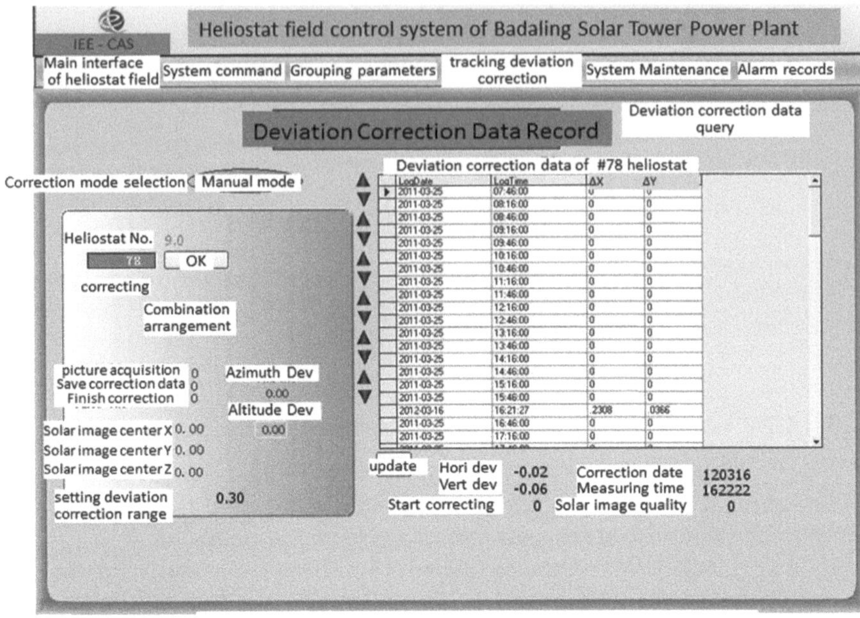

FIGURE 4.22 Data record sheet of tracking deviation correction at the Badaling solar thermal tower power plant.

and press the CONFIRM button to start the deviation correction process; at this time, the screen displays "deviation correcting." After deviation correction starts, the heliostat goes into the mode of "executing ... moving to deviation correction point" (refer to Fig. 4.23).

When the deviation correction point is reached, the current mode displays "deviation correction point reached"; the operator then initiates the CCD image acquisition camera while waiting for the image processing and analyzing system to return the deviation angle. As shown in Fig. 4.22, "horizontal deviation" and "vertical deviation" refer to the two deviation values from the white board center (unit: m).

To satisfy various working conditions, the range of deviation correction is adjustable by selecting the range through automatic or manual mode on the "System Maintenance" page. The manual mode can be modified on the "Deviation Correction Data Record" page; the range of the automatic mode can be automatically adjusted according to the solar field ring number where the heliostat has been located (distance from the heat-absorbing tower); the greater the distance, the greater the range for allowed deviation. Considering that deviation correction may not be completely adjusted to the white board center using the current setting, the maximum number

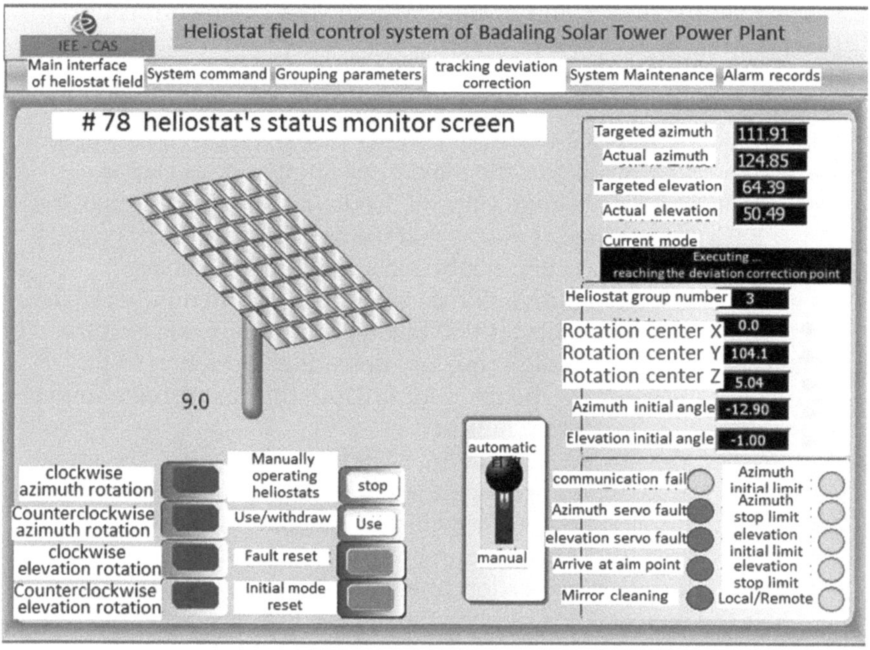

FIGURE 4.23 Heliostat tracking deviation correction control of the Badaling solar thermal power plant.

of times for deviation correction can be modified on the "System Maintenance" page.

Definition of range of deviation correction: when the absolute value of the deviation is less than the lower limit of this range, it means the respective deviation correction is qualified.

Once the heliostat has reached the deviation correction point, the deviation correction camera is initiated. In the case where the absolute value of the deviation is less than the lower limit of the deviation-initiated range, the deviation-initiated process finishes, and the azimuth deviation angle and elevation deviation angle are recorded as the corrected values of X and Y separately in the database while recording the completion date and time of deviation correction.

In the event that the absolute value of deviation is greater than the upper limit of the deviation correction range, and the number of deviation corrections is less than the maximum number of deviation corrections, new coordinates are calculated based on the current deviation and sent to the heliostat control system; when the heliostat once again reaches the deviation correction point, the camera is initiated. In the event that the absolute value of deviation is less than the lower limit of the range, the deviation correction process finishes, and the azimuth deviation angle and elevation deviation angle are recorded as the corrected values of X and Y separately in the database while recording the completion date and time of deviation correction. In the event that the number of deviation corrections is equivalent to the maximum number of deviation corrections, and the absolute value of the deviation is greater than the upper limit of the range, it means that the deviation rectification fails, and the respective data will not be recorded in the database.

After completion of deviation correction, the serial number of the deviation correction heliostat will be automatically reset, and the system returns back to the status of "deviation correction to be performed," waiting for the operator to input the next serial number of the deviation correction heliostat.

2. Automatic deviation correction. Under automatic deviation correction status, when pressing the "Initiate Automatic Deviation Correction" button, the system will automatically select the deviation correction heliostat number based on historical data and initiate the deviation-correction process (refer to Fig. 4.24).

Once the heliostat has reached the deviation-correction point, the camera is initiated. If the absolute value of deviation is less than the

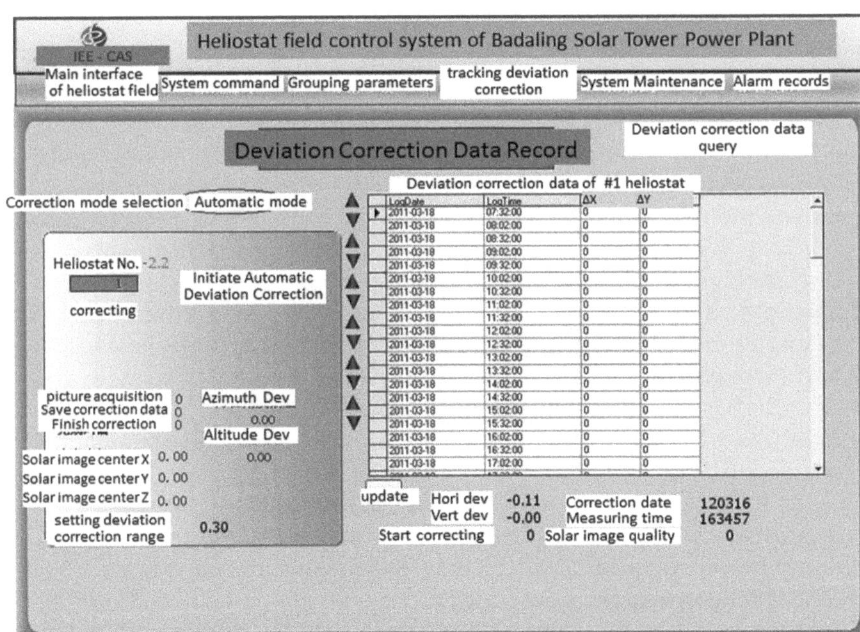

FIGURE 4.24 Picture of automatic deviation correction operational control system at the Badaling solar thermal power plant.

lower limit of the deviation correction range, the deviation-correction process finishes, and the azimuth deviation angle and elevation deviation angle are recorded as the corrected values of X and Y separately in the database while recording the completion date and time of deviation correction.

In the case that the absolute value of deviation is greater than the upper limit of the deviation correction range and the number of deviation corrections is less than the maximum number of deviation corrections, new coordinates are calculated based on the current deviation and sent to the heliostat control system; when the heliostat once again reaches the deviation-correction point, the camera is initiated. In the event that the absolute value of deviation is less than the lower limit of the range, the deviation correction process finishes, and the azimuth deviation angle and elevation deviation angle are recorded as the corrected values of X and Y separately in the database while recording the completion date and time of deviation rectification. In the event that the number of deviation corrections is equivalent to the maximum number of deviation corrections and the absolute value of deviation is still greater than the upper limit of the range, it means that the deviation correction fails, and the respective data will not be recorded in the database.

After completion of deviation correction, the deviation-correction heliostat number will be automatically reset, and the system returns to the status of "deviation correction to be performed." The control program will then select the next heliostat mirror number for deviation correction and initiate the next deviation correction process; the cycle repeats.

3. Special circumstances. If no tracking deviation angle is recorded in the respective daily time interval, the tracking deviation angle inspected in the respective time interval of the previous day, several days before, or from the previous month shall be used as the tracking deviation angle for that day. If there is no inspection result for the current time interval, the tracking deviation angle shall be 0. Then the tracking deviation angle inspected in the previous time interval shall be applied as the tracking deviation inspection result of the current time interval.

To achieve higher tracking precision, according to database records, the tracking deviation curve of the heliostat corresponding to any given daily time can be obtained through interpolation based on modifying the current angle of the heliostat so that the heliostat's concentrated solar image can be precisely projected to the target position.

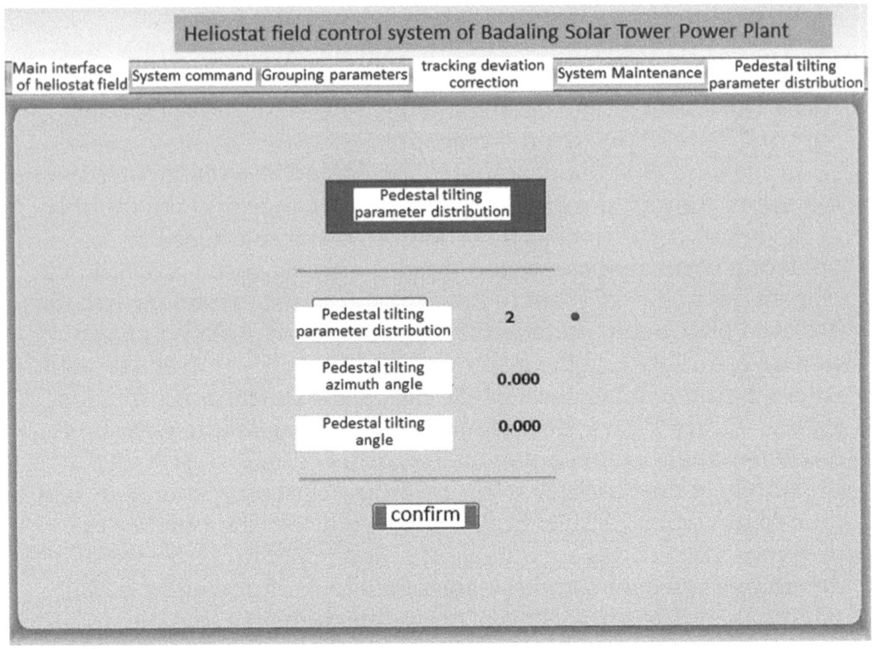

FIGURE 4.25 Interface of pedestal tilting parameter distribution.

Heliostat tracking errors are mainly caused by concentrated solar image shaking due to strong wind and regular deviation of the heliostat solar image due to pedestal tilting. Thus heliostat tracking errors are collected when the wind speed is low.

After heliostat deviation data are accumulated for a certain while, the angles of heliostat pedestal tilting can be inversely deduced according to tracking deviation features. By distributing these angles to the local controller of the heliostat (Fig. 4.25), satisfactory tracking effects still can be obtained even without introducing deviation-correction procedures.

4.5 SOLAR FIELD DESIGN OF PARABOLIC TROUGH POWER PLANT

The tracking rotation axis of a parabolic trough concentrator can be arranged in a north–south or west–east manner, which enables the concentrating surface to rotate around the pitch axis for single-axis tracking of the Sun and ensure that the incident beam can be contained in the plane determined by both the main normal and the focal line of the parabolic

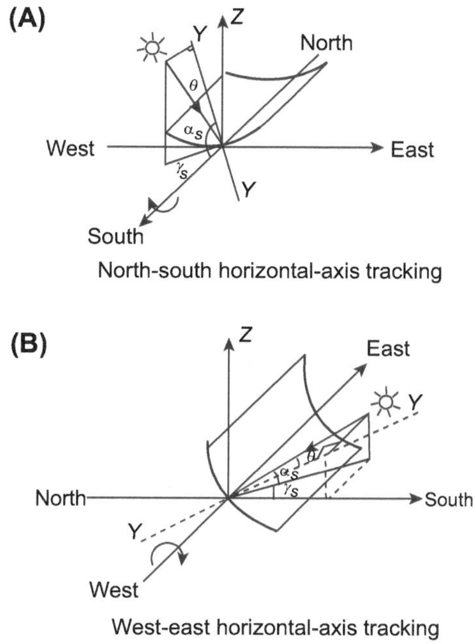

FIGURE 4.26 Different tracking modes of the parabolic trough concentrator.

trough reflective mirror. According to the different directions of the rotation axes, single-axis tracking can be divided into north–south and west–east axis tracking (Fig. 4.26). In terms of north–south axis tracking, the reflective mirror rotates around the long axis in the north–south direction tracking the azimuth of the Sun. In terms of the west–east axis tracking method, the reflective mirror rotates around the rotation axis in the west–east direction tracking the altitude angle of the Sun in the north–south direction (Fig. 4.26). The tracking formula is listed in Section 2.4 of this book.

The angle between the incident solar beam and main normal of the parabolic trough concentrator is called the solar incident angle. The smaller the incident angle, the higher the concentrator efficiency. The different tracking modes of a parabolic trough concentrator result in several different solar incident angles as well.

4.5.1 Axial Arrangement of the Concentrating Field

In order to explain the calculation process, Yanqing (40.4°N) in Beijing is taken as an example. The instantaneous solar irradiance is calculated by applying the Hottel model with atmospheric visibility of 5 km and an altitude less than 2.5 km. The DNI is:

$$G_{b,n} = 1367\left(1 + 0.033\cos\frac{2\pi n}{365}\right)\tau_b \tag{4.21}$$

in which $\tau_b = a_0 + a_1 \exp(-k/\cos\theta_z)$; this refers to the atmospheric transparency of direct radiation on a clear day. The values of relevant coefficients in the expression are:

$$a_0 = 0.97a_0^*,\ a_1 = 0.99a_1^*,\ k = 1.02k^*$$

$$a_0^* = 0.4237 - 0.00821(6 - A)^2 \tag{4.22}$$

$$a_1^* = 0.5055 + 0.00595(6.5 - A)^2 \tag{4.23}$$

$$k^* = 0.2711 + 0.01858(2.5 - A)^2 \tag{4.24}$$

in which A refers to the local altitude, m. The local altitude of Yanqing has a value of 525 m.

4.5.1.1 Calculation of Total Irradiation on a Typical Day

1. Spring equinox, March 21, $n = 59 + 21 = 80$

$$\delta = 23.45\sin\left(360 \times \frac{284+n}{365}\right)$$
$$= 23.45\sin\left(360 \times \frac{284+80}{365}\right) = -0.4$$

West–east horizontal-axis tracking:

$$\cos\theta = \sqrt{1 - \cos^2\delta\sin^2\omega}$$
$$= \sqrt{1 - \cos^2(-0.4)\sin^2\omega} = \sqrt{1 - \sin^2\omega}$$

North–south horizontal-axis tracking:

$$\cos\theta_z = \cos\phi\cos\delta\cos\omega + \sin\phi\sin\delta$$
$$= \cos 40.4\cos(-0.4)\cos\omega + \sin 40.4\sin(-0.4)$$
$$= 0.762\cos\omega - 0.0045$$

$$\cos\theta = \sqrt{\cos^2\theta_z + \cos^2\delta\sin^2\omega}$$
$$= \sqrt{(0.762\cos\omega - 0.0045)^2 + \cos^2(-0.4)\sin^2\omega}$$
$$= \sqrt{(0.762\cos\omega - 0.0045)^2 + \sin^2\omega}$$

2. Summer solstice, June 21, $n = 151 + 21 = 172$

$$\delta = 23.45\sin\left(360 \times \frac{284+172}{365}\right) = 23.45$$

West–east horizontal-axis tracking:

$$\cos\theta = \sqrt{1-\cos^2\delta\sin^2\omega}$$
$$= \sqrt{1-\cos^2(23.45)\sin^2\omega} = \sqrt{1-0.842\sin^2\omega}$$

North–south horizontal-axis tracking:

$$\cos\theta_z = \cos\phi\cos\delta\cos\omega + \sin\phi\sin\delta$$
$$= \cos(40.4)\cos(23.45)\cos\omega + \sin(40.4)\sin(23.45)$$
$$= 0.699\cos\omega - 0.258$$

$$\cos\theta = \sqrt{\cos^2\theta_z + \cos^2\delta\sin^2\omega}$$
$$= \sqrt{(0.699\cos\omega - 0.258)^2 + \cos^2(23.45)\sin^2\omega}$$
$$= \sqrt{(0.699\cos\omega - 0.258)^2 + 0.842\sin^2\omega}$$

3. Autumn equinox, September 23, $n = 243 + 23 = 266$

$$\delta = 23.45\sin\left(360 \times \frac{284+266}{365}\right) = -1.009$$

West–east horizontal-axis tracking:

$$\cos\theta = \sqrt{1-\cos^2\delta\sin^2\omega}$$
$$= \sqrt{1-\cos^2(-1.009)\sin^2\omega} = \sqrt{1-\sin^2\omega}$$

North–south horizontal-axis tracking:

$$\cos\theta_z = \cos\phi\cos\delta\cos\omega + \sin\phi\sin\delta$$
$$= \cos 40.4\cos(-1.009)\cos\omega + \sin 40.4\sin(-1.009)$$
$$= 0.76\cos\omega - 0.01$$

$$\cos\theta = \sqrt{\cos^2\theta_z + \cos^2\delta\sin^2\omega}$$
$$= \sqrt{(0.76\cos\omega - 0.01)^2 + \cos^2(-1.009)\sin^2\omega}$$
$$= \sqrt{(0.76\cos\omega - 0.01)^2 + \sin^2\omega}$$

4. Winter solstice, December 21, $n = 334 + 21 = 355$

$$\delta = 23.45\sin\left(360 \times \frac{284+355}{365}\right) = -23.45$$

West–east horizontal-axis tracking:

$$\begin{aligned}\cos\theta &= \sqrt{1-\cos^2\delta\,\sin^2\omega}\\ &= \sqrt{1-\cos^2(23.45)\sin^2\omega}\\ &= \sqrt{1-0.8417\,\sin^2\omega}\end{aligned}$$

North–south horizontal-axis tracking:

$$\begin{aligned}\cos\theta_z &= \cos\phi\,\cos\delta\,\cos\omega+\sin\phi\,\sin\delta\\ &= \cos 40.4\,\cos(-23.45)\cos\omega+\sin 40.4\,\sin(-23.45)\\ &= 0.699\,\cos\omega-0.2579\end{aligned}$$

$$\begin{aligned}\cos\theta &= \sqrt{\cos^2\theta_z+\cos^2\delta\,\sin^2\omega}\\ &= \sqrt{(0.699\,\cos\omega-0.2579)^2+\cos^2(-23.45)\sin^2\omega}\\ &= \sqrt{(0.699\,\cos\omega-0.2579)^2+\sin^2\omega}\end{aligned}$$

By applying the above mathematic models, solar DNIs from different axial directions projected on the collector at different time points can be calculated and are proportional to the cosine values of the incident angles of the solar beam. Fig. 4.27 shows cosine efficiencies proportional to direct normal irradiances on the unit area of the collector for different tracking modes corresponding to the four typical days of the spring equinox, summer solstice, autumn equinox, and winter solstice. Due to the characteristics of solar radiation, among the four tracking modes, the respective irradiances on the unit area of the collector are all comparatively low in the morning and in the afternoon, and reach maximum values around midday. Yet in the main working time interval of 8:00–16:00 for the west–east horizontal-axis tracking mode, irradiance on the collector varies to the largest extent; for the north–south horizontal-axis tracking mode, irradiance on the collector is most even and varies to the smallest extent. Furthermore, for west–east horizontal-axis tracking mode, there is only one peak value of instantaneous irradiance during the whole day that appears at 12:00; for the north–south horizontal-axis tracking mode, there are two peak values of instantaneous irradiance during the whole day that separately appear about 3 h before and after 12:00 noon.

Table 4.3 shows daily irradiations on the unit area of the collector during the whole day for different tracking modes corresponding to the

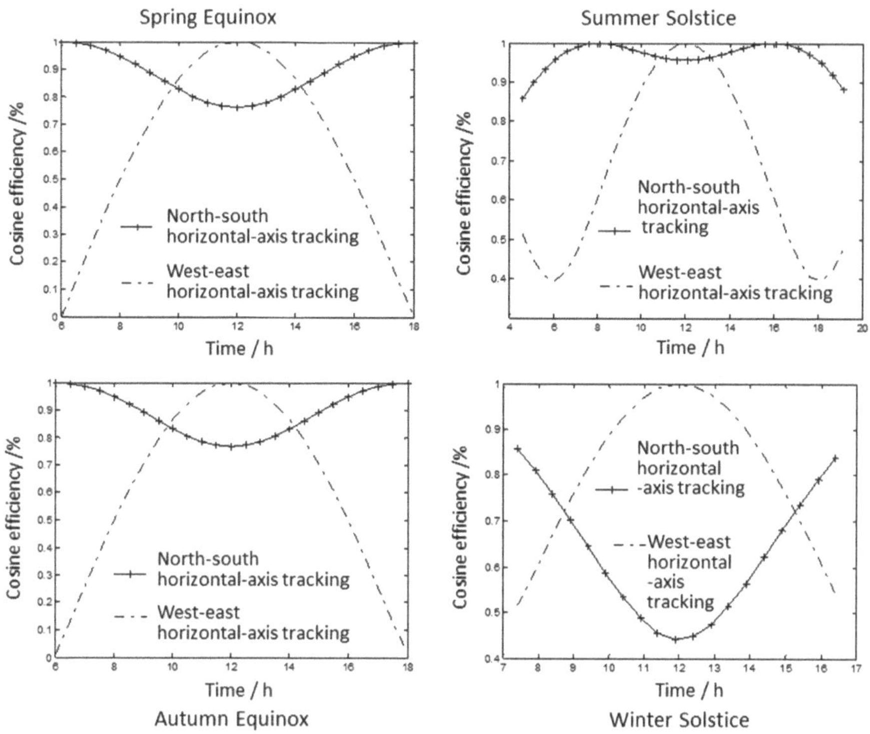

FIGURE 4.27 Cosine efficiency variations for different tracking modes.

spring equinox, summer solstice, autumn equinox, and winter solstice. As shown in Table 4.3, as daily irradiation on the unit area of the collector always reaches maximum value in the two-axis tracking mode, the received daily irradiation in this tracking mode is defined to be 100%. For the north–south horizontal-axis tracking mode, as the incident angles of the solar beam on the spring equinox, summer solstice, and autumn equinox are comparatively small, received daily irradiations of the concentrator are also significant, accounting for more than 86% of total irradiation; however, on the winter solstice the incident angle of the solar beam is comparatively large, instantaneous irradiance on the unit area is small, and daily irradiation only accounts for 58.19% of total irradiation. Conversely, for the west–east horizontal-axis tracking mode, daily irradiations on the spring equinox, summer solstice, and autumn equinox are comparatively small and only account for about 72% of total irradiation; yet on the winter solstice, irradiance on the unit area is comparatively large, and daily irradiation can account for up to 86.51% of total irradiation.

TABLE 4.3 Comparison of Received Daily Irradiations by Collector Under Different Tracking Modes

	Percentage of Daily Irradiation/%			Daily Irradiation/(MJ/m^2)		
Tracking Method	**Spring and Autumn Equinoxes**	**Summer Solstice**	**Winter Solstice**	**Spring and Autumn Equinoxes**	**Summer Solstice**	**Winter Solstice**
North–south horizontal-axis tracking	86.17	97.37	58.19	24.683	35.625	10.107
West–east horizontal-axis tracking	72.73	73.02	86.51	20.833	26.717	15.026

4.5.1.2 Annual Irradiation

As shown in Fig. 4.28 for all tracking modes during the whole year, daily irradiation peaks in summer and reaches its lowest value in winter. For the west–east horizontal-axis tracking mode, daily irradiation varies to the smallest extent, whereas under the north–south horizontal-axis tracking method, daily irradiation varies to the largest extent. In terms of monthly irradiation (Fig. 4.29), the highest value appears during May–July, whereas the lowest value appears during January and December; the former is almost two times that of the latter. Therefore, the optimal period for solar utilization is April–September.

Based on the calculation results of annual irradiation in Table 4.4, annual irradiation achieves its maximum value in the two-axis tracking mode, followed by that of the north–south polar axis tracking mode and

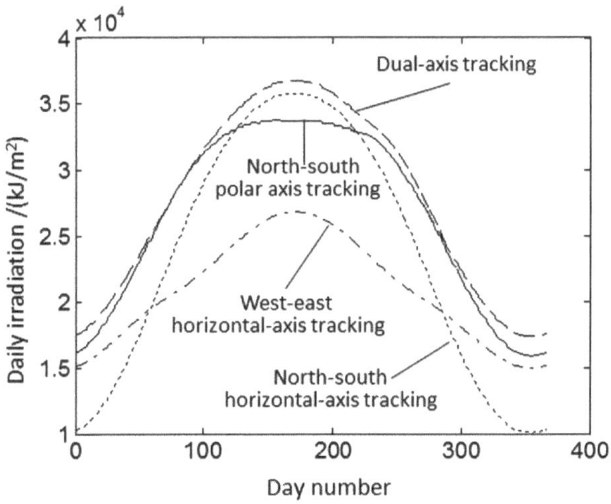

FIGURE 4.28 Daily irradiation for different tracking modes for the entire year.

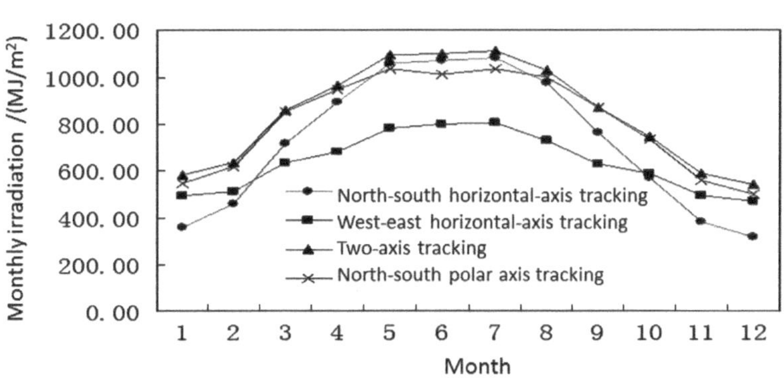

FIGURE 4.29 Monthly irradiation for different tracking modes.

TABLE 4.4 Annual Output of Collector for Different Tracking Methods

Tracking Method	North–South Horizontal Axis	West–East Horizontal Axis	Two-Axis	North–South Polar Axis
Annual irradiation/ (GJ/m^2)	8.64	7.62	10.12	9.71

that of the north–south horizontal-axis tracking mode; annual irradiation for the west–east horizontal-axis tracking mode is the lowest. Although annual irradiation for the two-axis tracking mode is the highest, as the system needs to track the Sun in directions of both solar altitude and azimuth angle, the equipment structure is more complex; in addition, it requires higher tracking control precision and thus manufacturing and service costs are higher, so it has been mostly applied in thermal power generation systems with higher temperatures. Normally a parabolic trough system adopts a single-axis tracking mode, which has not only a simple equipment structure, but also a lower requirement for tracking precision. Although annual irradiation for the north–south polar axis tracking mode is high, it is not easy to successfully drive the tilt-axis tracking system. Thus a parabolic trough collector is often driven by a horizontal axis. Compared with the west–east horizontal-axis tracking mode, collector output with north–south horizontal-axis tracking is higher; however, the output difference in summer versus winter is significant. Accordingly, in the case of requiring maximum collector energy output in winter, the west–east horizontal-axis tracking method should be applied; in the case of mainly focusing on summer utilization, the north–south horizontal-axis tracking mode should be adopted. As for Beijing, with its rainy summer and shorter sunshine days and durations, using the north–south horizontal-axis tracking mode does not make a significant difference.

Except for the factors just discussed, generally speaking, many parabolic trough solar collectors can be mutually connected in series and in parallel to reach the required heat collection temperature, and thus it is also necessary to consider the mutual shading of solar beam between collectors. For the polar-axis tracking mode when multiple collectors are used, the front collector will shade adjacent collectors. For the north–south horizontal-axis tracking mode, the respective shade influence is smaller and only appears during mornings and evenings. The shading influence for the west–east horizontal-axis tracking mode is minimal. Solar shading is mainly generated on winter solstice when the collector pitch angle reaches tis maximum value, and the collector shades other collectors located to the north.

To sum up, tracking modes of the collector shall be selected by considering not only the heat collection temperature requirement and purpose, but also processing, manufacturing, and tracking control system costs, as well as comprehensively considering many other factors such as efficiency reduction caused by solar shading and land coverage increase caused by shading avoidance.

4.5.2 Thermal Efficiency Evaluation of the Parabolic Trough Collector

The main thermal performance evaluation index of a parabolic trough solar concentrator is efficiency, which is defined as:

$$\begin{aligned}\eta &= \frac{P_{\mathrm{GAIN}}}{DNI \times \cos\theta \times A}\\ &= \frac{DNI \times \cos\theta \times A \times \rho \times \eta_{\mathrm{int}} - P_{\mathrm{LOSS}}}{DNI \times \cos\theta \times A}\\ &= \rho\eta_{\mathrm{int}} - \frac{P_{\mathrm{LOSS}}}{DNI \times \cos\theta \times A}\\ &= \rho\eta_{\mathrm{int}} - \frac{U_{\mathrm{L}}}{5.76 \times DNI \times \cos\theta}\end{aligned} \tag{4.25}$$

in which A refers to the width of concentrator aperture, which takes a value of 5.76 m here. As for an evacuated parabolic trough collector, when the difference between the internal and external temperature of the receiver reaches 400°C, $U_L = 220$ W/m; when $DNI = 800$ W/m^2, $\rho = 0.85$ (reflectivity of reflective mirror), $\eta_{\mathrm{int}} = 0.9$:

$$\eta = 0.765 - \frac{220}{5.76 \times 800 \times \cos\theta} = 0.765 - \frac{0.048}{\cos\theta} \tag{4.26}$$

The respective variation is shown in Fig. 4.30, according to which, when incident angle exceeds 60°, efficiency decreases at a faster rate.

The following test results of a parabolic trough concentrator are an example for explaining the concentrator evaluation method:

4.6 DESCRIPTION OF SOLAR CONCENTRATOR

4.6.1 Description of Concentrator

- Total area: 574.89 m^2
- Aperture area: 550.10 m^2
- Internal diameter of evacuated receiver: 70 mm
- External diameter of evacuated receiver: 120 mm
- Length of evacuated receiver: 97,200 mm

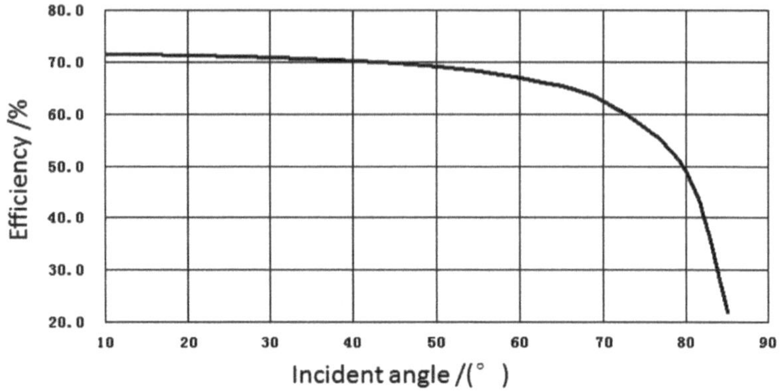

FIGURE 4.30 Variation of efficiency over incident angle.

4.6.2 Description of Heat-Transfer Medium

- Type: Heat-transfer oil
- Producer: Dow Chemical
- Model: Dowtherm A
- Description (additive, etc.): Null

4.6.3 Transparent Cover

- Type: Glass pipe
- Producer: Dow Chemical
- Material: Borosilicate glass
- Description (surface preparation, etc.): Null

4.6.4 Heat Absorber

- Material: Steel
- Surface preparation: Selective coating
- Structural type: Straight steel pipe
- Fluid volume: 0.384 m^3

4.6.5 Restrictive Conditions

- Maximum operating temperature: 400°C
- Maximum operating pressure: 1.6 Mpa

4.6.6 Schematic Diagram of Concentrator (Fig. 4.31)

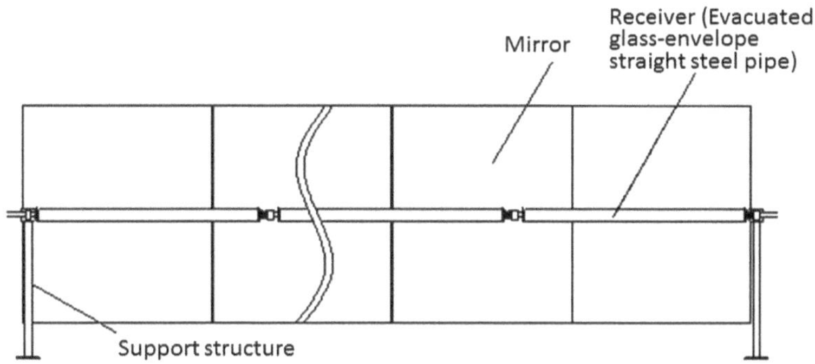

FIGURE 4.31 Concentrator.

4.6.7 Picture of Concentrator (Fig. 4.32)

FIGURE 4.32 Picture of concentrator to be tested.

4.7 INSTANTANEOUS EFFICIENCY

4.7.1 Schematic Diagram of Test Loop (Fig. 4.33)

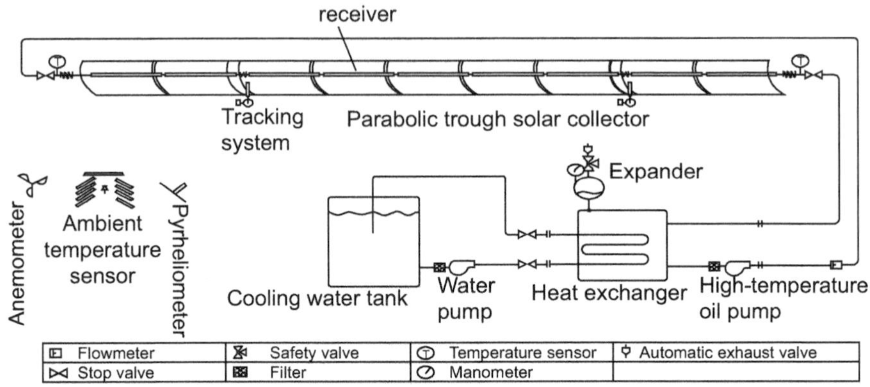

FIGURE 4.33 Test loop.

4.7.2 Test Results, Measurement, and Calculation Data

Latitude: 40.38° Longitude: 115.94°
Concentrator inclination angle: 0°
Concentrator rotation direction: Aperture moves from east to west
Local time corresponding to solar noon: 12:10 (17/09); 12:09 (19/09)
Test results are shown in Table 4.5.

Based on the total aperture area and the average temperature of heat-transfer fluid, the instantaneous efficiency curve can be derived by linear data fitting, with instantaneous efficiency $\overline{\eta}_g$ defined as:

$$\overline{\eta}_g = \frac{Q}{A_g G_{bpe}} \tag{4.27}$$

in which Q refers to the useful power output by the collector at each test data point; and G_{bpe} refers to the direct solar irradiance on the aperture area of collector, $G_{bpe} = G_{bp}\cos\theta$.

The flow rate used in the test is 0.59—1.50 m^3/h, and the total heat collection area is 574.89 m^2.

Data linear fitting (refer to Fig. 4.34):

$$\overline{\eta}_g = \overline{\eta}_{0g} - \overline{U}_g \frac{t_m - t_n}{G_{bpe}} \tag{4.28}$$

in which $\overline{\eta}_{0g}$ refers to the intercept of the collector efficiency equation, $\overline{\eta}_{0g} = 0.69$; and $\overline{U}_g$ refers to the slope of the equation, $\overline{U}_g = 0.67\ W/(m^2\ K)$.

TABLE 4.5 Test Results—Measurement Data

Time	G_{DN}/(W/m^2)	t_a/°C	u/(m/s)	$t_{out}-t_{in}$/°C	v/(m^3/h)
11:30–11:34	590.17	17.6	2.2	19.27	1.496
11:35–11:39	587.95	17.65	2.5	19.29	1.462
11:40–11:44	580.22	17.71	3.0	19.16	1.432
11:45–11:49	583.27	17.76	2.8	18.69	1.387
11:50–11:54	583.51	17.81	3.6	19.15	1.325
11:55–11:59	579.95	17.87	2.4	19.43	1.276
12:00–12:04	576.41	17.92	3.3	19.18	1.238
12:05–12:09	577.35	17.97	2.8	18.48	1.190
12:10–12:14	587.51	18.02	3.2	18.52	1.145
12:15–12:19	587.95	18.08	1.0	19.19	1.119
12:20–12:24	577.58	18.13	1.6	18.91	1.059
12:25–12:29	574.09	18.18	3.2	18.42	1.007

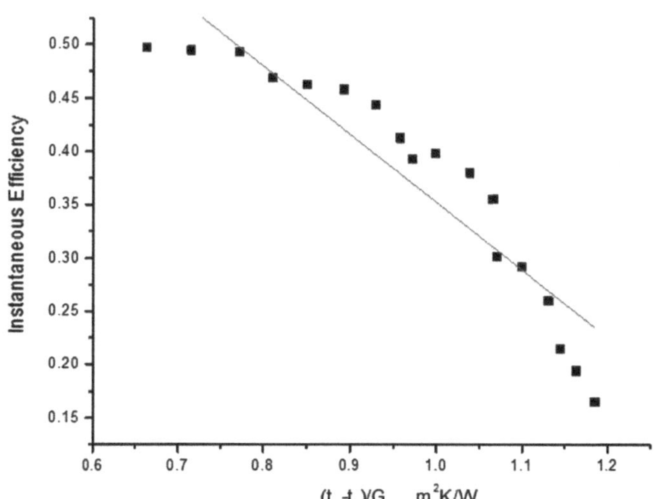

FIGURE 4.34 Linear fitting curve of efficiency.

4.7.3 Data Quadratic Fitting

Data quadratic fitting (refer to Fig. 4.35):

$$\overline{\eta}_g = \overline{\eta}_{0g} - \overline{\alpha}_{1g}\frac{t_m - t_a}{G_{bpe}} - \overline{\alpha}_{2g}G_{bpe}\left(\frac{t_m - t_a}{G_{bpe}}\right)^2 \tag{4.29}$$

$$\overline{\eta}_{0g} = 0.41$$

$\overline{\alpha_{1g}}$ refers to the first-order power term of the collector efficiency equation: $\overline{\alpha_{1g}} = -2.44$ W/(m^2 K).

$\overline{\alpha_{2g}}$ refers to the second-order power term coefficient of the collector efficiency equation: $\overline{\alpha_{2g}} = 0.004$ W/(m^2 K^2).

4.7.3.1 Experimental Formula of Solar Energy-Generating System (SEGS), United States

Heat collection efficiency equation [36] for the tested concentrator in LS-2 loop of SEGS power plant, United States:

$$\eta = k_\theta[73.3 - 0.007276(\Delta T)] - 0.496\left(\frac{\Delta T}{DNI}\right) - 0.0691\left(\frac{\Delta T^2}{DNI}\right) \tag{4.30}$$

in which DNI refers to direct normal solar irradiance, W/m^2; and k_θ refers to the IAM:

$$k_\theta = \cos\theta - c_1\theta - c_2\theta^2 \tag{4.31}$$

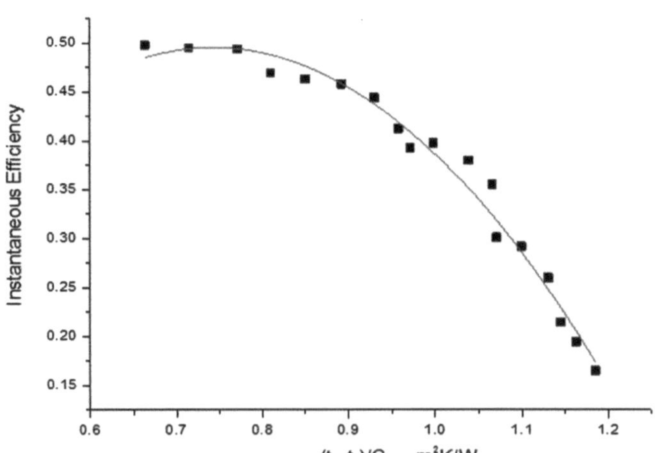

FIGURE 4.35 Quadratic fitting curve of efficiency.

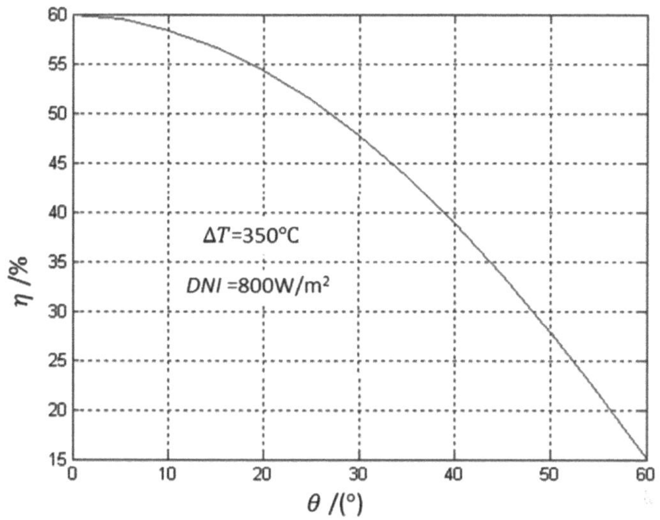

FIGURE 4.36 Variation of efficiency of solar energy-generating system parabolic trough concentrator along with incident angle.

As for the LS-2 loop of SEGS, $c_1 = 0.0003512$, $c_2 = 0.00003137$.

$$\Delta T = \frac{T_i + T_o}{2} - T_a \tag{4.32}$$

in which T_i refers to the concentrator inlet fluid temperature; T_o refers to the concentrator outlet fluid temperature; and T_a refers to the ambient temperature.

Fig. 4.36 shows the heat collection efficiency curve of the loop estimated according to Eqs. (4.30–4.32) when $DNI = 800\ \mathrm{W/m^2}$ and $\Delta T = 350°\mathrm{C}$.

Based on the foregoing, the relationship of heat collection efficiency to incident angle (the angle of direct solar radiation and the normal of the aperture area) of an SEGS parabolic trough concentrator is significant, as shown in Fig. 4.36.

4.8 DESIGN OF THE PARABOLIC TROUGH COLLECTOR FIELD

1. Analysis of the energy output characteristics of the parabolic trough concentrator field. Parabolic trough concentrator field energy output is an unsteady quantity that varies during the day along with variations in solar position, the variation rule of which is analyzed in Section 4.5.2.

2. **Extreme thermal boundary conditions of the heat-absorbing tuber.** The concentrator shall be able to concentrate 99.95% of the DNI onto the surface of the heat-absorbing tube, while the heat-absorbing tube shall be designed to satisfy the boundary conditions of the maximum concentrated flux density of the collector field (normally 100 kW/m^2).
3. **Boundary conditions of the heat-absorbing tube at work.** The normal working flux density on the heat-absorbing surface of an evacuated receiver is 35 kW/m^2.
4. **Interlock control of the concentrator and the receiver.** A parabolic trough concentrator can be calibrated by applying a closed-loop control; the concentrator's operational attitude can be interconnected with the outlet fluid temperature of the heat-absorbing tube. In the case of fluid overtemperature, the control system shall immediately move the concentrator in order to protect the heat-absorbing tube.

   Due to the angle between DNI and the normal of the aperture plane of a parabolic trough concentrator, it is not allowed to control the attitude of concentrator by measuring surface temperature at the inlet or outlet end of the heat-absorbing tube and then determining the temperature of the internal fluid.
5. **Receiver protection.** In the event that heat-transfer fluid inside the heat-absorbing tube is cut off due to power-off of the entire field, a pump fault, or pipeline leakage, the concentrator shall be immediately stopped. To avoid false alarms, there shall be two flow rate measuring points at the receiver outlet. There also shall be two main transfer pumps for the heat-transfer working fluid, one for operation and one for backup.
6. **Receiver heat-transfer fluid.** The heat-absorbing working fluid in a parabolic trough system is normally water, heat-transfer oil, or molten salt. In the event that a molten-salt receiver experiences freeze blockage in the fluid transmission pipeline, molten salt will be unable to flow into the heat-absorbing tube; concentrator operating conditions can be controlled based on the flow rate of the molten salt.
7. **Receiver initiation mode.** Before initiating a receiver, the pipeline shall be preheated. During initiation, before projection of solar beam onto the receiver, heat-transfer fluid shall be put into the heat-absorbing tube, the temperature of which shall approach that of the heat-absorbing tube.
   a. For a parabolic trough system that uses molten salt as the heat-absorbing fluid, 24-h circulation of molten salt is recommended for antifreezing and shall be conducted through temperature control in addition to power-generating hours. All transmission

pipelines shall be equipped with electric auxiliary heating facilities, especially at the elbow, flange, valve, pumping hole, etc., where antifreezing heating shall be conducted with care. At night, the circulation of molten salt in the pipe loop shall be continuous with no stops. During no-load thermal insulation operation, the fluid temperature shall exceed the melting point of molten salt by 50°C. Therefore, all receiver joints shall be strictly well insulated.

b. For a parabolic trough system that uses water as the heat-transfer fluid, the temperature of the heat-absorbing tube shall exceed 0°C before filling with water and starting the heat collection process; the receiver can be preheated through the injection of hot water. In the case where the ambient temperature is less than 0°C, whether there are any appropriate solar irradiation conditions, the circulation of heat-transfer fluid shall be continuous, and the water temperature shall exceed 8°C. Circulation can be controlled by associating the outlet fluid temperature of the heat-absorbing tube with that of the water pump.

4.9 CONCENTRATOR FIELD CONTROL DESIGN OF THE PARABOLIC TROUGH POWER PLANT

4.9.1 Technical Conditions for Concentrator Field Control System of a Parabolic Trough Power Plant

Control of the concentrator field in a parabolic trough power plant shall be determined according to solar irradiation conditions, wind speed, ambient temperature, and heat-transfer working fluid while satisfying the following requirements:

1. The concentrator field of the parabolic trough power plant consists of multiple rows of concentrators; each row is equipped with driving equipment. The control of the parabolic trough concentrator controls the actions of the driving equipment. The commonly used action modes of the driving equipment include hydraulic, gear drive, and electric push cylinder. The concentrator array controller controls driving equipment actions through local controllers.
2. The concentrator field controller determines and controls the working status of the concentrator through external meteorological conditions and heat-transfer pipe loop emergency alarm information, or determines the working status of the concentrator according to instructions from the power plant's main controller.

4.9.2 Control System Structure

Each concentrator is equipped with a set of local controllers to control the actions of the rotary axis. The concentrator field controller controls the rotation of the concentrator through local controllers.

1. The concentrator's local controller independently calculates the concentrator attitude that corresponds to each time point according to the astronomical formula of the Sun or obtains data from the host controller using more powerful computational capabilities. The main function of the local controller is to correct concentrator deviation so that concentrator precision during normal work can be maintained. Therefore, the local controller is connected to the deviation correction sensor of the concentrator.
2. The concentrator field controller is connected to the receiver's wind speed sensors and safety alarm facilities to provide concentrator attitude control in emergencies. The controller can also be connected to the power plant's main controller.
3. The method of control grounding is the same as that described in Section 4.4.1.
4. Logic connection between the concentrator field and the main control computer: the concentrator field controller sends concentrator driving position instructions to the local controller and receives ON/OFF instructions from the power plant's main controller.
   a. The power plant's main controller receives information from the concentrator field controller of the parabolic trough and gives information to the concentrator field controller, such as meteorological conditions, vacuum conditions of the receiver and thermal storage, as well as the working status of the steam turbine; however, it cannot directly input control instructions to the local controller mounted among the concentrators. All instructions on the actions of the concentrator are sent by the concentrator field controller, which means that individual concentrators only receive action instructions from the concentrator field controller.
   b. The concentrator field controller cannot be connected to the main controller of the power plant; accordingly, it directly receives safety control sensor and meteorological condition signals from the receivers to instruct actions in the concentrator field.
5. Open-loop control and closed-loop control: a parabolic trough concentrator can be controlled in a closed-loop manner through a solar sensor or in a semiclosed manner using a gravity inclinometer or a rotary encoder mounted on the rotation axis of the concentrator.

4.10 WIND LOAD CHARACTERISTICS OF THE CONCENTRATOR

Wind load is one of the most important issues encountered during concentrator design. A concentrator that has been poorly designed may influence precision or collapse or result in excessive costs due to large safety factor values. To carry out precise studies on wind load, experiments are normally required. There are two experimental measures, laboratory wind tunnel experiments and field measurements. This book separately uses examples of the heliostat and parabolic trough concentrator to demonstrate the respective research processes. Due to different wind speed conditions in different site locations, by understanding concentrator wind load characteristics, a power plant designer can achieve better scope during equipment type selection.

4.10.1 Wind Tunnel Test—Heliostat Wind Load Characteristics

Wind load may result in heliostat errors; in order to manufacture a highly efficient heliostat at the lowest cost, the respective wind-resistance structure of the heliostat must be optimized. Due to wind speed distribution, the complex structure of the heliostat, and the rotation feature of the heliostat, results through simple numerical simulations or engineering calculations have low precision. In heliostat design, wind load is normally determined through wind tunnel experiments, from which the details of wind load conditions under different working conditions can be understood. In this section, experiments on heliostat wind load at the Badaling Dahan solar tower power plant of the Chinese Academy of Sciences is used as an example to demonstrate the kind of data that can be obtained through a wind tunnel test and how to arrange such a test.

4.10.1.1 Brief Introduction

As stated above, heliostat precision is related to wind-resistance characteristics of the heliostat. It is very important for heliostat design to develop sufficient wind-resistance capability with low costs.

From 1986 to 1992, researchers carried out dozens of field measurement studies on the ATS and SPECO heliostats in Albuquerque, New Mexico, mainly focused on researching wind effects on the heliostats [53]. They proposed a theoretical method to define the structural wind load of a heliostat; however, as restricted by experimental conditions at the time, some issues were left pending: (1) Studies of heliostat mirror panel gaps have not been carried out, influences of gaps and the width of gaps on wind resistance of the heliostat are unknown; (2) the heliostat model in the wind tunnel test has an idealized flat pattern that does not consider the influences of the mirror surface support, rotation axis, and support arm components on overall wind resistance of the heliostat; (3) relevant studies on heliostat wind pressure distribution have not been carried out; (4) studies on the wind-induced stress and response of the heliostat structure have not been carried out; and (5) studies on wind-induced loads and responses corresponding to tornado, sandstorm, typhoon, thunderstorm, and other conditions have not been carried out.

Therefore, in terms of application and design of the heliostat system, many profound studies remain to be further carried out. This section focuses on analyzing data from wind tunnel tests and researching the variation rule of the heliostat wind load along with wind direction angle; it compares experimental analysis results with the report by Peterka (1992) and presents the fluctuating wind pressure and peak wind pressure distributions of the heliostat, which have laid a theoretical foundation for studies on wind load distribution regularities of reflective mirror panels and the prevention of mirror panel damage caused by local wind pressure.

4.10.1.2 *Overview of Experiments and Data Processing*

The dimension of the wind tunnel testing section is 5.5 m × 4.5 m; the wind speed of the testing section can be adjusted continuously in the range of 0–18 m/s. Commissioning and determination of the simulated atmospheric boundary layer wind field are conducted using a hot film anemometer; an electric pressure scanning system is applied as the pressure-measuring device. A six-component force-measuring balance is mounted at the bottom of the model, the scaling factor of which is one-tenth. A total of 144 (pressure) measuring points are designed on the front elevation and back elevation. The measuring point layout is shown in Fig. 4.37, and the heliostat model picture is shown in Fig. 4.38.

The heliostat of a solar tower power plant is located on an open flat ground subject to a class-B topography wind field. In this section, gratings, ladders, baffles, and the like are used to simulate the atmospheric boundary layer, wind speed, and turbulence intensity profile corresponding to B-class topography and are shown in Fig. 4.39. The overall wind load of the heliostat model is measured by using a six-component balance mounted at the bottom of the model with a sampling frequency of 100 Hz and a sampling period of 1 min. Coordinate labels of the balance are shown in Fig. 4.40.

For the horizontal wind angle a, tests and data acquisition shall be performed every 30° while the angle is increased from 0° on the north to 180° on the south in the anticlockwise direction.

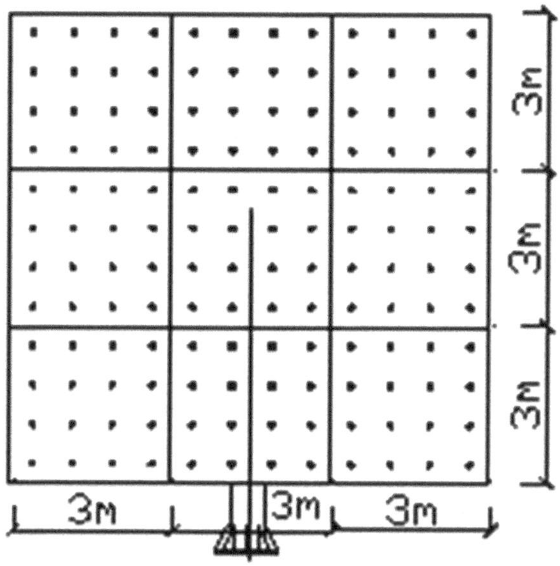

FIGURE 4.37 Measuring point layout of front elevation.

FIGURE 4.38 Heliostat model picture.

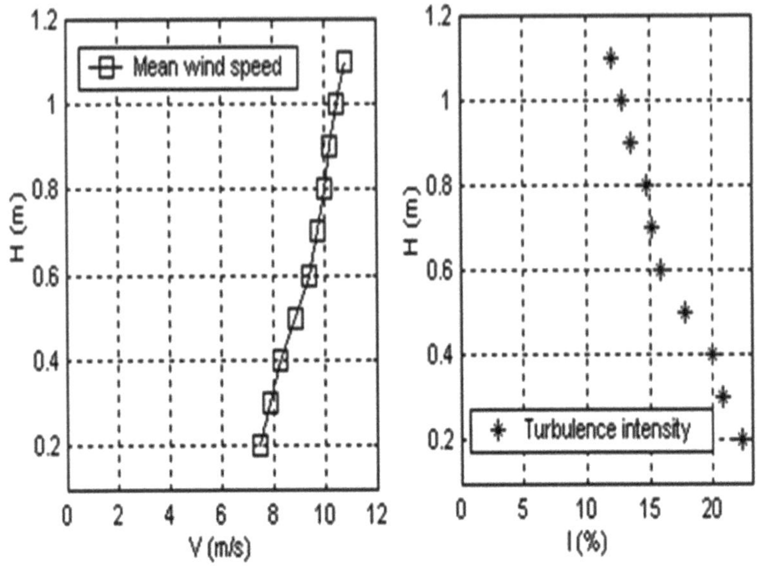

FIGURE 4.39 Profiles of wind speed and turbulence intensity.

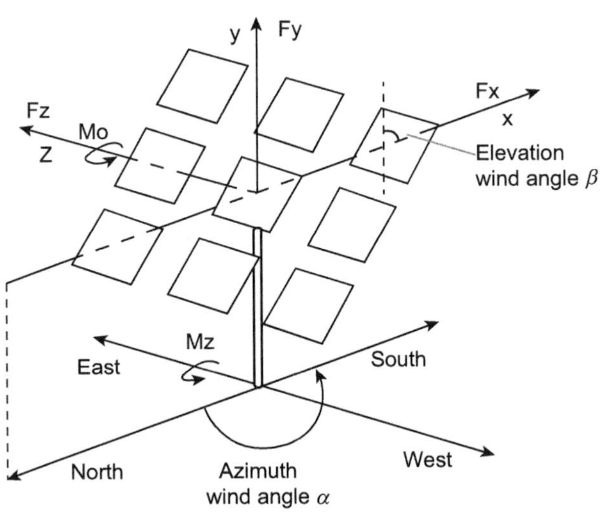

FIGURE 4.40 Coordinate label of balance.

As for the vertical wind angle β, tests and data acquisition shall be performed every 15° while the angle is increased from 0° in x axis direction to 90° in the y-axis direction by following the z-axis left-hand rule. Wind angle labels are shown in Fig. 4.40.

4.10.1.3 Data Processing

Measuring points on the front elevation of the heliostat are arranged corresponding to those on the back elevation. The superimposed net wind pressure coefficient at the measuring point is calculated as follows:

$$\Delta C_{p_i}(t) = \frac{p_i^f(t) - p_i^b(t)}{\frac{1}{2}\rho V_H^2} \tag{4.33}$$

in which $p_i^f(t)$ refers to the wind pressure value of the model at the measuring point on the front elevation; $p_i^b(t)$ refers to the wind pressure value of the model at the corresponding measuring point on the back elevation; ρ refers to the air density; and V_H refers to the incoming wind speed at the reference height inside the wind tunnel. For each measuring point, 20,000 data points of p_i shall be recorded. By analyzing $\Delta C_{p_i}(t)$, the mean wind pressure coefficient, fluctuating wind pressure coefficient, and peak wind pressure coefficient of the measuring point can be obtained.

Resistance coefficient C_F and moment coefficient C_M can be calculated as follows:

$$C_{F_x} = \frac{\mu_{F_x}}{\frac{1}{2}\rho V_H^2 A} \quad C_{F_y} = \frac{\mu_{F_y}}{\frac{1}{2}\rho V_H^2 A} \quad C_{F_z} = \frac{\mu_{F_z}}{\frac{1}{2}\rho V_H^2 A} \quad C_{M_x} = \frac{\mu_{M_x}}{\frac{1}{2}\rho V_H^2 Ah}$$

$$C_{M_y} = \frac{\mu_{M_y}}{\frac{1}{2}\rho V_H^2 Ah} \quad C_{M_x} = \frac{\mu_{M_z}}{\frac{1}{2}\rho V_H^2 Ah} \quad C_{M_0} = \frac{\mu_{M_0}}{\frac{1}{2}\rho V_H^2 Ah}$$

in which μ_{F_x}, μ_{F_y}, μ_{F_z}, μ_{M_x}, μ_{M_y}, μ_{M_z}, μ_{M_0} refer to mean values of balance data $F_x(t)$, $F_y(t)$, $F_z(t)$, $M_x(t)$, $M_y(t)$, and $M_z(t)$ respectively; A refers to the area of the model mirror surface; and h refers to the distance from center of the model mirror surface to the model pedestal base.

4.10.1.4 Analysis of Experimental Results

Component force and component moment: a heliostat is a reflective mirror that tracks the movement of the Sun and is adjusted by the control system in a 360° horizontal angle range and 90° vertical angle range; under different angles, all values of component force and moment of the heliostat in various directions (referred to as the resistance and moment coefficients) undergo significant changes, and the great resistance, lateral force, and lift force would all result in structural damage to different extents. Therefore, during heliostat analysis, it is necessary to extract the most adverse wind angle, comprehensively analyze component force and moment under different wind angles, and perform the calculation and design according to the most adverse working conditions. Fig. 4.41

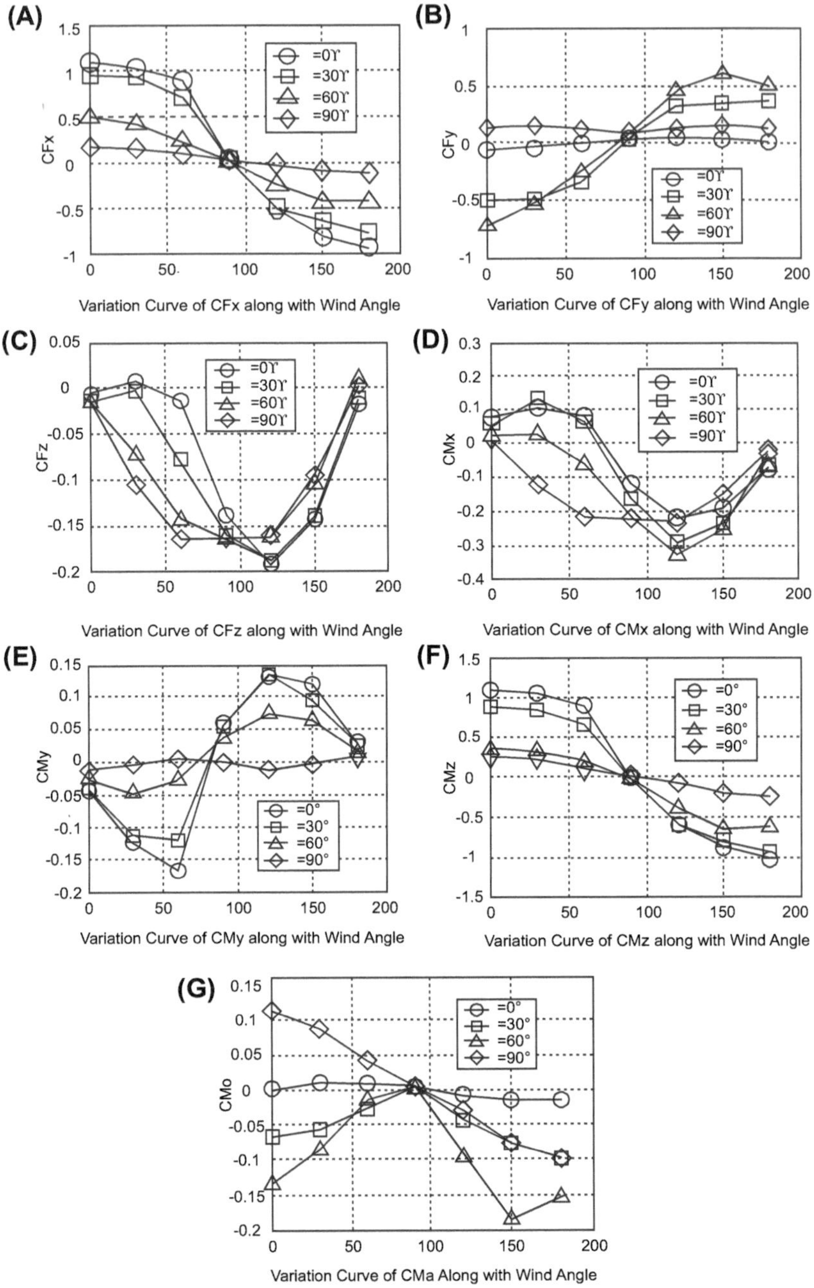

FIGURE 4.41 Variation curves of resistance coefficient and moment coefficient of the heliostat along with wind angle in various directions.

displays curves of variation of resistance and moment coefficients of the heliostat in various directions along with wind angle.

The experimental heliostat model in this section has been simulated completely according to the scaling factor of the real object, and thus the obtained values for the resistance and moment coefficients as well as the corresponding wind angles are precise. As for the heliostat, C_{Fy} with a downward force basically does not result in any structural damage, whereas C_{Fy} with an upward force easily results in damage to the connecting bolt at the bottom of the pedestal and the joint of the rotary axis. At a horizontal angle of 120°, lateral wind resistance is comparatively great; during structural design, it is necessary to perform design checking for the wind resistance corresponding to such a wind angle.

4.10.1.5 Peak Wind Pressure Distribution

Through experimental analysis, peak wind pressure distribution of the heliostat in the Badaling Dahan solar tower power plant is presented in this section, which serves as the theoretical foundation for studies of the wind load distribution rule on the mirror surface and the prevention of mirror surface damage from local wind pressure. Fig. 4.42 shows the peak wind pressure coefficient distribution on the mirror surface corresponding to the typical wind angle.

In Fig. 4.42, variation levels of wind pressure values are significant and have demonstrated excellent regularity; however, there also are smaller and bigger values in partial areas of the mirror surface that are different from others. According to the analysis, the reason is that the rotary shaft and support arm components have been designed at the back of the mirror surface; these components have certain blocking and interfering effects on the wind on the back surface, which lead to variation in the mean wind pressure and fluctuating wind pressure at partial areas of the back surface as well as increases and decreases in peak wind pressure.

When wind angle β is not less than 60°, the wind pressure distribution of the mirror panel is similar to that of slope and flat roofs of low-rise buildings. Airflow breakaway in the incoming flow direction, at the corner, and around the edge of the mirror panel makes the respective shear layer and cone vortex occur at the corner and around the edge of the mirror panel, which leads to significant peak negative pressure at the spot. Peak negative pressure reaches its maximum value when wind angle $a = 150°$, $\beta = 60°$, which occurs at the corner of the upper end of the mirror panel, with a magnitude of up to 3.5. In this section, it is suggested that necessary structural measures are taken while designing to intensify the stiffness of this weak section and prevent partial areas from being damaged by wind vibration.

When incoming wind is acting upon the back surface of the mirror panel (namely $a > 90°$), due to blocking and interference by the mirror

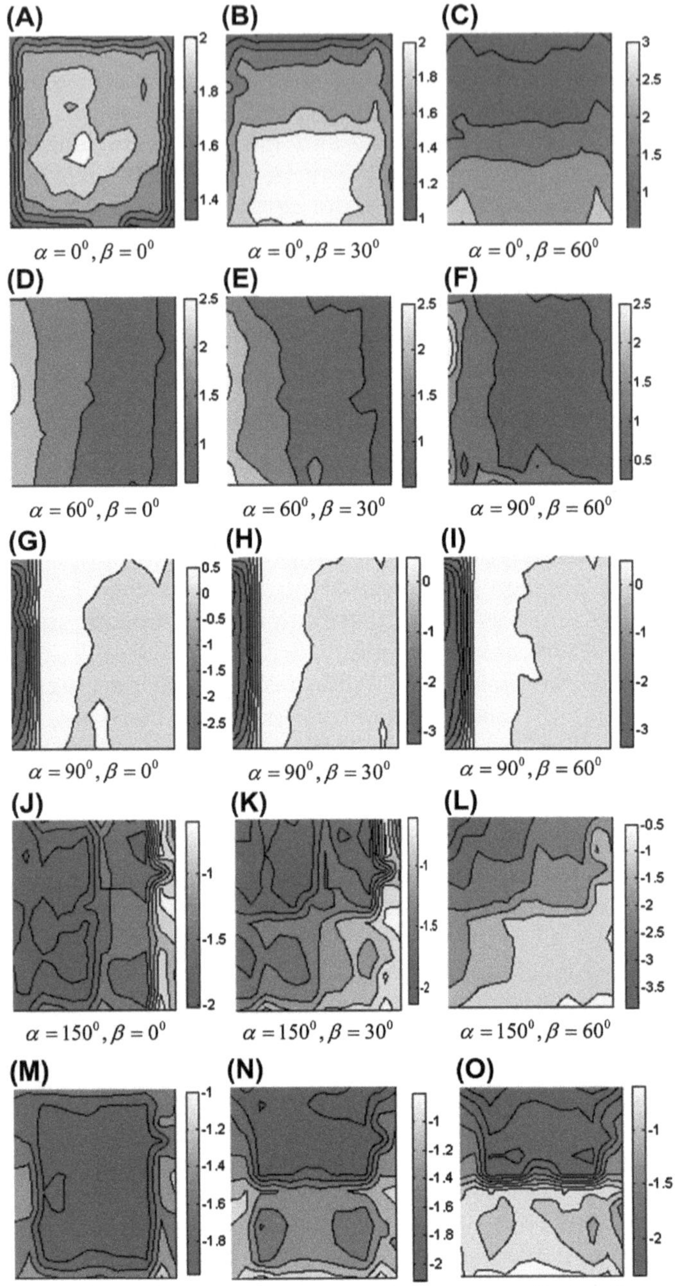

FIGURE 4.42 Mirror peak wind pressure coefficient distribution.

surface support, rotary shaft, and support arm component mounted on the back surface, the peak wind pressure distribution becomes more complex. Circumfluence on these components is more significant; especially when a certain vertical angle is generated by the tilting mirror panel, circumfluence on the rotary shaft is most significant; On the rotary shaft, peak wind pressure distribution demonstrates an intensive gradient change, and the maximum value of hinge moment M_0 around the rotary axis occurs when wind angle $a = 150°$, $\beta = 60°$, and $a = 180°$, $\beta = 60°$.

Mean wind resistance corresponding to wind angle $a = 0°$, $\beta = 0°$ reaches the maximum value; however, due to interferences from components on the back surface of the mirror panel, fluctuating wind pressure on the mirror panel corresponding to wind angle $a = 180°, \beta = 0°$ exceeds that corresponding to wind angle $a = 0°, \beta = 0°$, and mirror panel peak wind resistance corresponding to wind angle $a = 180°$, $\beta = 0°$ is greater than that corresponding to wind angle $a = 0°$, $\beta = 0°$.

4.10.1.6 Conclusions

Wind load on the heliostat structure has been analyzed in detail in this section, and the following conclusions have been made:

1. Rules for variations of component force and component moment obtained in this section and the extracted values corresponding to the most adverse wind angles are precise; in the successive heliostat structure design, it is suggested to check the lift force corresponding to wind angle $a = 150°$, $\beta = 60°$ and the lateral force corresponding to wind angle $a = 120°$, $\beta = 0°$.
2. Peak negative pressures at the four corners and four edges of the mirror surface are comparatively large. In this section, it is suggested to take necessary structural measures during design by intensifying the stiffness of the four corners and four edges and preventing partial areas of these weak spots from being damaged by wind vibration.
3. The rotary shaft of the heliostat is a weak spot likely for wind vibration damage, the maximum value of hinge moment M_0 occurs when wind angle $a = 150°, \beta = 60°$, and $a = 180°, \beta = 60°$. It is suggested to check and analyze the wind load and wind vibration response corresponding to these two pairs of wind angles.
4. Mean wind resistance corresponding to wind angle $a = 0°$, $\beta = 0°$ reaches the maximum value. However, due to the influences of fluctuating wind pressure, the peak value wind resistance corresponding to wind angle $a = 180°$, $\beta = 0°$ is larger than that corresponding to wind angle $a = 0°$, $\beta = 0°$. Therefore, when calculating the maximum equivalent wind resistance, the working

condition of wind angle $a = 180°$, $\beta = 0°$ shall be selected; namely in the case that the heliostat is placed perpendicular to the ground, the wind load reaches the maximum value under the working condition when wind blows from the back of the heliostat to the heliostat.

4.10.2 Field Measurement—Wind Load of Parabolic Trough Concentrator

In this section, based on actual field measurement and analysis of wind pressure and wind vibration of parabolic trough concentrators at the Badaling Solar Thermal Power Generation Test Base of the Institute of Electrical Engineering, Chinese Academy of Sciences, the method of actual field measurement of the concentrator [37] is explained. The field measurement includes the field prototype actual measurement, which acts as the most reliable way to achieve structural modal parameters, wind load, and wind-induced response. Results obtained through the wind tunnel experiment, flow field numerical simulation and structural finite element numerical simulation still must be verified in the end by field prototype actual measurement. Therefore it is of great significance to carry out wind field real-time monitoring and concentrator field prototype actual measurement throughout the year. In this section, the conditions of conducting field prototype actual measurement works by utilizing parabolic trough concentrators at the Badaling Solar Thermal Power Generation Experimental Base of the Institute of Electrical Engineering, Chinese Academy of Sciences are introduced.

4.10.2.1 Introduction to Actual Measurement

1. Parabolic trough concentrator. The dimension of the aperture of the parabolic trough concentrator is 5.76 m, and the height of the support pedestal is 3.55 m. Total length of a single row of concentrators is 96 m and consists of eight groups of single-span concentrators supported by the moment tube, and the length of each group of single-span concentrators is 12 m.
2. Ambient environment of the parabolic trough concentrator. The neighborhood picture and site description of the concentrator are shown in Fig. 4.43. Location of the concentrator is under the influence of wintertime winds, which mainly come from the west and northwest. Looking over the neighborhood, the western and northwestern areas are large areas of low crops that can be deemed open flat ground. The wind speed tower is located northwest of the concentrator and used for real-time monitoring and measuring of wind speed and wind direction data of the concentrator field; these

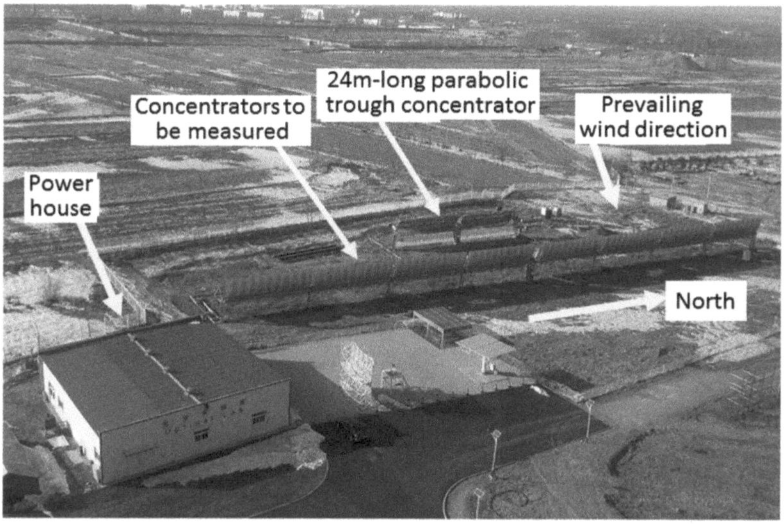

FIGURE 4.43 Ambient environment of solar trough concentrators to be measured.

data are synchronous with the wind load sampling data of the concentrator. The heliostat field is located 100 m to the east of the solar trough concentrator with a height of about 12 m. A single-story house is located south of the concentrator. West of the concentrator, another 24-m long parabolic trough concentrator has been built.

The respective test periods are selected from March 2011 and November 2011 to January 2012. During the test, wind speed, wind direction, wind pressure, and wind-induced deformation data were collected in a synchronous manner. Prototype testing studies on the concentrator mainly aim at testing and analyzing the wind characteristics of the boundary layer of the concentrator field, boundary conditions for wind load of the concentrator, wind-induced deformation and the respective rules and characteristics of the concentrator, and applying the analytical conclusions to the actual wind resistance design of the concentrator.

4.10.2.2 Test Instruments

Four types of test instruments are involved in actual measurement.

The first type is mechanical wind speed anemometer 09101, with a total quantity of four. As shown in Fig. 4.44, the four are separately mounted at heights of 1.7, 4.7, 8.0, and 9.5 m on the wind speed tower; it is normally used for testing two-dimensional wind speed and wind direction data at different heights.

The second type is the wind pressure sensor CY2000, which is normally used for testing the mirror panel local wind pressure of the concentrator,

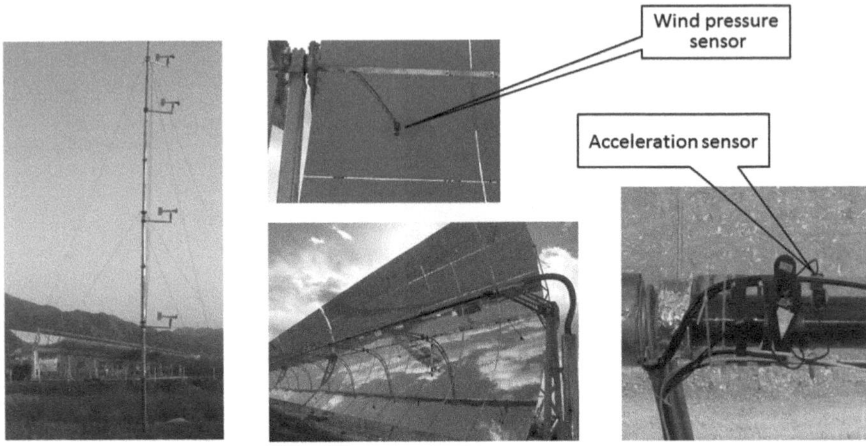

FIGURE 4.44 Test instruments.

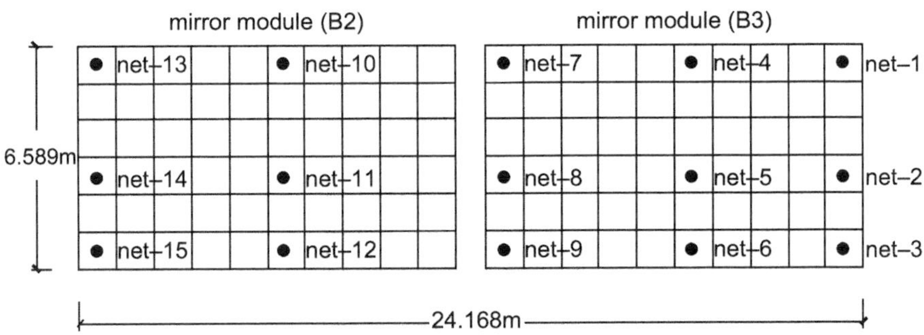

FIGURE 4.45 Wind pressure measuring point layout.

with a testing range of ±1,500 Pa and testing precision of ±0.2%. Thirty wind pressure sensors have been evenly mounted on two cross-mirror panels as shown in Fig. 4.45, with fifteen wind pressure sensors used for testing local wind pressure on the frontal surface of the concentrator and the other 15 used for testing the local wind pressure of the corresponding rear surface; all wind pressure sensors are mounted at the center of the curved mirror panel.

The third type is the acceleration sensor MSI4000, which is used for testing the vibration of the vacuum heat-absorbing tube of the concentrator. Eight acceleration sensors have been mounted on the cross-middle and cross-end sections of the evaluated receiver at B3 and attached on the exterior glass wall of the evaluated receiver with high-strength adhesive. They are arranged at four measuring points in an orthogonal manner, with four sensors used for testing F_x direction vibration and four sensors used for testing orthogonal F_z direction vibration.

4.10.2.3 Working Conditions of Actual Measurement

The parabolic trough concentrator performs a single-axis rotation while tracking the movement of the Sun. Each rotation angle corresponds to a pitch angle working condition of the concentrator, and wind loads on the concentrator under pitch angle working conditions are different from each other. Furthermore, different incoming wind directions may result in the variation of concentrator wind load.

The concentrator location is under the influence of wintertime winds mainly coming from the west and northwest. After months of testing, it is believed that the fluctuation amplitude of the incoming wind direction is very small. Based on this, according to the differences in the pitch angle of the concentrator, this test is divided into 12 testing working conditions, and the sampling frequency is selected to be 30 Hz; all channel data are collected in a synchronous manner. In addition, according to relevant China regulations, a 10-min sampling data segment is used each time for analysis.

4.10.2.4 Wind Field Actual Measurement

1. Wind speed and wind direction. A mechanical anemometer has recorded the time history data of the collected wind speed $U(t)$ and wind angle $a(t)$, in which wind speed components in x direction and y direction are defined as:

$$u_x(t) = U(t)\cos[\alpha(t)]$$
$$u_y(t) = U(t)\sin[\alpha(t)]$$

in which when the incoming wind direction is north, wind angle $a(t) = 0°$; and when the incoming wind direction is east, wind angle $a(t) = 90°$.

Within a unit time interval, mean wind speed $\overline{U}(t)$ and mean wind angle a can be defined as:

$$\overline{U}(t) = \sqrt{\overline{u}_x^2(t) + \overline{u}_y^2(t)}$$

$$\alpha = \arccos\left[\frac{\overline{u}_x(t)}{\overline{U}(t)}\right]$$

Within a unit time interval, time history data of along-wind direction wind speed $u(t)$ and across-wind direction wind speed $v(t)$ can be defined as:

$$u(t) = u_x(t)\cos(\alpha) + u_y(t)\sin(\alpha) - \overline{U}(t)$$
$$v(t) = -u_z(t)\sin(\alpha) + u_y(t)\cos(\alpha)$$

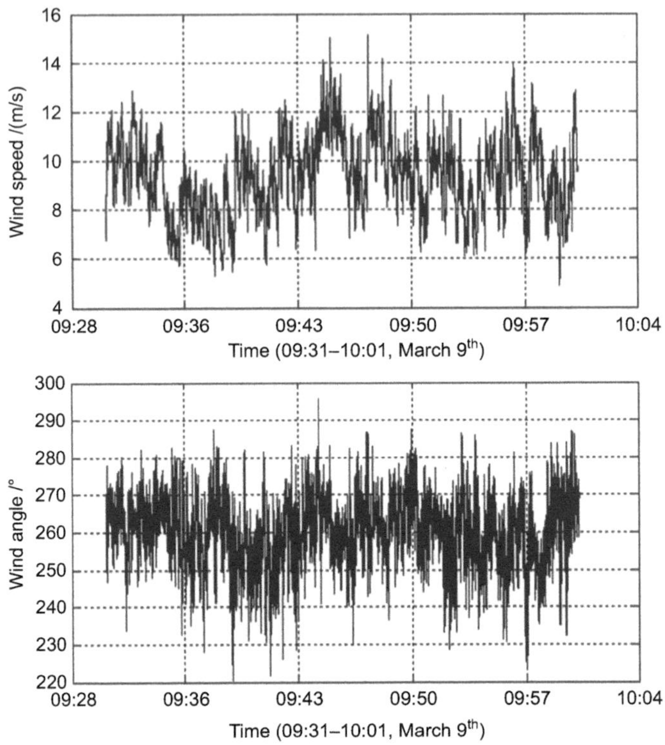

FIGURE 4.46 Time histories of wind speed and wind direction.

In this section, statistics and analysis are performed on the measured wind speed and wind direction data of the parabolic trough concentrator. Fig. 4.46 shows the time history true curve of wind speed and wind direction of the concentrator field under strong wind conditions within 30 min. Fig. 4.47 displays the statistics of wind speed range and wind direction range corresponding to the entire test period.

2. Turbulence intensity and gust factor. The atmospheric boundary layer close to the ground is a high-turbulence wind field, based on which respective low-rise buildings and structures have been under significant influences from fluctuating winds. Turbulence wind field parameters mainly include turbulence intensity and gust factor. In this section, turbulence wind field parameters corresponding to near-ground heights of 1.7, 4.7, 8.0 and 9.5 m during the test period are analyzed and researched.

   Turbulence intensity refers to the ratio of the standard deviation of fluctuating wind speed to the mean wind speed, which is used to

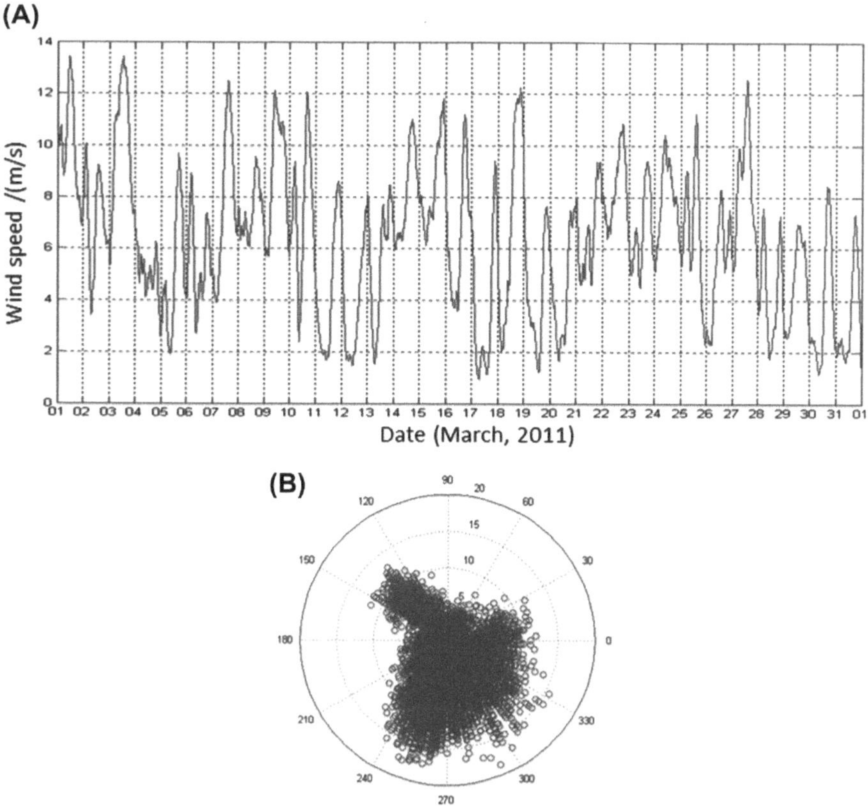

FIGURE 4.47 Statistics of wind speed and wind direction (March 2011).

describe the intensity of fluctuating wind speed. The respective calculation formula is as follows:

$$I_u = \frac{\sigma_u}{\overline{U}(t)}$$

$$I_v = \frac{\sigma_v}{\overline{U}(t)}$$

in which σ_u and σ_v refer to the standard deviations of fluctuating wind speed in the along-wind and across-wind directions; $\overline{U}(t)$ refers to the mean wind speed; and I_u and I_v respectively refer to turbulence intensities in the along-wind and across-wind directions.

Gust factor is used to describe the intensity of the fluctuating component of wind speed, which is defined as the ratio of the mean wind speed during gust period (the gust duration is normally 3 s) to mean wind speed in the unit time interval:

$$G_u(t_g) = 1 + \frac{\max\left[\overline{u}(t_g)\right]}{\overline{U}(t)}$$

$$G_v(t_g) = \frac{\max\left[\overline{v}(t_g)\right]}{\overline{U}(t)}$$

in which t_g refers to gust duration (which is selected to be 3 s during this test); $\max\left[\overline{u}(t_g)\right]$ and $\max\left[\overline{v}(t_g)\right]$ refer to mean wind speeds in the along-wind and across-wind directions during the maximum gust period within a unit time period (which is selected to be 10 min during this test); $\overline{U}(t)$ refers to the mean wind speed; and $G_u(t_g)$ and $G_v(t_g)$ separately refer to gust factors corresponding to the along-wind and across-wind directions.

Statistical characteristics of turbulence intensities in the along-wind and across-wind directions are shown in Table 4.6, in which the wind speed test range is 4–15 m/s and the ground clearances of wind speed testing include 1.7 , 4.7, 8.0, and 9.5 m. Based on the analysis in Table 4.6, turbulence intensities in both directions decrease along with increases in ground clearance. At the heights of 1.7, 4.7, 8.0, and 9.5 m, the statistical mean values of turbulence intensity in the along-wind direction are 0.27, 0.23, 0.2, and 0.2, and the along-wind direction turbulence intensities obtained through tests are obviously larger than those in the concentrator wind tunnel experiment applied by Hosoya and Peterka. The range of turbulence intensity ratios of across-wind and along-wind directions corresponding to the four heights in the test is 0.95–1.06, which is different from the test result $I_v{:}I_u = 0.75$ of Solari and Piccardo. A possible reason for such a difference is that the test height applied by Solari and Piccardo is obviously larger than that in this test.

Statistical characteristics of gust factors in the along-wind and across-wind directions are shown in Table 4.7, in which the wind speed test range is 4–15 m/s and the ground clearances of wind speed testing include 1.7, 4.7, 8.0, and 9.5 m. Based on the analysis on Table 4.7, values of the gust factor in both directions decrease along with increased ground clearance. At heights of 1.7, 4.7, 8.0, and 9.5 m, statistical mean values of the gust factor in the along-wind direction are 1.7, 1.54, 1.47, and 1.46. According to the analysis, it is believed that gust factors in the down-wind direction obtained through tests basically coincide with the specified value of 1.53 in the American Standard Code for Loads on Structures.

TABLE 4.6 Statistical Characteristics of Turbulence Intensity

Sampling Time (March 8, 2011)	**Along-Wind Direction**	**Along-Wind Direction**	**Along-Wind Direction**	**Along-Wind Direction**	**Across-Wind Direction**	**Across-Wind Direction**	**Across-Wind Direction**	**Across-Wind Direction**
Item	(9.5 m)	(8.0 m)	(4.7 m)	(1.7 m)	(9.5 m)	(8.0 m)	(4.7 m)	(1.7 m)
Maximum Value	0.30	0.31	0.33	0.35	0.24	0.26	0.27	0.35
Minimum Value	0.14	0.15	0.18	0.21	0.15	0.16	0.17	0.19
Mean Value	0.20	0.20	0.23	0.27	0.19	0.20	0.22	0.27

TABLE 4.7 Statistical Characteristics of Gust Factor

Sampling Time (March 8, 2011)	Along-Wind Direction	Along-Wind Direction	Along-Wind Direction	Along-Wind Direction	Across-Wind Direction	Across-Wind Direction	Across-Wind Direction	Across-Wind Direction
Item	(9.5 m)	(8.0 m)	(4.7 m)	(1.7 m)	(9.5 m)	(8.0 m)	(4.7 m)	(1.7 m)
Maximum Value	1.66	1.66	1.68	1.94	0.71	0.71	0.75	0.84
Minimum Value	1.33	1.33	1.34	1.49	0.25	0.26	0.27	0.33
Mean Value	1.46	1.47	1.54	1.70	0.42	0.43	0.45	0.53

3. Wind speed and turbulence intensity profile. The atmospheric boundary layer wind speed profile and turbulence intensity profile are both major parameters for wind engineering research. Many experts and scholars have carried out highly fruitful analysis and studies while obtaining the logarithmic law model, exponential law model, and Deaves-Harris model, which can be used to describe the wind speed profile. Due to the simplicity and effectiveness of the exponential law, it has been widely applied by main structural load codes at home and abroad. The calculation formula for wind speed profile under the exponential law is:

$$V_Z/V_o = (Z/Z_o)^a$$

in which V_z refers to the mean wind speed at the place with a ground clearance of Z; V_0 refers to the mean wind speed at the reference height of Z_0; and a refers to the ground roughness exponent.

The wind speed profile obtained through the field test is shown in Fig. 4.48, which basically coincides with the wind speed profile formula under exponential law when the ground roughness exponent $a = 0.22$.

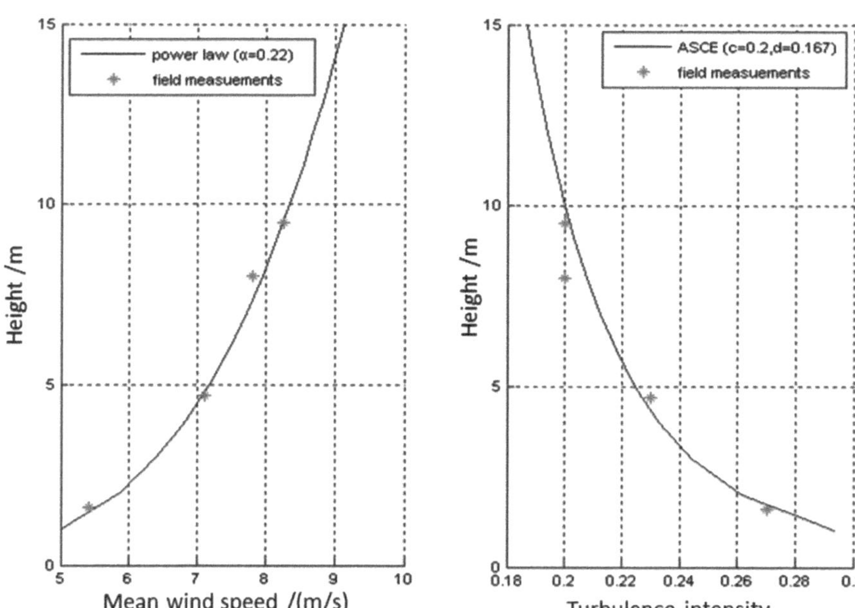

FIGURE 4.48 Profile of wind speed and turbulence intensity.

According to the American Standard Code for Loads on Structures, the theoretical formula for the along-wind direction turbulence intensity profile [38] is as follows:

$$I_u = c(Z/10)^{-d}$$

in which according to the regulations of ASCE (American Society of Civil Engineers) for open flat ground, coefficients c and d take values of 0.2 and 0.167 respectively.

The turbulence intensity profile obtained through the field test is shown in Fig. 4.48, which basically coincides with the profile formula and value of coefficient corresponding to the specified open flat ground. The height of the concentrator is normally within 10 m. Wind field parameters obtained through testing can well describe the atmospheric wind field within the height range of the concentrator.

4.10.2.5 Wind Load of Concentrator

1. Definition of formula. Concentrator wind load is mainly generated under the combined action of wind pressures on the frontal surface and back surface of the concentrator mirror. It is hereby defined as the time history value of the net wind pressure coefficient and refers to the ratio of net wind pressure value to atmospheric incoming wind pressure acting upon the mirror; it belongs to a nondimensional quantity and can be calculated as follows:

$$\Delta C_{p_i}(t) = \frac{p_i^f(t) - p_i^b(t)}{\frac{1}{2}\rho V_0^2}$$

in which $\Delta C_{p_i}(t)$ refers to the time history value of the net wind pressure coefficient of the mirror at measuring point i; $p_i^f(t)$ and $p_i^b(t)$ respectively refer to time history values of frontal surface wind pressure and back surface wind pressure of the mirror at measuring point i; ρ refers to the air density; and V_0 refers to the time history value of wind speed corresponding to the reference height (in this section, reference height is 9.5 m, where the anemometer is mounted).

By performing the mathematical statistical analysis toward the time history value of the net wind pressure coefficient, the mean value C_{p_i}, mean and standard deviation value C_{p_i}, rms can be obtained separately, and the peak positive pressure coefficient and peak negative pressure coefficient can be further calculated through the formula below:

$$C_{p_i,\max} = C_{p_i}, \text{mean} + gC_{p_i}, \text{rms}$$
$$C_{p_i,\min} = C_{p_i}, \text{mean} - gC_{p_i}, \text{rms}$$

in which $C_{p_i,\max}$ and $C_{p_i,\min}$ respectively refer to the peak positive pressure coefficient and peak negative pressure coefficient; $C_{p_i,\text{mean}}$ and $C_{p_i,\text{rms}}$ respectively refer to the mean wind pressure coefficient (namely the mean value of net wind pressure coefficient) and fluctuating wind pressure coefficient (namely the standard deviation of net wind pressure coefficient); and g refers to the peak factor.

In this section, according to analysis of wind pressure data, it is believed that the probability density distribution of fluctuating wind pressure basically coincides with characteristics of Gaussian distribution; when the peak factor takes the value of 2.5, the guarantee rate can be as high as 99% and meet the requirement (in this section, g takes the value of 2.5).

Through the integration of wind pressures at various measuring points on the mirror surface, the resultant force of wind loads acting upon the concentrator can be calculated (including wind load resistance and wind load lift force) and can be calculated as follows:

$$Q = \sum_{i-1}^{N} w_i p_i$$

in which Q refers to the resultant force of wind loads; p_i refers to the net wind pressure at measuring point i; N refers to the quantity of wind pressure measuring points; and w_i refers to the weighted area of measuring point i.

The wind load coefficients include the wind load resistance coefficient and wind load elevating force coefficient, which can be calculated as follows:

$$C_{F_z} = \frac{F_x}{\frac{1}{2}\rho V_0^2 LW}$$

$$C_{F_z} = \frac{F_z}{\frac{1}{2}\rho V_0^2 LW}$$

in which F_x and F_z respectively refer to the resistance and lift force along x and z directions, which are shown in Fig. 4.40; L refers to the concentrator single-span length; W refers to the aperture dimension of the concentrator; ρ refers to the air density; and V_0 refers to the time history value of wind speed corresponding to the reference height (in this section, the reference height is 9.5 m, where the anemometer is mounted).

2. Resistance and lift force. Operational attitude of the parabolic trough concentrator: Single-axis rotation, for tracking movement of the Sun,

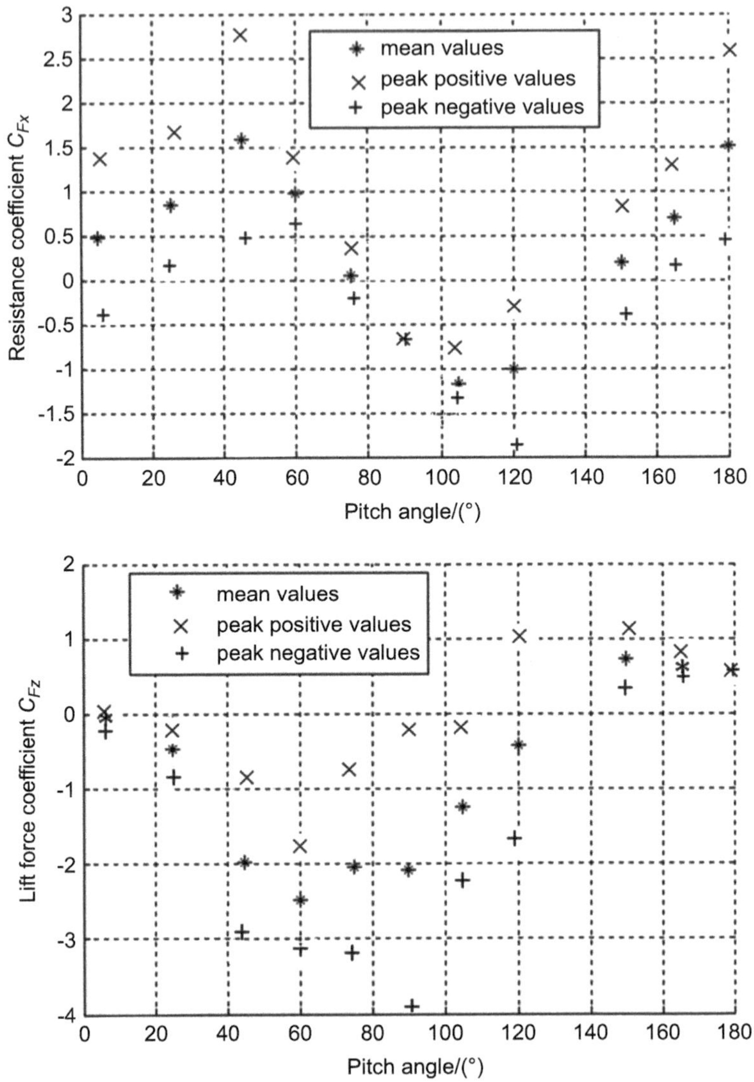

FIGURE 4.49 Variation curves of resistance coefficient and lift force coefficient.

revolving from east to west with a revolving range of about 180°. In this section, wind loads corresponding to various operational angles of the concentrator are tested and analyzed. As shown in Fig. 4.49, they respectively refer to the variation curves of concentrator wind load resistance coefficient and lift force coefficient along with the variation of the pitch revolving angle, in which the testing range for pitch angle is 5°–180°.

By analyzing Fig. 4.49, the maximum resistance obtained in this section occurs when the pitch angle is 45°, while the minimum resistance occurs when the pitch angle is 75°; based on wind tunnel experiment results by Hosoya, the maximum resistance occurred when the pitch angle was 30°, while the minimum resistance occurred when the pitch angle was 90°. According to experiment results obtained in this section, the maximum elevating force occurs when the pitch angle is 60°, while the minimum lift force occurs when the pitch angle is 5°; based on wind tunnel experiment results by Hosoya, the maximum lift force also occurred when the pitch angle was 60°, while the minimum lift force occurred when the pitch angle was 0° and 180°.

3. Wind pressure distribution and wind pressure spectrum characteristics. The trough concentrator is parabolic and cylindrical-shaped; at different pitch angles, the airflow mechanism and respective wind pressure distribution law on the concentrator surface may vary significantly. Fifteen wind pressure measuring points are set on the frontal and back surfaces of the concentrator for testing, and the wind pressure distribution laws of the concentrator at different pitch angles are tested and analyzed in order to conduct scientific research on the airflow mechanism.

   The wind pressure distribution law on the surface of concentrator is shown in Fig. 4.50, which is subject to the mean value distribution of net wind pressure coefficient obtained through combined wind pressure acting upon the frontal and back surfaces of the concentrator.

In the case that the pitch angle is within the range 0° to 60° (namely aperture of the concentrator facing toward the incoming flow direction), the frontal surface of the mirror is influenced by the positive effect of the incoming wind pressure, and the back surface is influenced by the negative pressure effect of airflow vortex, the mean value of the net wind pressure coefficient falls into the range of 0–5, in which MDD area has a smaller wind pressure and a coefficient value falling into the range of 0–2.5; MUU area has a larger wind pressure and a coefficient value falling into the range of 3–5; and wind pressure on the entire mirror surface of the concentrator demonstrates a gradual increase from MDD area to MUU area while following an obvious hierarchical distribution. Centered on a plane containing the line of the receiver, MUU represents the upper section of the upper half of the mirror, while MDD represents the lower section of the lower half of the mirror.

In the case that the pitch angle is within the range of 75° to 150°(namely aperture of the concentrator facing upward or toward the upper inclined side), airflow breakaway occurs at the frontal edge of incoming airflow in

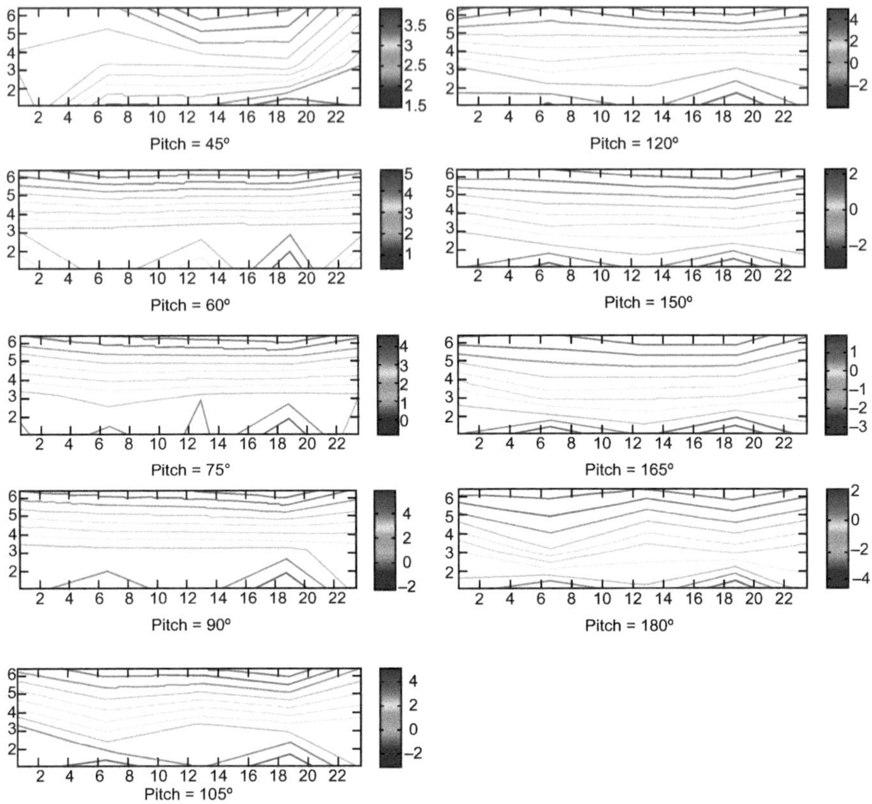

FIGURE 4.50 Distribution of mean value of net wind pressure coefficient on the concentrator surface.

the along-wind direction of the mirror and is subject to a large negative-pressure area; the airflow is attached to the rear edge and is subject to a large positive-pressure area. The mean value of net wind pressure coefficients at various measuring points on the frontal edge (including measuring points 1, 4, 7, 10, and 13) is in the range of −5 to −3, while at various measuring points on the rear edge (including measuring point 6, 9, and 12), it falls in the range of 5 to 7. Similarly, wind pressure on the entire mirror surface of the concentrator demonstrates a gradual increase from MDD area with a negative wind pressure to MUU area with a positive wind pressure while following an obvious hierarchical distribution.

In the case that pitch angle is within the range of 165° to 180° (namely aperture of the concentrator facing toward the opposite direction of incoming flow), the back surface of the mirror is influenced by the positive effect of the incoming wind pressure, and the frontal surface is influenced by the negative pressure effect of the airflow vortex, and most of the mirror areas are subject to combined negative-pressure areas with the coefficient value falling into the range of −5 to -3; however, the MUU area is subject to a combined positive-pressure area with the coefficient value falling into the range of 0 to 2. In this section, the respective reason is inferred to be the MUU area being influenced by the aerodynamic interference effect of the ground and the airflow vortex on the frontal surface of the MUU area having a positive-pressure effect on the frontal surface.

By analyzing the wind pressure acting on the middle area of the mirror, it is believed that wind pressure gradients on both sides of the middle area vary significantly, indicating that when incoming wind passes through, airflow breakaway around the cylinder occurs on the moment tube in the middle of the concentrator. In this section, it is believed that the peak wind pressure distribution law is consistent with the mean wind pressure distribution law; as the value of the design load parameter, the peak wind pressure coefficient is capable of better ensuring reliability of the concentrator structure during usage.

Spectrum analysis is an effective method for analyzing energy distribution of different frequency ranges. In this section, the fluctuating wind pressure power spectrum density is analyzed by applying the spectrum analysis method to carry out studies on the characteristics of fluctuating wind pressure on the surface of the concentrator. Power spectrum density is a probability statistical parameter of the random signal within the frequency domain that can be used to describe the distribution condition of wind pressure random signal power in the frequency domain and reflect the value of signal power on a unit frequency section. As the random signal is an infinite time domain signal and not qualified to be integrated, a respective Fourier transformation does not exist, and it can only be described through statistical methods instead of using mathematical expressions for precise description. In this section, the power spectrum density function of the wind pressure signal corresponding to each measuring point is estimated by applying the Welch power spectrum method:

$$S(n) = \frac{1}{MUL} \sum_{i-1}^{L} \left| \sum_{n=0}^{M-1} X_N^i(n) w(n) W_N^{-kn} \right|$$

in which $S(n)$ is the power spectrum density function of a random wind pressure signal; L refers to the data segment number of the random wind pressure signal; M refers to the quantity of data samples in each data segment; $U = \frac{1}{M} \sum_{n=0}^{M-1} w^2(n)$ is the normalization factor, which ensures that the power spectrum estimate obtained through the Welch method is an unbiased estimate; $X_N^i(n)$ is the time history sequence of the random wind

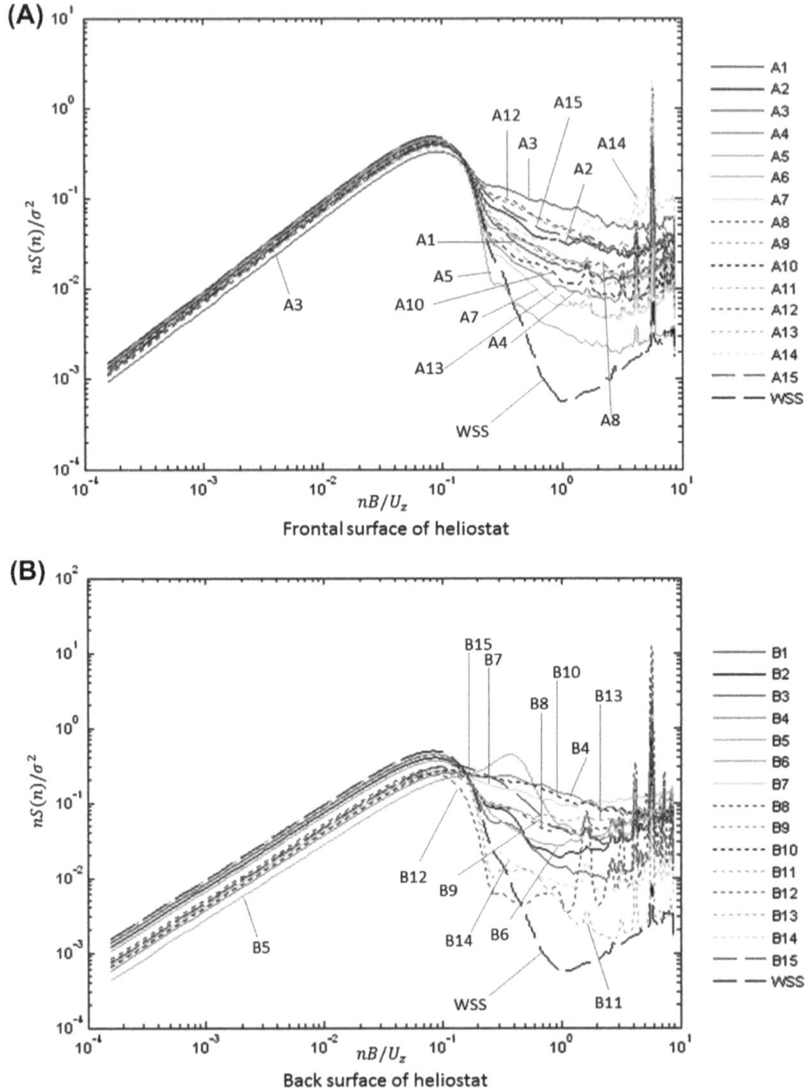

FIGURE 4.51 Power spectrum of fluctuating wind pressure.

pressure signal; $w(n)$ refers to the window function for reducing "spectrum leakage" of the signal and improving the respective spectral resolution; and W_N^{-kn} is the Fourier transformation of the signal sequence.

Fig. 4.51 shows the fluctuating wind pressure power spectrum obtained through analysis in this section, which uses log-log coordinates. The horizontal coordinate is the nondimensional reduction frequency

nB/U_z, in which n refers to the observation frequency, B refers to the characteristic length of the structure, and U_z refers to the referential wind speed, while the vertical coordinate is the nondimensional power spectrum density function $nS(n)/\sigma^2$, in which n refers to the observation frequency, $S(n)$ is the power spectrum density function of the random wind pressure signal, and σ^2 refers to the variance of wind pressure corresponding to the measuring point.

By analyzing Fig. 4.51 within the low-frequency section (namely the section less than a nondimensional reduction frequency of 0.1), the incoming wind speed spectrum is basically consistent with the fluctuating wind pressure spectrum. Within the intermediate- and high-frequency section (namely the section with a nondimensional reduction frequency being subject to 0.1–10) with increased reduction frequency, the incoming wind speed spectrum and fluctuating wind pressure spectrum demonstrate a decrease in tendency, in which the fluctuating wind pressure spectrum is obviously larger than the incoming wind speed spectrum. Within the high-frequency section (namely the section with a nondimensional reduction frequency subject to 1–10), the fluctuating wind pressure spectrum demonstrates several minor and major peaks, demonstrating that the larger wind pressure spectrum energy has concentrated here. A possible reason is that the respective frequency coincides with the mirror natural vibration frequency, which results in mirror resonance and leads to a sharp increase in energy.

4.10.2.6 Wind-Induced Vibration of Evacuated Receiver Tube

1. Definition of formula. The peak wind-induced vibration of the evacuated receiver tube of the concentrator tested and analyzed in this section can be used to describe the receiver's structural reliability under wind environment effects, which can be expressed as follows:

$$A = \max\left[\sqrt{\left(A_{F_x,\max}\right)^2 + \left(A_{F_z,\mathrm{rms}}\right)^2},\ \sqrt{\left(A_{F_x,\max}\right)^2 + \left(A_{F_z,\mathrm{rms}}\right)^2}\right]$$

   in which A refers to the peak receiver tube vibration; $A_{F_x,\max}$ and $A_{F_z,\max}$ respectively refer to the maximum values of the accelerated vibration of receiver tube on F_x and F_z within the unit time interval (which is selected to be 3 s in this section); and $A_{F_z,\mathrm{rms}}$ and $A_{F_x,\mathrm{rms}}$ respectively refer to standard deviations of the accelerated vibration of receiver on F_x and F_z within the unit time interval (which is selected to be 3 s in this section). The accelerated vibration of the receiver tube and the respective vibration damage are mainly caused by gust. Therefore, a gust duration of 3 s is selected in this section as the time interval of vibration analysis for the vacuum heat-absorbing tube.

2. Vibration analysis. Under strong wind conditions, the receiver may result in wind-induced vibration. Such long-term vibration may reduce receiver service life and solar concentration precision. When the vibration amplitude is too large, it may result in collision of the receiver tube's glass exterior and metal interior, further breaking the glass exterior. Based on this, vibration of the glass exterior under strong wind conditions is tested and analyzed in this section, and conclusions are drawn for reference in design and improvement.

Fig. 4.52 shows the time history curve of accelerated receiver vibration and the relationship between wind speed and wind direction within 30 min under strong wind conditions in the concentrator field. Fig. 4.53 shows the variation relationship curve of peak receiver vibration along with gust wind speed variation. Through the analysis, four conclusions can be drawn: (1) Gust wind speed increases from 4 to 13 m/s and the peak vibration basically stays unchanged, with no tendency of variation accordingly. In this section, the reason is inferred to be that along with the increase in gust wind speed, peak vibration may surge; as 13 m/s is slower than the boundary wind speed for the surging of peak vibration, no obvious variation rule between peak vibration and gust wind speed has been obtained in this test. (2) Within the section of 4–13 m/s for gust wind speed, peak vibration values corresponding to each pitch angle at various measuring points are basically stable. For example, for a pitch angle of 180°, the variation values at measuring point 1 concentrate within 0.4–0.6, while values at measuring points 2, 3, and 4 concentrate within 0.2–0.4. Based on this, peak vibration values of various measuring points under different pitch angles can be obtained through statistical analysis, and the respective rules can be summarized. (3) Under the condition that wind speed is fixed, different pitch angles may severely influence the peak receiver vibration amplitude. For example, when the pitch angle is 25°, peak vibration at measuring point 3 exceeds that in 180° by 45%. (4) By focusing on different pitch angles, position points of the maximum peak vibration amplitudes of receiver tube are different from each other. For example, when the pitch angle is small, the position point of the maximum amplitude is at measuring point 3; when the pitch angle is large, it is at measuring point 1.

Fig. 4.54 shows the relationship curve of peak vibration and pitch angle corresponding to different measuring points under the condition that the gust wind speed is 8 m/s. When the pitch angle falls within the range of 45°–120°, the respective peak vibration is comparatively small, and the peak acceleration value falls within a range of 0.2–0.35; when the pitch angle is between 150° and 180°, the respective peak vibration is comparatively large, and the peak

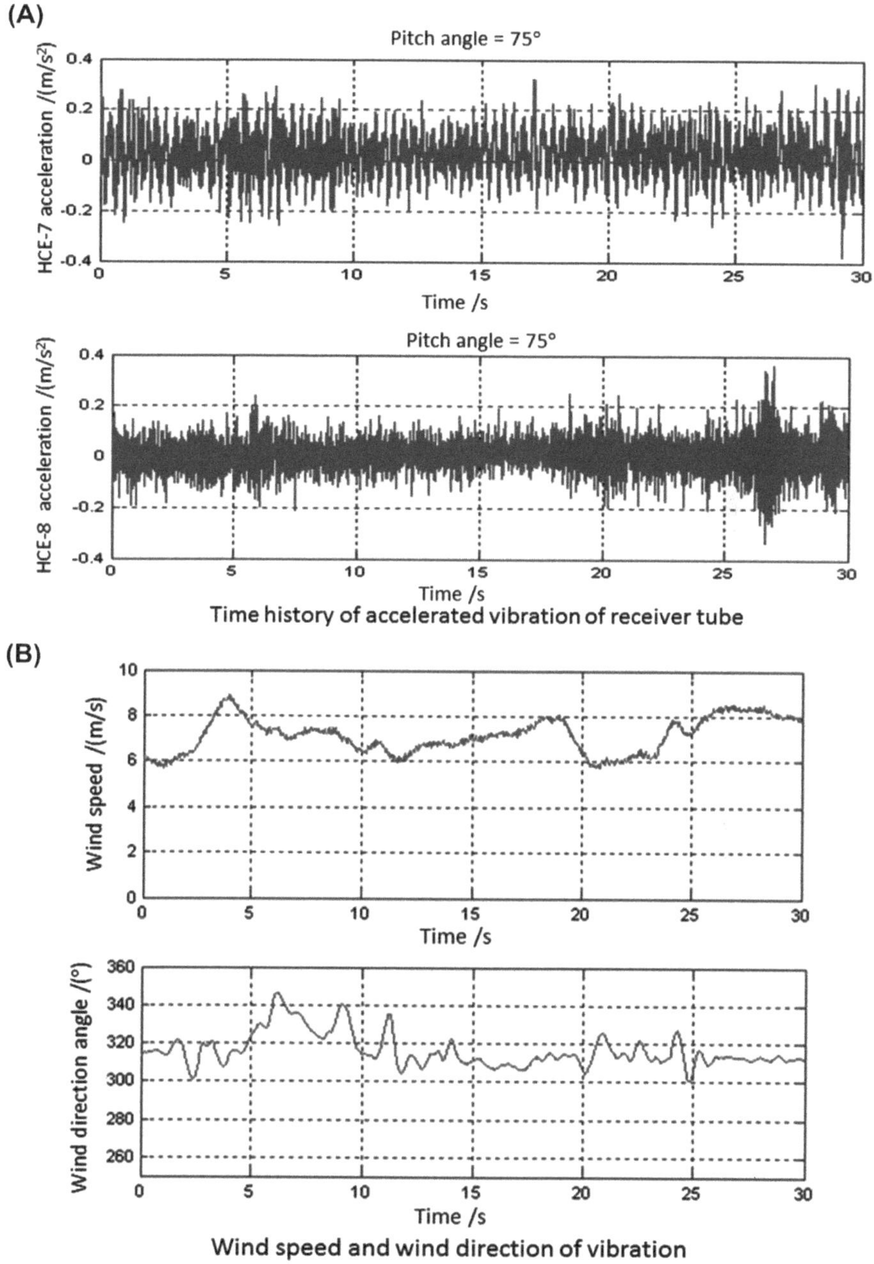

FIGURE 4.52 Relationship between the time history of accelerated vibration of receiver and the corresponding wind speed and wind direction.

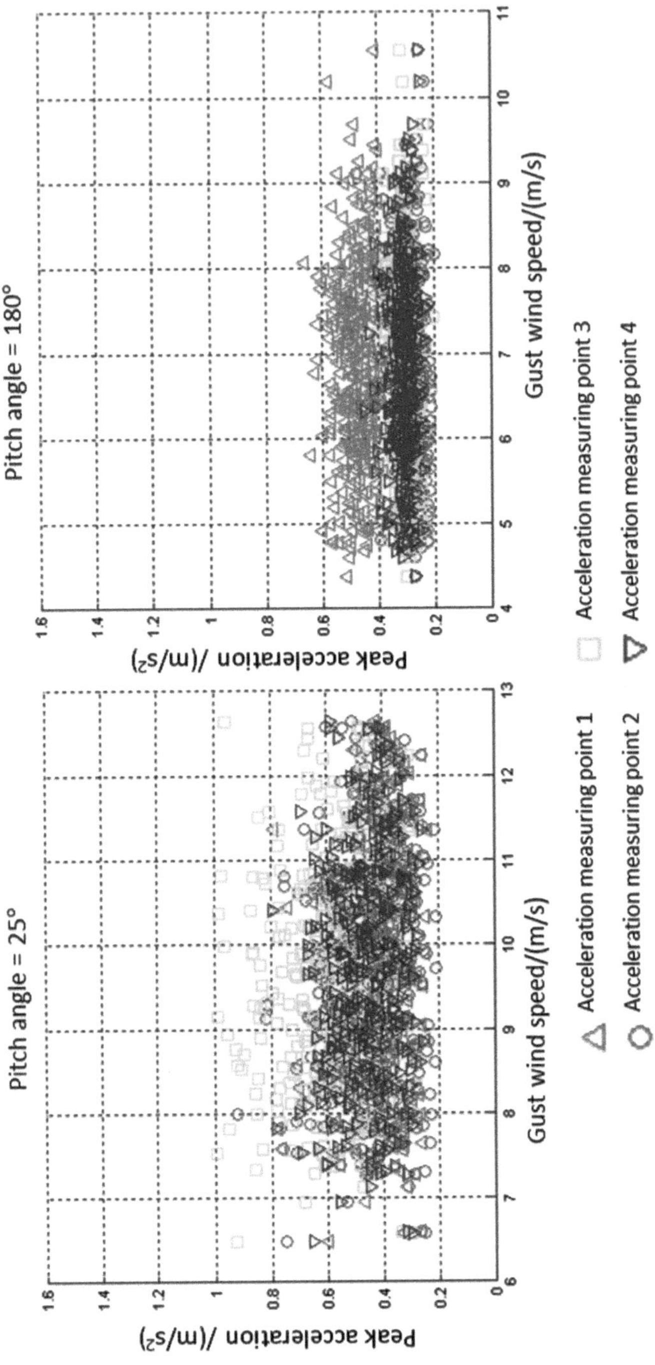

FIGURE 4.53 Relationship between peak vibration of receiver and wind speed.

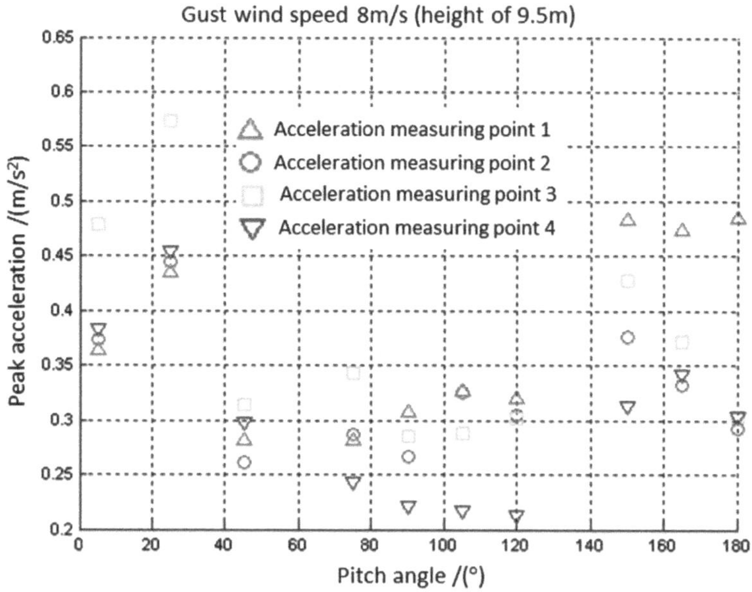

FIGURE 4.54 Relationship between peak vibration and pitch angle.

acceleration value is between 0.3 and 0.5; when the pitch angle is between 0° and 25°, the peak vibration reaches the maximum value, and the peak acceleration value is between 0.35 and 0.6. This is attributable to the direct impact of the incoming wind on the receiver; in addition, the mirror is located at the leeward direction of receiver, which may further lead to increased wind speed on the receiver tube.

HCE-7 and HCE-8 refer to the serial numbers of receiver tubes in a parabolic trough collector.

3. Natural vibration frequency. Modal parameter identification research means to identify the inherent modal parameters of the structure based on experimental vibration test results, including natural vibration frequency, vibration mode, and damping ratio, which aims at solving relevant issues in terms of structural dynamics by using identification results, such as vibration control, dynamic response analysis, and fault diagnosis. The conventional modal identification methods have already been widely applied in aviation, aerospace, automotive, and many other fields. These methods require the utilization of excitation and response signals simultaneously in order to achieve frequency response function or pulse response function and infer the system frequency, vibration mode, and damping ratio based on these functions. However, in terms of a large-scale civil structure, it is very difficult to perform

excitation. Thus, modal parameter identification techniques on the basis of environmental excitation (namely wind load excitation) have been enjoying significant development. The power spectrum peak value method is applied in this section to analyze vibration test data under wind load excitation and identify the receiver's modal parameters.

Power spectrum peak value method is a kind of frequency domain analyzing method based on structural modal parameters of environmental excitation rapid identification, the fundamental principle of which is to obtain natural vibration frequency of the structure through power spectrum peak values of a random response. For a structure with low damping and discrete natural vibration frequency, the natural vibration frequency can be easily identified. As this method is easy to operate and convenient to use (only needing to take advantage of the Fourier transformation and transform time history data into a power spectrum), it has been widely applied in civil engineering.

Fig. 4.55 offers acceleration power spectrum densities in both F_x and F_z directions and separately compares and analyzes the working conditions corresponding to pitch angles of 5° and 165°, which can be used to research the influencing effects of different pitch angles on the receiver's modal parameters. By applying the power spectrum peak value method and focusing on the power spectrum peak value indicated in Fig. 4.55, the respective frequency values are extracted as the natural vibration frequencies of the structure. Table 4.8 has listed the first five order values for the receiver tube's natural vibration frequency.

4. Vibration modes of collector receiver. For framed structure equipment supported on both ends, wind-induced vibrations can be divided into three categories: flutter, buffeting, and vortex-excited vibration. Flutter is a kind of aerodynamic instability occurring under certain wind speeds in which aerodynamics and the vibration structure jointly create a dynamic system with an interaction feedback mechanism. In this case, aerodynamics mainly represent a self-excited force. Under continuous interactive feedback effects, in the case of the positive damping value of the vibration system created by the combined actions of the structure and circumfluent airflow that generates self-excited aerodynamic force approaching a negative value, energy absorbed by the vibration system will exceed its own energy consumption capability and result in vibration system and motion divergence that further leads to the damage of the framed structure. Buffeting refers to the stochastic forced vibration of the structure under effects of natural wind fluctuation components. It is subject to an amplitude-limited vibration; unlike the divergence

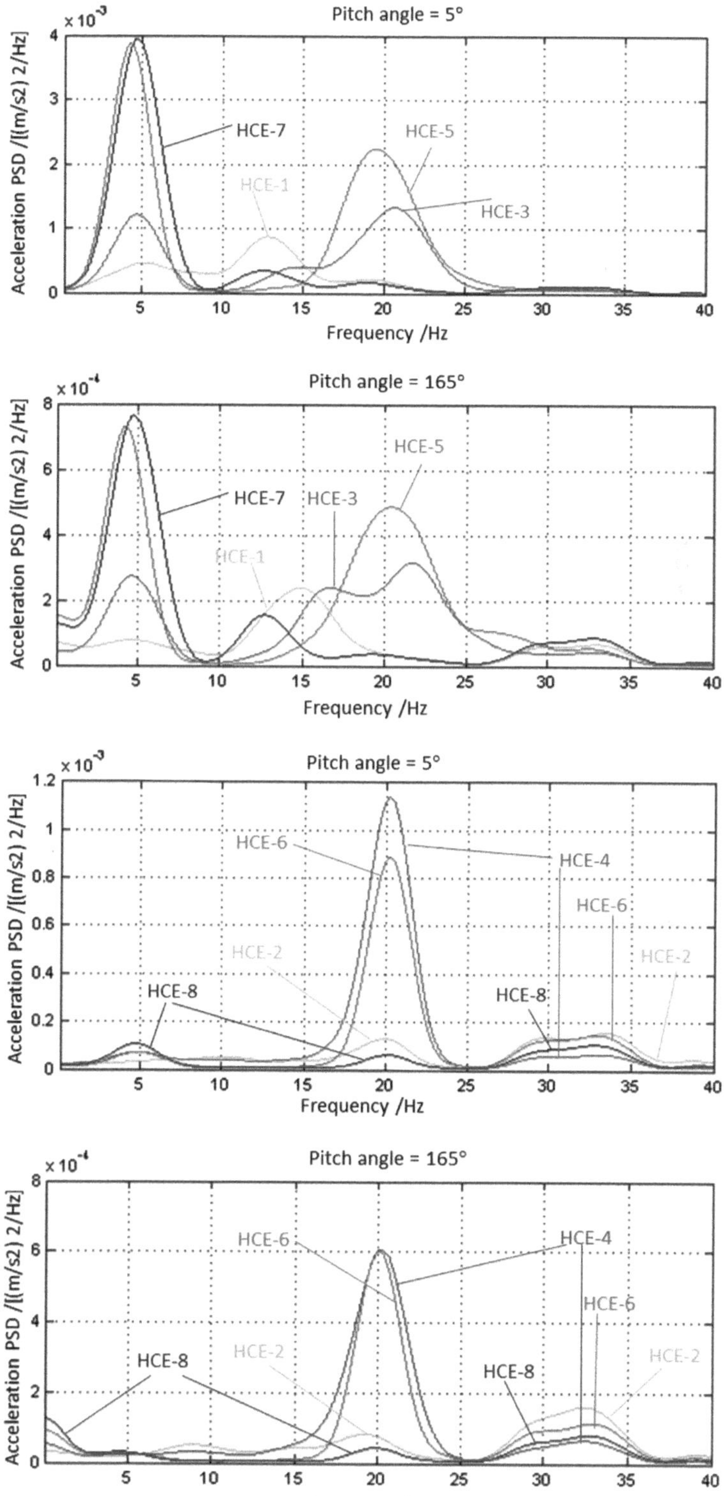

FIGURE 4.55 Acceleration power spectrum density. *PSD*, power spectrum density.

TABLE 4.8 Natural Vibration Frequency of Receiver

Working Condition (Pitch Angle)		First-Order Modal/Hz	Second-Order Modal/Hz	Third-Order Modal/Hz	Fourth-Order Modal/Hz	Fifth-Order Modal/Hz
F_z	5°	4.7	12.9	19.7	29.1	32.4
	165°	4.7	14.9	20.3	28.9	33.5
F_x	5°	4.8	9.1	20.2	29.6	33.6
	165°	4.6	8.8	20.2	29.5	32.8

nature of flutter, it normally does not result in any catastrophic instability damage to the structure. The current framed structure buffeting analysis is mainly focused on the structural buffeting caused by characteristic turbulence in the atmospheric boundary layer.

Frame structure vortex-excited vibration is subject to an important aeroelasticity phenomenon that may easily occur under low wind speeds. When a blunt-body structure is influenced by airflow, a vortex will be generated at the back end. In the case of periodic shedding of vortex from both sides of the structure, a Karman vortex is generated. In this case, the periodically shedding vortex will generate an alternative and periodic excited-vibration force (a vortex-excited force) to the structure, which leads to the periodic vibration of the structure. Such a vibration is referred to as a vortex-excited vibration. When the vortex shedding frequency approaches the natural vibration frequency, the structure generates a significant amplitude vibration; in addition, vortex-excited resonance often occurs on bridge components, such as stayed-cable. Vortex-excited vibration mainly has five features: (1) a limited amplitude vibration occurring under low wind speeds; (2) occurring only within a certain wind speed section; (3) significant dependence of the maximum amplitude on damping; (4) vortex-excited response being very sensitive to the subtle changes of section configuration; and (5) vortex-excited vibration being able to excite flexural vibration and torsional vibration.

By analyzing wind vibration damage and section configuration and the receiver's physical properties, buffeting and vortex-excited vibration are two major sources of wind-induced receiver vibration; vortex-excited vibration easily occurs under low wind speeds, which influences intensity and fatigue of the structure. Receiver tube vortex-excited vibration is discussed to a certain extent in this section in order to lay a solid foundation for subsequent works.

During the flowing process, inertial and viscous forces play leading roles on fluid particles. The ratio of inertial force to viscous force is referred to as the Reynolds number, the expression of which against air is as follows:

$$Re = \frac{\rho v l}{\mu} = 69000 v l$$

in which ρ refers to the air density; v refers to the wind speed; l refers to the characteristic dimension of the structure; and μ refers to the dynamic viscosity.

Based on the value of the Reynolds number, three critical ranges have been divided, including a subcritical range, which normally takes the values of $3 \times 10^2 < Re < 3 \times 10^5$; a supercritical range, which normally takes the values of $3 \times 10^5 \leq Re < 3 \times 10^6$; and a transcritical range, which normally takes the values of $Re \geq 10^6$.

Within subcritical and transcritical ranges, vortex shedding follows a quite definite frequency on a periodic basis; whereas within the supercritical range, vortex shedding is disordered and ruleless. Therefore, only subcritical and transcritical ranges are checked for vortex-excited resonance. Vortex shedding frequency refers to the quantity of vortices shedding from the flow per second. When vortex shedding frequency approaches the natural vibration frequency of the structure, the structure generates large amplitude vibrations, which can be expressed as follows:

$$n_s = \frac{v Sr}{l}$$

in which v refers to the wind speed; l refers to the characteristic dimension of the structure, namely receiver diameter, which represents the maximum scale of the object section in the direction perpendicular to the flow; S_r refers to the Strouhal number, which is a function of the geometrical shape of the object and the Reynolds number; the Strouhal number for a cylindrical structure like the receiver tube is selected to be 0.2. Table 4.9 shows vortex shedding frequencies of the vacuum tube

TABLE 4.9 Vortex Shedding Frequencies Corresponding to Receiver Tube

	$l = 0.12$ m	
Wind Speed (m/s)	***Re***	**n_s/Hz**
1	8.28×10^3	1.67
5	4.14×10^4	8.33
10	8.28×10^4	16.7
15	1.242×10^5	25.0
20	1.656×10^5	33.4

corresponding to different wind speed conditions. The first-order frequency of the vacuum tube approximates 5 Hz, whereas the second-order frequency is 10 Hz; both remain in the subcritical range. It is inferred that vortex-excited resonance conditions exist under low wind speeds.

CHAPTER

5

Design of the Receiver System

The receiver system consists of the solar receiver and the respective heat-transfer fluid power source (pump or fan) as well as equipment for flow rate measurement, temperature measurement, fluid pressure measurement, and safety monitoring. Receivers mentioned in this chapter include water/steam receivers, molten-salt and air receivers for solar tower power plants, parabolic trough receiver tubes, and linear Fresnel receivers.

5.1 GENERAL RECEIVER SYSTEM DESCRIPTION

The function of the receiver system is to transform solar energy concentrated by the concentration field into heat under certain parameters. The heat is safely transferred to the thermal energy storage unit and the steam turbine in the next step in the form of high-temperature fluid.

5.1.1 Receiver System Configuration

1. Receiver system components include receivers, pumps or fans, heat-transfer fluid pipeline valves, measurement and data collection equipment for recording temperature, pressure, and flow rate, and equipment for circuit fluid antifreezing and thawing safety, circuit fluid resistance measurement, receiver system control, and safety alarms and protection.
2. Receiver drive pump. The receiver uses heat-transfer fluid loop pumps to drive the heat-transfer fluid (water, oil, salt, liquid metal, etc.) to circulate inside the receiver. Normally two pumps are equipped to ensure continuous work by the receiver. Because fluid temperature changes continuously from startup to shutdown with a temperature difference that can be as high as 400°C, pumps or fans shall be selected while considering the viscosities and resistances of the fluid at different temperatures. For a fan-driven air receiver, the

Design of Solar Thermal Power Plants
https://doi.org/10.1016/B978-0-12-815613-1.00005-5

increased resistance caused by air viscosity variation at high temperature should be considered.

3. Thermal parameter measurement device of the receiver loop. Parameters to be measured include inlet and outlet fluid temperature, pipeline pressure, and flow rate. Temperature measuring points are designed on the frontal or back surface of the heat receiver in a pattern of one temperature measuring point in each "2m × 2m" area. In the case where both the fluid inlet temperature and the outlet temperature are less than 500°C, it is better to measure the platinum resistance with each test spot corresponding to two measuring points. It is better to use two flowmeters working in series to measure the flow rate of the pipeline. Voltage, current, and resistance signals from all measuring points are connected to the data collection system in the control room.
4. System equipment type selection principles. Receiver type selection shall be firstly based on the working parameters (temperature, pressure, and power) of the generator unit as well as the concentration ratio, concentrated solar flux distribution, and heat-transfer medium categories of the concentrator.

   Types of fluid pumps or fans should be selected in accordance with the pump head, medium temperature, medium pressure, pipeline caliber, continuous working period, and whether there are is frequent startup and shutdown etc.

   Heat-transfer medium shall be selected while considering ambient temperature, working temperature, freezing points of medium, flammability point, saturation temperature, and toxicity. For heat-transfer oil, attention shall be paid to the relationship between its freezing point and the ambient temperature, as well as to antifreezing in winter. The saturation temperature and circuit pressure of synthetic oil are connected to safety system design. In areas where the wintertime temperature is lower than the freezing point of the heat-transfer oil, the antifreezing design for synthetic oil and various circuit instruments shall be fully considered.

   For heat-transfer oil pipelines, heat sources required by the synthetic oil producer shall be used for the antifreezing and thawing of pipelines and containers. Electrical heating is forbidden in a regular oil loop.

   For molten salt, pipeline temperature protection in the heat-transfer loop is of great significance; the temperature configuration shall enable the pipeline wall surface temperature to exceed the freezing point of molten salt by 50°C. Pipeline materials shall be selected while considering molten-salt corrosion. Special attention shall be paid to pipeline fatigue from thermal stress circulation. For a molten-salt receiver, it is necessary to consider the convenience and thoroughness of salt elimination so as to prevent thorough salt elimination from causing freeze blockage.

The types of receiver temperature and fluid flow rate measurement instruments shall be selected in accordance with receiver working parameters. Flow rate measurement is related to receiver safety and thus a redundant design is required.

Receiver signals during surface temperature measurement are sent to the control room. Thus interference or selection methods that shield interference or apply trunking schemes should be avoided.

5. Connection between the receiver system and thermal energy storage. For parameters under different heat-transfer and storage medium, receivers are connected through the heat exchanger to the thermal energy storage units. The receiver outlet is connected to the heat exchanger inlet. Before hot fluid enters the heat exchanger, the heat exchanger shall be preheated and thermal storage warmed in advance to avoid thermal impact generated during the entry of high-temperature fluid.

   For an immersion heat exchanger, the thermal storage tank shall be completely filled with thermal energy storage materials in advance.

5.1.2 Receiver Structure and Principles of Receiver Adjustment and Control

A receiver is a device that absorbs solar radiation and transforms it into heat. Receivers in a tower power plant can be of cavity (Fig. 5.1) or

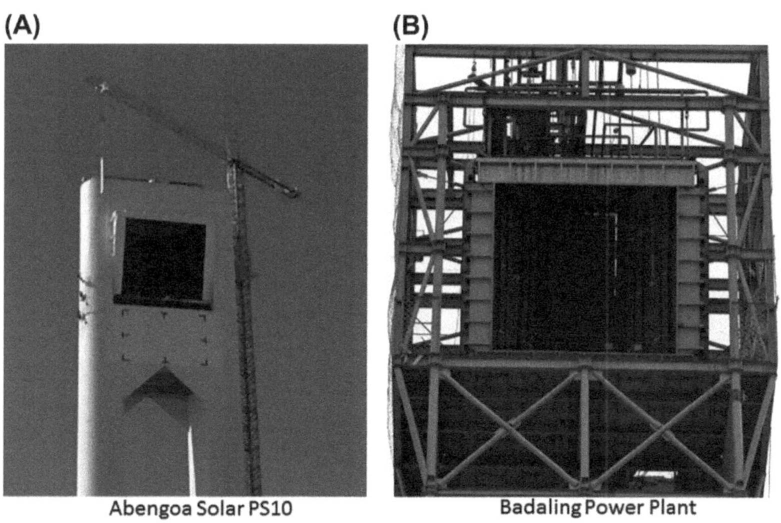

FIGURE 5.1 Cavity-type receiver.

cylinder type (Fig. 5.2); receivers in a parabolic trough power plant are normally vacuum tube type; a Fresnel receiver (Fig. 5.3) can be either a vacuum or nonvacuum heat collection tube; a solar dish receiver is normally cavity type.

FIGURE 5.2 External cylinder receiver (© SENER. The Gemasolar Plant is owned by Torresol Energy and developed by SENER).

FIGURE 5.3 Fresnel tube receiver (Himin Solar Energy Corp., China).

5.1.2.1 *Receiver Constitution*

A receiver consists of the following parts: absorber, solar selective coating, insulating layer, casing, high-temperature protection and fire-fighting facilities, and pump or fan.

A water/steam receiver shall also be equipped with a drum and the necessary primary and secondary instruments for temperature, pressure, and flow rate measurement.

A molten-salt receiver includes a gas protection system, salt buffer tank, hot salt expansion tank, and necessary primary and secondary instruments for temperature, pressure, and flow rate measurement as well as an electric tracer heating system and a receiver night-protective door.

A liquid metal receiver is equipped with a gas leakage inspection system to prevent explosions, especially for a liquid sodium circuit.

A nonpressure air receiver is normally equipped with an induced fan, whereas a pressure-type receiver is normally equipped with a blowing machine or compressor as well as a quartz glass cover and secondary concentrator.

5.1.2.2 *Receiver Heat-Transfer Loop Constitution*

The receiver heat-transfer loop includes the receiver, pump or fan, pipeline, temperature sensor, flow rate sensor, controller, receiver night-protective door, and heat receiver high-temperature protection system.

5.1.2.3 *Receiver Adjustment and Control Principles*

Receiver adjustment and control are conducted mainly according to thermal parameters. The thermal parameter measurement system mainly measures the outlet temperature and flow rate of heat-transfer fluid inside the receiver so that the steam turbine or heat accumulator can function regularly and steadily.

For a superheating type water/steam receiver in a tower power plant, measurement parameters normally include evaporation section surface temperature, evaporation section steam inlet temperature, evaporation section steam outlet temperature, and evaporation section steam outlet pressure; drum pressure (remote-transfer and local), drum water level (remote-transfer and local), and upper, intermediate, and lower temperatures inside the drum; drum cylinder body external surface temperature; superheating section surface temperature, superheating section steam inlet temperature, superheating section steam outlet temperature, and superheating section steam outlet pressure; main steam pipeline steam temperature, main steam pipeline steam pressure, and main steam pipeline steam flow rate; main pipeline wall surface temperature; and receiver daylight opening peripheral wall temperature.

For a saturated type water/steam receiver in tower-type power plant, measurement parameters normally include evaporation section surface temperature, evaporation section steam inlet temperature, evaporation section steam outlet temperature, and evaporation section steam outlet pressure; drum pressure (remote-transmission and local), drum water level (remote-transmission and local), and upper, intermediate, and lower temperatures inside the drum; drum cylinder body external surface temperature; main steam pipeline steam temperature, main steam pipeline steam pressure, and main steam pipeline steam flow rate; main pipeline wall surface temperature; and receiver aperture peripheral wall temperature.

For a molten-salt receiver, measurement parameters normally include absorber surface temperature, molten-salt inlet temperature and receiver outlet temperature, aperture wall temperature, inlet molten-salt flow rate, and molten-salt gas protection system pressure.

The quantity of temperature-measuring points for the receiver can be increased appropriately according to the spatial distribution of energy flux projected to the receiver surface, provided that there is not less than one measuring point within a "2m × 2m" area. In the case where the temperature exceeds the rated value, the concentration field must be removed. The fluid flow rate acts as a major parameter for receiver and pump control.

It is not possible to measure the absorber surface temperature for a parabolic trough evacuated tube, which shall instead be determined by measuring the outlet temperature of the collection tube so as to control the flow rate of the pump. A thermal infrared imager and other noncontact measurement methods can be adopted to roughly measure the receiver tube's surface temperature distribution.

For an air receiver, measurement of the absorber's surface temperature becomes quite difficult as the temperature approaches 1000°C; however, it can be determined by measuring its outlet air temperature. The receiver surface temperature must not be allowed to exceed its oxidation temperature. It is also possible to adopt a thermal infrared imager and other noncontact measurement methods to roughly measure the receiver's surface temperature distribution.

5.1.2.4 Receiver Protection System

Receiver protection is firstly about absorber protection, which normally adopts the method of measuring absorber surface temperature. However, as it has certain difficulties with fixing the thermocouple or thermal resistance on some absorber surfaces, it is suggested that the

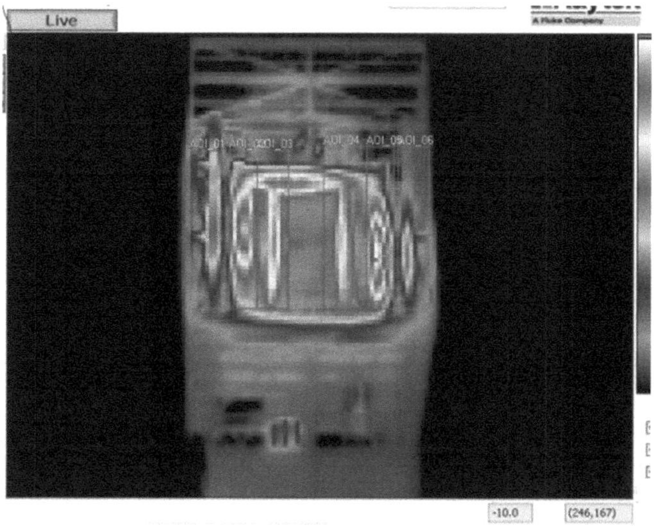

FIGURE 5.4 Infrared monitoring image [20] of receiver and tower.

thermal infrared imager be integrated with thermocouple verification to evaluate the temperature distribution on the absorber surface. Cutoff of heat-transfer fluid inside the receiver is the largest factor for receiver damage. Fig. 5.4 shows the infrared monitoring image of a receiver and tower in which the absorber's surface temperature has reached its maximum. High-temperature points on the heat-absorbing tower are located in the middle area of the receiver's left section and the middle area of the right section of the heat-absorbing tower, but the areas of high temperature are not large.

5.1.3 Principles for Determining the Receiver's Rated Power

The receiver is the most critical component of a tower-type concentrating solar power (CSP, also known as solar thermal power) generation system utilizes energy concentrated by the concentration field to heat the heat-transfer working medium inside the receiver tubes mounted on the receiver's interior wall, which can then be turned into high- grade (temperature and pressure) fluid to absorb and transform solar energy concentrated by the concentration field. Because this concentrated energy has highly non-uniform time and spatial distributions, it may result in concentration temperatures of partial areas inside the receiver reaching more than 1000°C and further lead to receiver tube explosion or melting as well as other severe accidents. In solar tower CSP generation, receiver energy comes from the concentration field. Based on receiver features in terms of nonuniform heating on partial areas, no coordination with the concentration field and a certain control logic are likely to result in receiver damage due to nonuniform heating or thermal shock. Initiating the receiver is even more important; the receiver must follow a certain control logical sequence in order to be initiated. In addition, power plant operation is often under the influences of weather conditions, such as occurrences of obnubilation and obnubilation duration that result in strict requirements for the control logic of the entire system. When designing a receiver system, in addition to determining the rated power, the startup regular operation, standby, shutdown emergency, and various receiver operational conditions under normal and abnormal working circumstances must be considered.

The heat-transfer process inside the receiver has several features: unstable energy distribution, nonuniform and unstable space, nonuniform working temperature and hot fluid density, and an energy-transmission process that involves mutual coupling of radiation, transmission, and convection. During receiver system design, calculation models shall be established for the entire receiver that integrate the concentrated energy of the concentration field, typically including preheating, evaporation, and superheating sections, drum models, and circulating pumps and evaporators. According to the logical connection sequence of the system, procedures for the entire receiver system can be established and used to create modes for receiver convection loss, radiation loss, and transmission loss. This research should also be used to study the variation rules for receiver efficiency and monitor variations in metal temperature on the heat-absorbing section's surface for the entire receiver. This also helps in designing the control logic of the receiver, the concentration field, and the concentrator to ensure that the receiver can function steadily under rated working conditions to the greatest extent.

The rated power and quantity of CSP plant receivers shall be determined according to the following requirements:

1. Receiver power. Receiver power shall depend on the value that corresponds to the design point. At the design point, power shall match the air inflow of the steam turbine under maximum working conditions. When thermal energy storage is incorporated into the thermodynamic system, capacity must be increased to satisfy the requirements of thermal energy storage and the steam turbine. Determination of the design point is quite important. A poorly determined design point may result in the steam turbine functioning under nonrated low efficiency for a long period or wasting energy.

2. Receiver quantity. In a solar tower power plant, a turbine generator may be equipped with multiple receivers, whereas in a parabolic trough power plant, a turbine-generator may correspond to a respective quantity of receivers, but the fluid-resistance balance during variations in working conditions must be fully considered.
3. Steam pipeline. If a power plant expansion and header system are applied to the main steam pipeline, a new receiver capacity shall be selected that incorporates the capacity of the original. Joint positions for various receivers on the header shall be designed to balance the pressures at various joints; otherwise, fluid backflow may occur, especially under temperature variations caused by environmental and solar irradiation changes.

5.2 SELECTION OF MATERIALS FOR THE RECEIVER SYSTEM

5.2.1 Heat-Transfer Medium

Heat-transfer fluid is the key for transforming solar energy into heat. Currently used heat-transfer medium are typically fluids, mainly including water/steam, heat-transfer oil, molten salt, air, and the like. Furthermore, ceramic solid particles can be used as a heat-transfer medium for the fluidized bed receiver. Except for chemical stability under high temperatures, heat-transfer medium shall be determined while considering the following aspects:

1. High heat conductivity, low viscosity, and high density, especially in the high-temperature section; For nonpressure fluid, the heat resistance limit must exceed the working temperature by more than 50°C, and the freezing point shall be as low as possible. Water, steam, air, heat-transfer oil, molten-salt, liquid metal, etc. are commonly used.
2. Good low-temperature properties. Under low temperatures, fluid viscidity variation shall be as high as possible. A fluid with high variation can be used to facilitate heat transfer and reduce pump resistance during circuit startup on cold winter mornings.
3. Good safety: high flammability point, nontoxic, nonexplosive, easy replacement, and easy degradation and cleaning in case of leakage.

5.2.2 Materials of Absorber

Selection of absorber materials depends on the surface thermal flux density and heat-absorbing surface temperature of the receiver as well as the pressure and some chemical properties of the absorber.

According to the request for heat-work conversion, normally the receiver's outlet-fluid temperature has already been given. When selecting materials, the greatest difficulty is determining the thermal load (W/m^2) of the heat-absorbing surface. The mean thermal load of the receiver can be found by dividing the power concentrated onto the receiver from the solar concentration field by the area of the receiver aperture. For a cavity-type receiver, the area of the aperture varies significantly from the area of the absorber. Therefore the mean thermal load of the heat-absorbing surface and the mean thermal load of the receiver vary significantly from each other. The thermal boundary condition of the heat-absorbing surface is normally heat radiation, which includes the peak and mean thermal loads. Due to variations in solar position over time, cosine values of the concentration field change accordingly; concentrator errors are also a function related to time. Combination of these two quantities makes energy flux distribution a function related to time, which is also extremely complex. And the rules for the variation of these two values over time are also quite complicated.

For a solar tower water receiver, the value of the receiver's mean thermal load is normally designed to be 400 kW/m^2; the mean thermal loads of the water-cooled wall and the superheating surface are 200–300 kW/m^2. For a solar tower molten-salt receiver, the mean thermal load is 500 kW/m^2. A liquid absorber's limit thermal load shall exceed 1000 kW/m^2. The limit thermal load of an air receiver shall exceed 1200 kW/m^2.

For a parabolic trough receiver, because its aperture is the same surface the heat-absorbing surface, the thermal load of the receiver is consistent with the mean thermal load of the heat-absorbing surface. In the case of using oil, water, or molten salt as the heat-transfer medium, the mean thermal load is normally 35 kW/m^2. Parabolic trough collectors with large focal ratios have been developed, and their concentration ratios can be higher than 90.

Absorber materials can be selected based on the limit thermal load, receiver heat-transfer coefficient design, and working parameters of the steam turbine.

Absorber materials of a tower receiver shall have excellent properties against oxidation and thermal shock. Thermal-insulating materials shall be determined according to the absorber's working temperature as well as being refractory to spontaneous combustion under high-temperature baking. The absorber of a parabolic trough vacuum tube must have heat resistance that exceeds the working temperature of tube fluid by 100°C so that the vacuum can be considered on an overall scale. Normally, 316 stainless steel shall be selected as the material for the receiver tube.

Absorber materials must have high heat resistance. Heat resistance of materials during long-term work shall exceed the normal outlet temperature of the heat-transfer fluid by at least 100°C.

For a molten-salt receiver, pipe materials shall be selected while considering their respective corrosion resistance under high temperature. When using nitrates at 360°C, chemical corrosivity is insignificant; it is suggested that pipelines and containers made of carbon steel are selected. Chloride has strong corrosivity; it is normally suggested that a titanium alloy with strong corrosion resistance is selected. In terms of sulfate and carbonate, special attention shall be paid to its properties under high temperature.

For a liquid metal receiver, pipe materials shall be processed while paying special attention to preventing hydrogen from overflowing and leaking from the material and accessing the liquid metal while causing explosion.

The heat resistance of absorber materials under pressure and the thermal circulation status of daily startup and shutdown shall be specially analyzed and considered when discussing service life.

5.2.3 Surface Coating of Receiver

For a tower receiver, the solar energy selective coating must have weather and high-temperature oxidation resistance. For a vacuum tube, the heat-absorbing coating must feature low emittance and high absorptance under high temperature.

The bonding force of coating and substrate materials shall be of great importance when discussing coating performance.

5.2.4 The Thermal-Insulating Material of a Tower Receiver

To reduce heat loss, a cavity-type absorber shall be completely covered with thermal-insulating material on all sides. Such thermal-insulating material shall have high fire resistance and high temperature resistance; its fire point shall be not less than 900°C, and it shall not decompose or emit toxic gases under high temperature.

Fire-resistant thermal-insulating materials shall be selected to be attached closely to the absorber. The back surface of fire-resistant thermal-insulating material shall be attached with other types of thermal-insulating materials having excellent thermal-insulating performance. Thermal-insulating materials are normally provided by the receiver manufacturer.

Thermal-insulating materials are normally not used in external receivers, yet the receiver's top and bottom must be attached with fire-resistant and thermal-insulating materials to prevent structural damage to the heat-absorbing tower. This section can be painted white or another highly reflective color; thermal-insulating materials of this section are an integral part of the heat-absorbing tower.

5.2.5 Tubing Material

The pipeline material that will contact heat-transfer fluid shall meet the following requirements: high temperature resistance, corrosion resistance, high-temperature stress corrosion resistance, and high heat conductivity.

5.3 SELECTION OF PIPES AND PUMPS FOR RECEIVER SYSTEM

1. Feed pump. The maximum transmission capacity of the receiver's heat-transfer fluid electric feed pump shall be designed according to the following requirements:
   a. Normally, two electric feed pumps are required, one for operation and one for backup, and shall be controlled through a local controller.
   b. Electric fluid pump head shall be calculated according to the most adverse temperature. Fluid viscidity normally increases along with temperature reduction. Thus pump head shall be designed while considering low-temperature viscidity.
   c. The pump's explosion-proof characteristics shall be designed according to the flammability point and explosion-proof characteristics of the fluid;
   d. Due to variations in solar radiation, a receiver's working condition points vary over time, and thus the working condition points of a pump also vary accordingly. The selected pump must be able to adapt to such variations in terms of flow rate, pump head, temperature resistance, and pressure resistance.
2. Receiver pipeline and valve
   a. The receiver valve shall have certain temperature- and pressure-resistance capabilities. The valve's working temperature shall be determined by the system's designed temperature; it is not necessary for the maximum working temperature to be set greater than the receiver's outlet fluid temperature. For a pipeline using oil as its heat-transfer medium, attention shall be paid to pipeline material welding requirements. The flow rate allowance of the circulating pump shall be 5%–10%; pump head allowance shall be 10%–20%; and temperature allowance shall be 15%. For oil in the pipeline, explosion-proofness must be considered. In such a case, pneumatic valves can be used.

b. The fluid transmission pipeline shall be selected by considering pressure and temperature as core parameters. Furthermore, wall surface material and chemical compatibility of the tube fluid are also areas of emphasis.

Pipeline overheating protection shall be achieved using the multipoint temperature monitoring method. Measuring points shall be set on dangerous temperature zones and points, such as the receiver outlet.

An CSP plant does not function continuously, 24 hours a day, so startup and shutdown of pipelines occurs frequently, and the fatigue and damage caused by thermal stress are significant. Stress monitoring and measuring points shall be set on pipelines, especially at joints, including the flange, the valve, the sampling point, and various primary instrument interfaces.

c. Receiver cleaning in a solar tower power plant. In addition to the exterior wall surface of the receiver's heat-transfer tube must be free of oxidation and there shall be no scaling of suspended matter on the tube's inner wall; before startup the tube's inner wall surface shall be mounted with chemical cleaning devices. Because the receiver and connecting pipeline are comparatively long, various elbows are needed. During cleaning, the cleaning of all pipelines shall be verified, with cleaning inspection a necessary step in the process.

Cleaning of the tube's exterior wall surface can be conducted using the oxidation film chemical removal method.

5.4 RECEIVER SYSTEM CONTROL

5.4.1 Logic of the Receiver Control System

1. Receiver control design is firstly about guaranteeing absorber safety, the major parameters of which include tube fluid temperature, pressure, tube wall surface temperature, and thermal stress. Because a solar receiver is under the impact of thermal stress for long periods, materials can be easily damaged. Therefore, regular material stress inspections and checks are emphases of control.
2. Because solar irradiation and wind speeds change quickly, the energy flux density of the heat-absorbing surface also undergoes fast changes, and emergencies may occur at any time. The layers in the receiver control system should not be excessive but must include redundancy.
3. Receivers can be controlled by a central computer to directly collect data and to output control information. Computational complexity

is not difficult because the number of measuring points is not large, so an independently designed local control cabinet for the receiver's programmable logic controller is not necessary.

4. Sampling data points include the multipoint temperature on the absorber surface, exterior surface temperature and temperature distribution of the absorber, internal pressure of the absorber, inlet fluid temperature of the receiver, receiver pump, and electric motor, and fluid flow rate inside the receiver.
5. Control interlock: absorber surface temperature, temperature distribution, and internal pressure are keys for safe operation of a receiver. These measuring points shall be interlocked with the receiver's fluid pump in order to maintain parameters within the safety range. Surface temperature distribution of the absorber is normally collected with an infrared imager. Sectional image information shall be appropriately processed so that information can be submitted to the master distributed control system (DCS) in time. The DCS sets algorithms to collect and analyze infrared images, extracts information about dangerous temperature areas, and sends this information to the concentration field control unit; the processor of the unit determines the geometric positions of the receiver's high-temperature points and calculates the serial number of the heliostat associated with the high temperature by integrating the corresponding solar angle. The concentration field controller instructs heliostats to move from this section, which is also designed with a feedback step that continues until the dangerous high-temperature area has been eliminated. During this process, the heliostat shall be moved as slowly as possible to prevent the absorber temperature from dropping too fast. When a heliostat is moved too quickly, the temperature and pressure fluctuations at various points will be significant. As shown in Fig. 5.5, during the period from 12:00 noon to 2:20 p.m., all nine steam pressure fluctuations inside the tube were caused by heliostat adjustment.

5.4.2 Designed Scope of Control

1. A physical control layer including the concentration field, receiver, and auxiliary boiler and its auxiliary system and equipment
2. A water system including a generator unit condensing water processing system that consists of desalination, chemical dosing, and deoxidization
3. A chemical water processing system, mainly including sampling and dosing. It is quite difficult to take samples in a tower power plant because its tower is so high. Thus a design incorporating automatic field sampling and analysis should be considered.

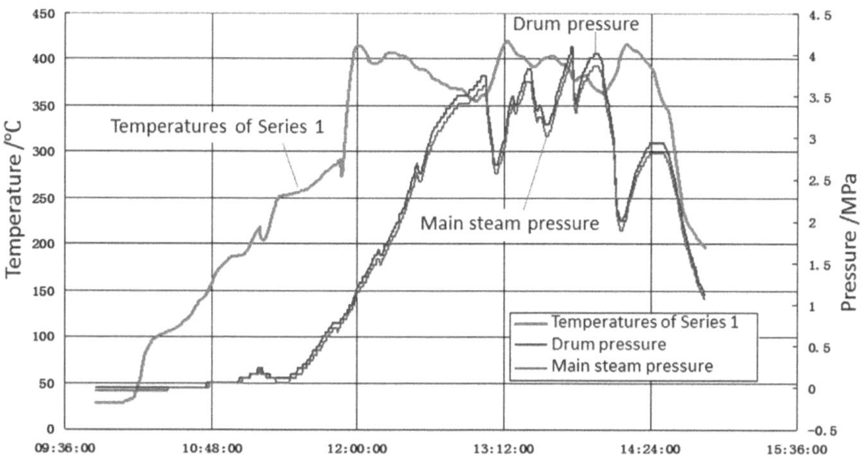

FIGURE 5.5 Pressure fluctuation caused by heliostat adjustment [20].

4. A comprehensive pump house, which serves as the thermal power plant's core; pump medium include water, oil, salt, and so on. Pumps are mainly used for antifreezing and frost resistance and mounted below the frozen layer.
5. Whole-plant fire hazard inspection alarm and fire control system. The whole-plant regional closed-circuit television monitoring system's main control level consists of a host computer and data collection modules. The risk of concentration field fire is related to the leakage of synthetic oil as well as fire on the cable sheath. The use of flame-retardant cables is recommended.

5.4.3 Control Mode

The design principles of control function distribution and information centralized management are applied:

1. The automatic network containing the DCS is normally applied.
2. The centralized control of the concentrators, receiver, thermal storage, turbine-generator unit, and auxiliary workshop is normally applied. The whole plant is designed with one central control room; the concentrators shall be equipped with an independent host computer for monitoring. This host computer shall be connected to other master control computers throughout the plant. Inside the central control room, the operator monitors turbine-generator-unit startup/shutdown control, regular operation surveillance and adjustment, and the treatment of abnormal turbine-generator-unit operations and emergency

working conditions through the LCD operator station and large-screen LCDs.

The heliostat control host computer in a tower power plant can be either connected or disconnected to the DCS, especially at the initial stage when the mode of direct connection to the DCS and heliostat is normally not applied. However, the distance between two operators shall be short as possible to guarantee smooth and safe communication as well as a high level of security. A one-button reset mode for the heliostats shall be designed under which all heliostats can be defocused using one button in the event of receiver problems.

3. Network control shall be integrated into DCS unit monitoring and control; no independent network control building or network monitoring system shall be designed. Backup monitoring and control equipment and regular display instruments in the central control room are not needed; all that is needed is a small number of hard-wired control switches and buttons independent from the DCS and used for emergency treatment. Two displays and a closed-circuit television monitoring system for major unattended regions shall be designed.

5.5 DESIGN OF THE OPERATION MODES OF THE RECEIVER SYSTEM

The receiver's operation modes depend especially on the respective working medium, the concentration ratio, and weather conditions.

5.5.1 General Principles

The operation of various receiver types shall center on safety while considering efficiency as the target.

5.5.2 Startup of the Receiver

1. For a molten-salt tower receiver, the entire pipeline must first be preheated to a temperature 50°C greater than the freezing point of molten salt before initiating the molten-salt pump and filling the pipeline with molten salt from the storage tank to form a stable closed-circuit cycle. Then the concentrators shall be moved so that solar radiation can be concentrated on the absorber, during which the user shall refer to the startup instruction manual from the receiver manufacturer.
2. A molten-salt parabolic trough receiver tube runs continuously with molten salt even when the concentrator is not running so that the

pipeline temperature always exceeds the melting point of molten salt by 50°C. While running, the operator shall rotate the concentrator, move the facula to the receiver tube, and gradually adjust the flow rate of tube fluid in order to control fluid temperature in response to solar radiation and the meteorological environment, referring as needed to the startup instruction manual of the receiver manufacturer. Freezing of molten salt in a parabolic trough receiver tube may result in catastrophic destruction of the system and should definitely be avoided. In addition, this process normally occurs during system startup and long-time shutdown at night, and thus temperature detection and prevention are core issues for molten-salt heat collection systems in molten-salt parabolic trough systems.

3. For water/steam and oil receivers, the absorber pipeline must be preheated to a temperature greater than the freezing point of water before the absorber is filled with water. After the operation is stable, the concentrator shall be moved so that solar radiation can be concentrated onto the absorber. The operator shall gradually adjust the flow rate of tube fluid in order to control the fluid temperature to adapt to the solar radiation and meteorological environment, which shall be referred to the startup instruction manual of receiver manufacturer. Such kind of receiver is the same with the molten-salt receiver, antifreezing is also a very important aspect. Antifreezing of a synthetic oil pipeline is normally conducted without applying the electric heating mode; the pipeline initiation preheating mode normally applies steam pipe external preheating mode.

5.5.3 The Working Mode of the Receiver

In the basic heat-absorbing process of the receiver, pumps/fans are used to drive heat-transfer fluid to transform the solar energy concentrated on the absorber surface into heat. The absorber's surface temperature is higher than the fluid temperature and is difficult to measure. Control of the absorber's surface temperature is the key to controlling the safety and efficiency of receiver.

5.5.4 Technical Improvement of the Receiver

The influences of system technical innovations mainly derive from receiver tube improvements; by improving the absorption coating and joints, the temperature and pressure level of the process can be raised (to reduce pressure drop). Receiver improvement requires circulation innovations, such as adding the reheat section or using an organic Rankine steam turbine. All technical innovations (Table 5.1) can raise the power output level.

TABLE 5.1 Technical Innovation Approaches of the Parabolic Trough System With Synthetic Oil as the Heat-Transfer Medium

Technical Innovation Approaches [32]		Potential Technical Benefit	Potential Economic Benefit
Receiver tube innovation	Advanced coating (applicable for a higher temperature range) Removal of corrugated tube Increase in working temperature Advanced joints Increase in pressure Reduction of pressure drop and pipe length Advanced working media	Improvements in receiver tube performance and service life	Reduction in operation and maintenance, assembly and installation, and certain parts costs
Circulation innovation	Increase of temperature of superheated steam Introduction of reheat section Apply the organic Rankine cycle (minor system)	Increase in system efficiency	Reduction in system installation and levelized electricity costs

All the improvements mentioned in Table 5.1 are related to parabolic trough systems using water/steam as the heat-transfer medium to eliminate the influences of multiphase flow instability inside the receiver tube. A direct steam generator can be used to improve system efficiency. By considering the accumulated experiences of PSA on the DISS testing platform and R&D on advanced selective coating for the DSG receiver tube in direct superheating of steam, 450°C seems to be the upper limit for feasible temperatures. Increases in temperature lead to significant increases in heat loss as well as a significant reduction in receiver tube durability. Technical innovation approaches in a parabolic trough system using the water/steam heat-transfer mechanism are shown in Table 5.2.

5.6 THE DISCHARGE SYSTEM AND EQUIPMENT OF THE RECEIVER

5.6.1 Discharge Range

In order to control parabolic trough and tower receivers that use water/steam as the heat-transfer media, so that water quality satisfies the specified standard and in order to restrict receiver water impurities to a

TABLE 5.2 Technical Innovation Approaches of Parabolic Trough System With Water/Steam as the Heat-Transfer Medium [36]

Technical Innovation Approaches		Potential Technical Benefit	Potential Economic Benefit
Receiver tube Innovation	Advanced coating (applicable for a higher temperature range) Removal of corrugated tube Increase in working temperature Advanced joints Increase in pressure Reduction of pressure drop and pipe length Eliminate influences of multiphase flow instability of the DSG receiver tube	Improvements in receiver tube performance and service life	Reduction in operation and maintenance, assembly and installation, and certain parts costs
Circulation innovation	Increase of working temperature	Increase in system efficiency	Reduction in system installation and levelized electricity costs

certain range, it is necessary to continuously remove receiver water that contains a large quantity of salts and alkalis as well as deposited slags and loose sediment from the receiver, which are referred to as the discharge of the receiver.

5.6.2 Discharge Mode

Receiver discharge can be divided into continuous discharge and regular discharge. Continuous discharge is also known as surface discharge, which needs to continuously drain off partial circulating water from the part of the receiver with the highest salt and alkali concentrations in order to reduce the levels of salts, alkalis, silicates, and suspended slags in the circulating water. Thus continuous tubes are mounted 80−100 mm below the receiver's normal water level. Regular discharge mainly removes slags, sludge, and other sediment from the receiver. Thus its discharge outlet is normally at the lower section of the drum and the bottom of the header. The operational process of regular discharge does not last long. It should be conducted when the receiver water level is high and with a low load, or with low load in the output state. Normally, an MW-level minor receiver is only designed with regular discharge.

5.6.3 Discharge Devices

Discharge devices refer to discharge short pipes, valves, and conduits inside the drum and within range of the receiver. The discharge conduit must have sufficient length and shall be mounted horizontally with one end sealed. The discharge pipe shall be mounted with the minimum number of elbows necessary to ensure that discharge is smooth and connected to a safe location. Joints between the discharge pipe and the boiler drum, header, and discharge valve shall be firm, reliable, and corrosion-free. Gate, sector, and inclined stop valves may be used as discharge valves. The nominal diameter of a discharge valve is $\Phi 20 \sim 65$ mm. For a receiver with a rated evaporation capacity ≥ 1 t/h or a working pressure ≥ 0.7 MPa, the discharge pipe shall be mounted with two discharge valves connected in series. During discharge, the discharge valve undertakes the washing of high-temperature fluid and dirt from abrasion; after the discharge ends, the valve cools to room temperature. To lessen frequent pressure differences (great pressure drop), fouling corrosion wear, vibration, thermal shock, and other adverse discharge value working conditions, discharge valves shall be connected in series and follow a certain operational sequence including the following connection sequence: boiler drum (or lower header), valve 1 (slow valve), valve 2 (quick valve).

5.6.4 Discharge Principles and Methods

1. Principles. Principles for receiver discharge are high frequency, small quantity, and evenness. Discharge must be conducted in a timely manner according to receiver boiler water test results while maintaining receiver water quality in line with the requirements of the applicable standard. The time interval for each discharge should be basically balanced, and all discharge valves shall be used for discharge. Because the receiver is located about 100 m above the ground, discharge operations should avoid times of strong wind as much as possible. When conducting discharge and sampling during winter, attention should be paid to antifreezing measures.
2. Correct operational approaches. Receiver discharge shall be conducted under low loads because receiver water impurities can easily precipitate. Discharge in a water-cooled wall system is not recommended under high loads because the water circulation system can be damaged; however, it is possible to discharge boiler water in the receiver drum in an appropriate manner.

Discharge shall be conducted intermittently in short intervals; the discharge duration for each discharge valve group normally lasts 20–30 s. During discharge, discharge valves shall be closed immediately after opening, and opened immediately after closing; the process shall be repeated for two to three times to attract slags to rapidly flow toward the discharge outlet, form vibrations with water flow, and intensify discharge effects.

Previously opened valves shall be closed successively during discharge operations, and successively opened valves shall be closed in order; previously opened and closed valves shall be specially protected.

For a receiver mounted with surface discharge devices, the degree of the discharge valve's opening shall be adjusted in an appropriate manner according to boiler water quality test results; in addition, discharge shall be conducted on a regular basis based on specific circumstances, especially the pH values of circulating water, so that discharge operations are conducted in a timely manner.

The application of correct discharge operations is a necessity and an important means of ensuring receiver operational safety when located high above. In order to prevent accidents, operating instructions shall be strictly executed to ensure the receiver's safe, reliable and long-term operation.

5.7 VACUUM PERFORMANCE OF THE PARABOLIC TROUGH RECEIVER TUBE

5.7.1 Brief Introduction to Parabolic Trough Receiver Tube Structure

As shown in Fig. 5.6, a parabolic trough receiver tube mainly consists of the glass cover tube, the metal absorber tube, the corrugated tube, the glass–metal sealing adapting piece and getter, and other elements and components.

1. Glass cover tube. Parabolic trough power plants are normally located in wild Gobi Desert areas and heavily influenced by wind and rain erosion. Thus, glass materials must have comparatively higher hardness and stronger chemical resistance than would be typically expected. Glass cover tube temperature fluctuations, which are caused by frequently occurring factors such as alternating day and night conditions cloud cover, require that glass materials have thermal stability. High-borosilicate glass such as Pyrex is normally used. By considering the matching of the sealing between glass and metal, the SCHOTT Corporation independently developed a special glass with an expansion coefficient of 5.5×10^{-6}/K that can be

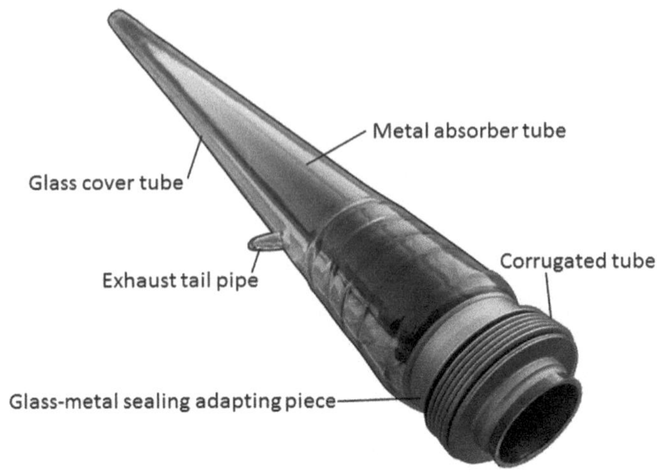

FIGURE 5.6 Parabolic trough receiver tube. *Picture provided by Himin Solar Energy Co., Ltd., CHINA, 2010.*

matched and sealed with Kovar alloy; the respective hardness, corrosion resistance, and thermal stability meets operating requirements. Because the metal absorber tube bends when heated unevenly, the diameter of the glass cover tube must not be too small. Based on production costs, thermal efficiency, and other factors, external diameters of commercial parabolic trough evacuated tubes are normally 115–120 mm with a wall thickness of 3 mm.
Solar radiation transmittance of the glass cover tube serves as a key factor in solar conversion efficiency. Solar transmittance of borosilicate glass normally approximates 92%. In order to improve transmittance, the gel–sol process can be used to make antireflecting film for glass tube interiors and exteriors that can increase the solar transmittance level to more than 97% while maintaining excellent stability.

2. Metal absorber tube. The absorber tube is the thermal absorption component of a parabolic trough tube. Its external diameter shall be determined to simultaneously meet both the optical and the thermal requirements of the concentrator while further striving to reduce material usage and manufacturing costs. The metal absorber tube's external diameter tube shall exceed the width of the facula band of the parabolic trough concentrator. Under ideal conditions, the parallel incident radiation with an angle of 0° will be concentrated onto the axis line of the metal absorber tube; however, due to the limits of processing precision and tracking precision as well as hot–cold concentrator deformation, the intercept rate of the metal absorber tube is normally around 95%.

A metal absorber tube with a larger diameter can better adapt to the optical errors of the entire concentrator, but increasing the external diameter results in an increased radiation area, which influences the receiver tube's maximum temperature. Meanwhile, the increased diameter also increases production costs. With these factors in mind, a parabolic trough concentrator with an opening width of 5 to 6 m typically has metal absorber tube radius of around 70 mm.

Metal absorber tube materials must have superior thermal conductivity, which depends on the type, thickness, and diameter of metal materials as well as the type and state of the heat-transfer fluid. Heat transfer can be improved by reducing tube diameter and wall thickness; however, reductions in diameter may introduce additional concentration optical errors, whereas reductions in wall thickness make the tube incapable of withstanding the high pressure of heat-transfer fluid inside the tube. When selecting organic synthetic oil (such as Therminol VP-1 or Dowtherm A) as the heat-transfer fluid, the resulting internal pressure in the tube is small, so wall thickness can be 2–3 mm; when selecting water as the heat-transfer fluid, the tube must withstand pressure of 10 MPa, so the wall thickness must be increased to about 6 mm.

When selecting metal absorber tube materials for a parabolic trough receiver tube, additional factors must be considered including structural strength, corrosion resistance, vacuum performance, welding and installation convenience, and materials costs. Commonly used materials include 321H, 316L, 304L, and so on.

A solar-selective absorption coating is applied to the exterior surfaces of metal absorber tubes in parabolic trough receiver tubes to improve receiver tube heat absorption efficiency. The merits and demerits of selective absorption coating directly influence the efficiency of parabolic trough CSP generation, which have high absorptivity and low emissivity under high temperature.

3. Glass-metal sealing adapting piece. One end of the glass–metal sealing adapting piece is connected to the glass cover tube through heat sealing, while the other end is welded to the corrugated tube. This mainly aims to solve problems of inconsistency in the corrugated tube and glass cover tube in terms of the expansion coefficient and selecting a sealing alloy. The expansion coefficient must be as consistent as possible with that of the glass cover tube before conducting matching and sealing; otherwise, the nonmatching housekeeper sealing (also known as knife-edge sealing) method may be used, but the resulting sealing strength and reliability will be greatly weakened. The adapting piece shall be easily welded to the corrugated tube and have satisfactory corrosion resistance.

4. Corrugated tube. A corrugated tube is mainly used for compensating the thermal expansion by the metal absorber tube and the glass cover tube. Thus a corrugated tube shall have excellent flexibility and high tensile fatigue strength as well as high-temperature, acid, and alkaline resistance. The corrugated tube's length shall be minimized in order to increase the receiver tube's concentration length and improve heat absorption efficiency. The corrugated tube's wall thickness after forming is small, normally around 0.3 mm.

   Corrugated tubes can normally be divided into seamless tubes and welded tubes. Based on the sectional shape, seamless corrugated tubes can be divided into U-shaped, C-shaped, Ω-shaped, V-shaped, and echelon-shaped tubes, of which U-shaped and C-shaped corrugated tubes can be used after hydraulic forming without reshaping, or can be slightly reshaped while maintaining great stiffness and high sensitivity [27]. Welded corrugated tubes feature satisfactory flexibility, but with many welding seams and a high cost, and thus they are rarely used in parabolic trough receiver tubes. To ensure the vacuum and reliability of a parabolic trough receiver tube, the 316L stainless steel hydraulic-forming corrugated tube with high-temperature and corrosion resistance has been most widely used. The service life of corrugated tubes is related to work displacement, pressure, temperature, impacts, shocks, and other operating conditions.

5. Gas getter. In order to maintain the vacuum capacity inside parabolic trough receiver tubes, gas getters are installed in the vacuum jacket to absorb residual gases inside the tube's vacuum gap, which is generally divided into two series, evapotranspiration getters and nonevapotranspiration getters. An evapotranspiration getter is a metal material that inhales air during evapotranspiration and after evapotranspiration forming, and refers to materials mainly made of barium, strontium, magnesium, and calcium; barium–aluminum–nickel, nitrate, and similar getters are the most commonly used. Evapotranspiration of the inhaling metal is not necessary in a nonevapotranspiration getter. A getter that requires activation of the air-inhaling metal surface to enable it to inhale air is mainly made from zirconium, titanium, thorium, and their aluminum alloys. ZrAl 16, ZrVFe, and ZrG are commonly used.

   A tube is normally filled with both nonevapotranspiration-type and evapotranspiration-type getters. A large quantity of nonevapotranspiration-type getters are used mainly to absorb residual gases inside the vacuum jacket, whereas fewer evapotranspiration-type getters are filled and are mainly used to determine whether the receiver vacuum tube has become invalid by observing the color of the evapotranspiration film. When the evapotranspiration film turns from silver to white, it means the receiver vacuum tube has become invalid.

5.7.2 Current Status of Parabolic Trough Receiver Tube Vacuum Reliability

Currently, commercially operating parabolic trough CSP plants use mainly heat-transfer oil as the heat-transfer medium. Moens and Blake et al. from the National Renewable Energy Laboratory of the US Department of Energy (NREL) analyzed and carried out studies on the thermal decomposition of organic synthetic oil in parabolic trough power plants. Organic synthetic oil is 73.5% biphenyl ether and 26.5% biphenyl. DOW Chemical's Dowtherm A and Solutia Inc.'s Therminol VP-1 are commonly used at present. The thermal decomposition rate of such a synthetic oil at temperatures of less than 400°C is quite low, and thus it can be used on a long-term basis; once the temperature exceeds 400°C, synthetic oil decomposes quickly. Taking Dowtherm A as an example, when heated at 425°C for 120 h, about 8% of the synthetic oil is thermally decomposed, with products including hydrogen, micromolecular hydrocarbon, and aromatic compounds in which hydrogen accounts for about 44% of the total gas volume.

Research has also found that organic impurities inside synthetic oil can accelerate and catalyze thermal decomposition of synthetic oil while producing highly active hydrogen atoms. These impurities may be introduced while synthetic oil is being produced. The oxide layer inside the stainless-steel pipe also may accelerate and catalyze thermal decomposition of synthetic oil; this conclusion, however, must be further studied and demonstrated. Because the chemical bonds of biphenyl ether and biphenyl are unstable, they cannot prevent the production of hydrogen through chemical means. Currently, the method applied in the SEGS power plant in the United States is the removal of high boiling point matters by periodic exhalation of pipeline gases.

Hydrogen produced during synthetic oil decomposition can permeate the parabolic trough receiver tube. If an insufficient number of getters are placed in the receiver tube, hydrogen will accumulate inside the receiver tube. According to the calculation model for receiver tube thermal properties proposed by Forristall, even a small amount of hydrogen may significantly increase receiver tube thermal losses, as shown in Fig. 5.7.

Gretzmaier (from the NREL) studied the predicted partial pressure of hydrogen inside a parabolic trough receiver tube and determined solutions for hydrogen permeation using a steady-state assumption, namely that pressure inside the vacuum jacket of the tube is a fixed value that makes the hydrogen-permeation rates of various elements and components of the parabolic trough power plant near the atmosphere equivalent to the production rate of hydrogen decomposed from the synthetic oil in the power plant. A hydrogen-permeation model was also established.

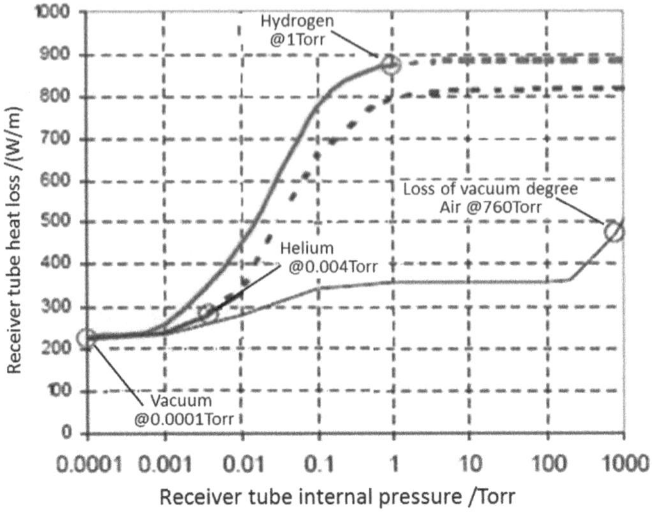

FIGURE 5.7 Heat loss of a parabolic trough receiver tube.
Operating conditions: LS-2 collector; Solel UVAC receiver tube; Therminol VP-1 heat-transfer oil, 350°C, 140gal/min; DNI 950 W/m^2; wind speed 1 m/s; ambient temperature 35°C. 1 Torr = 133.322 Pa; 1 gal (US) = 3.78541 dm^3.

To verify the calculation results of this model, Gretzmaier tested the permeation rate of hydrogen from inside to outside the receiver tube. During the test, he first heated and vacuumized the receiver tube manufactured by SCHOTT without getters, then filled it with a certain amount of hydrogen and maintained hydrogen pressure inside the tube within a certain range (37 and 55,600 Pa); then heated the metal absorber tube at 400°C for 14 h while observing whether variation of hydrogen pressure in the vacuum jacket was consistent with the calculation results of the model. However, during the test period, hydrogen pressure in the tube did not change, which was inconsistent with the model. Gretzmaier believed that this error was caused by the oxide layer on the stainless-steel surface or the hydrogen barrier, which might result in the hydrogen-permeation coefficients of the stainless-steel absorber tube and corrugated tube being less than the hydrogen-permeation coefficient of pure stainless steel during the test; plus model calculations were based on the hydrogen-permeation coefficient of the pure stainless steel, and thus experimental results were inconsistent with calculation results.

In order to reduce the hydrogen-permeation rate of a parabolic trough receiver tube, the interior surface of the stainless-steel pipe was processed to introduce a hydrogen barrier layer. This process oxidized the stainless-steel absorber tube within the steam containing free hydrogen at a temperature of 500–700°C to create a rich chromium oxide layer on the interior surface of the steel pipe with a thickness of 0.5–10 μm, namely a hydrogen barrier with a chromium oxide mass fraction of 20%–60%. Such a hydrogen barrier could reduce the permeation rate of hydrogen by 50 times.

To solve the hydrogen-permeation problem, Luz Corp. designed a hydrogen pump made of palladium and palladium alloy that was connected to the metal corrugated tube close to the glass–metal sealing. However, the temperature of this device rose sharply when it received concentrated solar radiation, which resulted in overheating of partial areas of the glass cover tube and increased stress, leading to further fracturing of the glass tube. Furthermore, such a device could be easily corroded, which could let rainwater in and result in absorber tube invalidity. Therefore such a device has not been used since.

The getter's installation position influences its inhaling performance. In the parabolic trough receiver tube manufactured by Siemens, the getter was installed inside the bridge groove on the stainless-steel absorber tube as shown in Fig. 5.8. Axial motion freedom existed between the groove and supporting foot and could relieve the stress caused by the different thermal expansion coefficients of the stainless-steel absorber tube and groove. Solar radiation shields under and on both sides of the groove blocked radiation heat sent from the stainless-steel absorber tube and solar radiation reflected by the concentrator, thus reducing the getter temperature and improving its inhaling performance.

SCHOTT installed the getter inside the annular space between the metal adapting piece and corrugated tube as shown in Fig. 5.9. This installation structure prevented the getter from being directly influenced

FIGURE 5.8 Installation position of parabolic trough receiver tube getter (from Siemens).

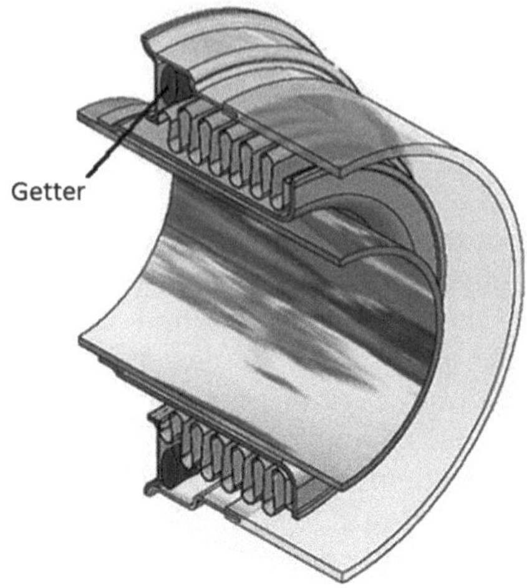

FIGURE 5.9 Installation position of parabolic trough receiver tube getter from SCHOTT.

by solar irradiation, reduced its temperature, and further improved its inhaling performance.

SCHOTT also tested the getter's inhaling performance. Test results indicated that the getter's inhaling capability corresponding to depressurization conditions was stronger than that corresponding to constant pressure conditions. The relationship between the getter's inhaling capability and the temperature was also tested and is shown in Fig. 5.10.

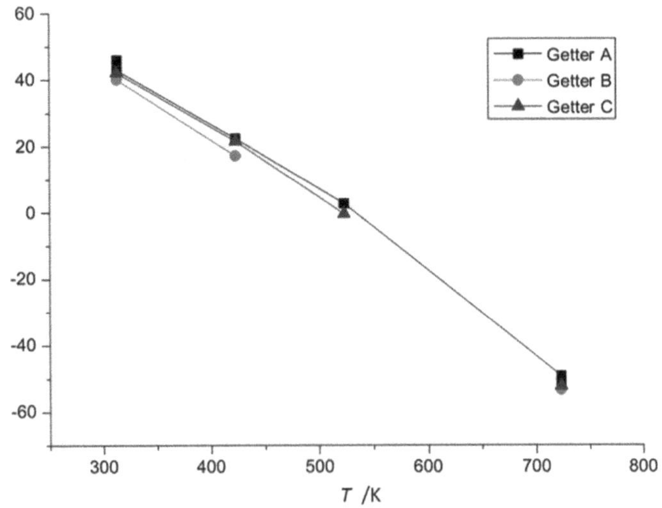

FIGURE 5.10 Variation curve of getter capacity along with temperature.

Using the getter's inhaling capability at 200°C as the reference point, it increased by 40% at 27°C, and decreased by 40% at 400°C. SCHOTT also conducted finite element analyses, laboratory testing, and field testing on getter temperature; the results proved that the temperature of the getter installed inside the pipe was below 200°C.

In order to verify the vacuum duration of a parabolic trough receiver tube, SCHOTT conducted a field test at the SEGS V power plant. SCHOTT installed absorbing tubes with different getter quantities on the end of the loop with an oil temperature of 352°C and determined whether the vacuum of these receiver tubes was invalid by measuring the glass surface temperature of the tube absorber; then they acquired the service lives of tubes through derivation of the test results, according to which, under the operating conditions of the SEGS V power plant, the getter quantities installed by SCHOTT were capable of maintaining a receiver tube service life of 50 years.

Whether reducing the permeation rate of hydrogen by utilizing a hydrogen barrier or by increasing the quantity of getters installed, the basic target is to prolong the parabolic trough receiver tube's vacuum life. However, due to the restrictions of installation space and production costs for getters inside the receiver tube, in unlimited increase in getters is not possible. In recent years, scientific researchers and receiver tube manufacturers have changed their minds and have carried out relevant studies of thermal loss reduction methods after invalidation of the receiver tube vacuum.

Frank Burkholder et al. from the NREL discovered from research that by filling the receiver tube with inert gases and mixing them with hydrogen that has permeated the receiver tube, receiver tube thermal losses can be significantly reduced. They established the gas heat conduction model of the parabolic trough receiver tube by utilizing the Sherman interpolation formula and direct Monte Carlo simulation to carry out experimental analysis on the thermal loss experiment platform through the NREL. According to their research, the heat conduction experiment and simulation results of the mixed gas of xenon and hydrogen are shown in Fig. 5.11.

According to Fig. 5.11, along with an increase in the proportion of inert gases in mixed gas, receiver tube thermal losses caused by heat conduction of gas increase at a faster rate. According to research results by Frank Burkholder, when the receiver tube's inner tube temperature reaches 350°C with only hydrogen in the tube, the additional thermal losses caused by the heat conduction of gas are greater than 500 W/m; when the proportion of inert gases in a mixture with hydrogen exceeds 95%, the additional thermal losses caused by heat conduction are only 50–100 W/m.

Based on the heat conduction characteristics of mixed gas in a mixture of inert gases and hydrogen, SCHOTT developed a special "xenon

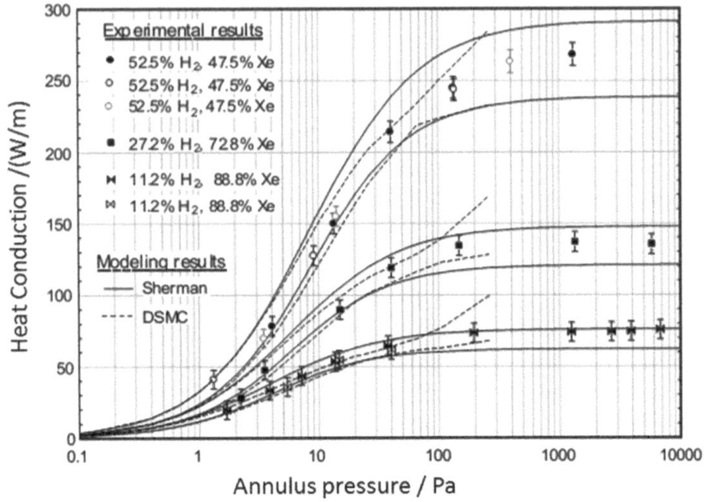

FIGURE 5.11 Heat conduction experiment and simulation results of xenon/hydrogen mixed gas.

capsule" and mounted it inside the parabolic trough receiver tube as shown in Fig. 5.12. When getters inside the parabolic trough receiver tube are saturated and cannot inhale gas, hydrogen permeating the receiver tube results in a sharp increase in thermal losses inside the receiver tube, and the temperature of the exterior glass pipe also increases accordingly. In this case, the operations and maintenance personnel of the power plant can open the "xenon capsule" and release the inert gases inside the capsule so they mix with hydrogen to reduce thermal losses in the parabolic trough receiver tube and prolong the receiver tube's service life.

Along with the development of the CSP generation industry, various scientific research units and companies in China have made active efforts

FIGURE 5.12 "Xenon Capsule" inside the SCHOTT parabolic trough receiver tube.

at carrying out relevant studies and the manufacturing of key techniques and equipment. Currently, an increasing number of studies on parabolic trough receiver tubes have been seen around the world, especially in China.

Some experimental analysis has been made of the outgassing conditions of borosilicate glass and stainless steel and have discovered that the gas components of borosilicate glass mainly include steam, CO_2, N_2, and CO, H_2; and the peak value of outgassing fell into a range of 160–230°C, which was caused by the desorption of gases absorbed on the glass surface. As the borosilicate glass surface had a comparatively stronger capability for water absorption, steam desorption was dominant, the outgassing quantity of which was far more than it was for other gases; CO_2, N_2, and CO started to become outgassed significantly within a temperature range of 300–400°C as shown in Fig. 5.13 [39].

The outgassing experiment on stainless steel [39] indicated that outgassing components mainly included steam, N_2, CO, H_2, and CO_2; below 300°C, outgassed steam was significant in quantity; over 300°C, N_2 and CO became the major outgassing components; in a range of 300–350°C, H_2 was the major outgassing component. Over 200°C, the outgassing quantities of N_2, CO, and CO_2 were significant; inferring from tendency, continuously increasing the temperature could lead to even greater outgassing of stainless steel, and the temperature of outgassing was estimated to exceed 430°C as shown in Fig. 5.14 [39]. Although the outgassing components of glass material and stainless-steel material were tested for during the experiment, the researchers did not carry out profound studies on the outgassing rules of materials and did not offer quantity variation rules for the outgassing rates or outgassing of these two materials at a certain temperature.

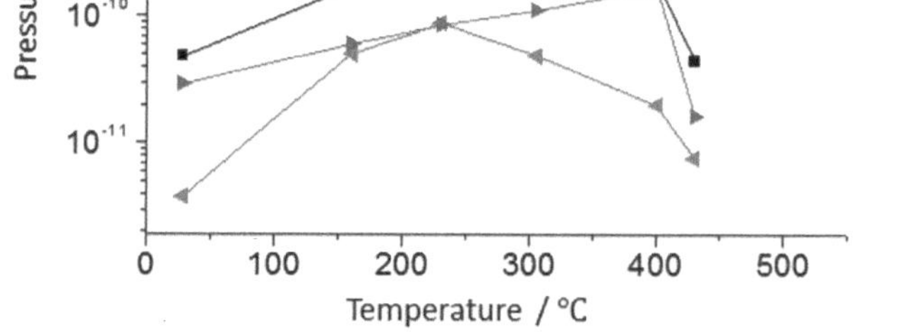

FIGURE 5.13 Borosilicate glass outgassing conditions.

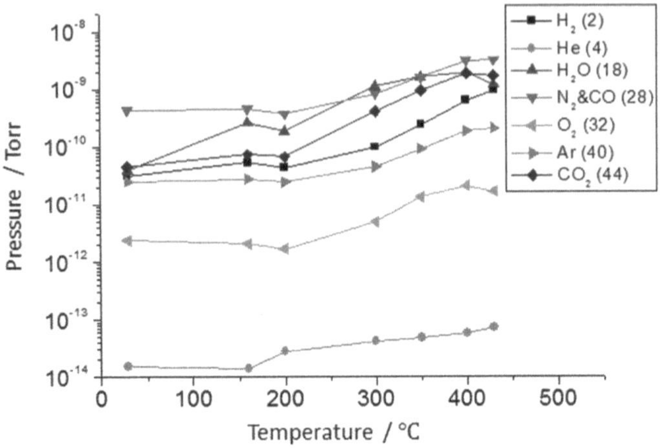

FIGURE 5.14 Stainless-steel outgassing conditions.

Some experiments simulated actual operating conditions of a parabolic trough solar power plant by heating the receiver tube's absorber tube to 450°C during daytime and allowing natural environmental cooldown at night. After heating and cooling the receiver tube constantly for half a year, the internal pressure of the tube showed maintenance at the 10^{-2} Pa level; the experimental testing platform is shown in Fig. 5.15.

Some coating of preferential oxidation is on the interior surface of a parabolic trough receiver tube's stainless-steel absorber tube and achieved a dense and rich Cr_2O_3 coating that prevents hydrogen from permeating the receiver tube as shown in Fig. 5.16.

Installing a gas exhaust pipe on the receiver tube's glass–metal adapting piece is an interesting option. If the receiver tube vacuum is invalid, the metal exhaust pipe could be opened to reinstate the vacuum as shown in Fig. 5.17.

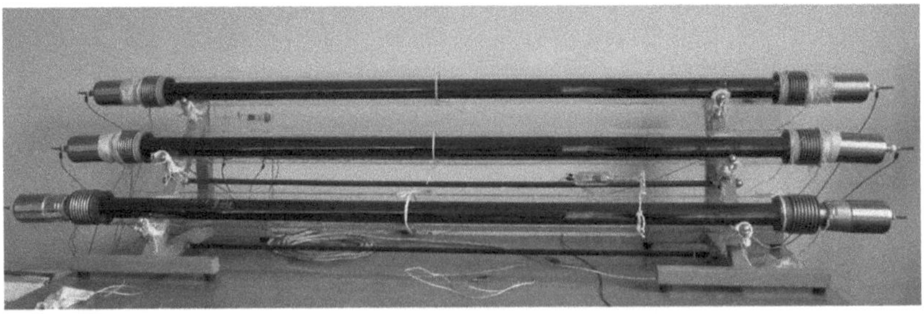

FIGURE 5.15 Parabolic trough heat-absorbing sample tube vacuum experimental testing platform. *From Tsinghua University, China.*

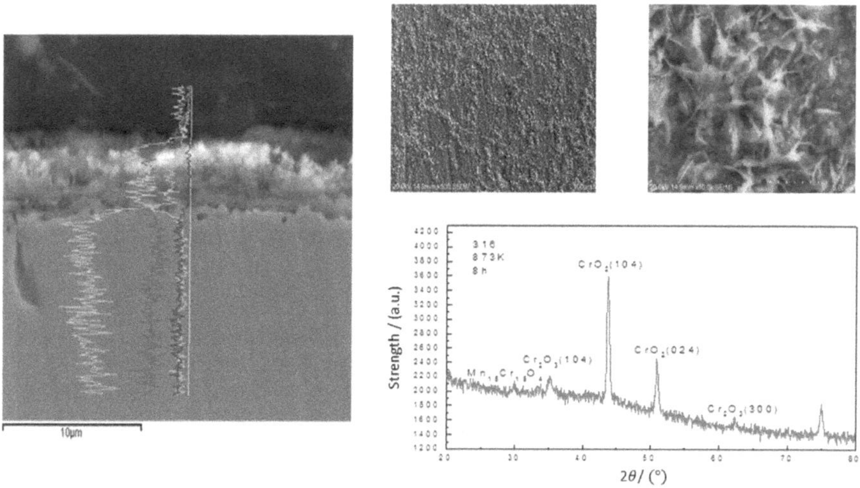

FIGURE 5.16 Hydrogen barrier development. *a.u.*, a random unit. *From CHINA General Research Institute for Nonferrous Metals.*

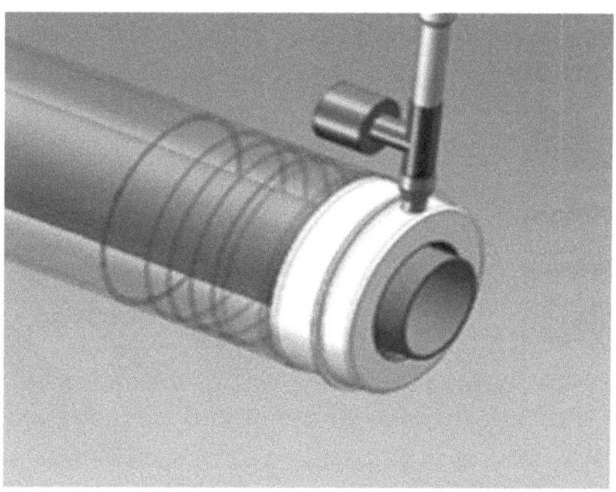

FIGURE 5.17 Parabolic trough receiver tube.

5.7.3 Parabolic Trough Receiver Tube Gas Release Performance

Material outgassing of the parabolic trough receiver tube is a key factor that influences vacuum reliability and life. In order to maintain a receiver tube's degree of vacuum, material outgassing must be reduced. Many scholars have carried out studies and analyses on the outgassing performance of vacuum system materials; however, the factors that influence the outgassing performance of materials are numerous and are not only related to material type, but also closely associated with the production, processing, manufacturing, pretreatment, and operating environment of materials as well as many other factors. These studies have important referential values for analyzing the outgassing performance of materials in a parabolic trough receiver tube, but it is quite difficult to carry out quantitative studies and analyze outgassing rules in parabolic trough receiver tubes because they must be analyzed specifically case by case and solved through experimental measurement. Here, based on the theoretical analysis of material outgassing and testing the outgassing rate and outgassing composition of the parabolic trough receiver tube, material outgassing rules of the receiver tube have been found that can provide theoretical instructions for preparing the exhaust process and improving the vacuum reliability of the parabolic trough receiver tube.

Material outgassing includes diffusion dissolution of dissolved gases inside the material and desorption of surface absorption gases. Calder and Lewin pointed out in their article that the diffusion of gasses inside the material was the key step influencing the material outgassing rate. They also proposed a diffusion calculation model for the material outgassing rate, the diffusion limited model (DLM); namely, when gas is diffused to the material surface, it should be immediately desorbed to vacuum, and the material outgassing rate is therefore equivalent to the diffusion rate. Diffusion of gases inside materials follows Fick's law:

$$q = -D\mathrm{grad}C \tag{5.1}$$

$$\frac{\partial C}{\partial t} = \mathrm{div}(D\mathrm{grad}C) \tag{5.2}$$

in which q refers to the outgassing rate of material, $\mathrm{Pa}\cdot\mathrm{m}^3/(\mathrm{s}\cdot\mathrm{m}^2)$; D refers to the diffusion coefficient, m^2/s; and C refers to the gas concentration of interior material, $\mathrm{m}^3/\mathrm{m}^3$.

Taking a flat plate with a thickness of l as an example, the outgassing rate is obtained through the one-dimensional diffusion equation:

$$\frac{\partial C}{\partial t} = D\frac{\partial^2 C}{\partial x^2} \tag{5.3}$$

The internal initial gas density of the entire flat plate is assumed to be C_0; when $t = 0$, it is placed in the vacuum environment, and then the initial and boundary conditions are:

when $t = 0$, $0 < x < l$, $C = C_0$; and

when $t > 0$, $x = 0$ and $x = l$, $C = 0$:

$$C(x,t) = C_0\frac{4}{\pi}\sum_{0}^{\infty}(2n+1)^{-1}\sin\frac{\pi(2n+1)x}{l}\exp\left\{-\left[\frac{\pi(2n+1)}{l}\right]^2 Dt\right\} \tag{5.4}$$

can be solved.

Thus, the outgassing rate of the flat plate is:

$$q = -D\frac{\partial C}{\partial x} = \frac{4C_0D}{l}\sum_0^\infty \exp\left\{-\left[\frac{\pi(2n+1)}{l}\right]^2 Dt\right\} \tag{5.5}$$

In many actual applications, the outgassing rate of the flat plate can be approximated as:

$$q = \frac{4C_0D}{d}\exp\left(-\left[\frac{\pi^2 D\tau}{d^2}\right]\right) \tag{5.6}$$

in which d refers to the thickness of the outgassing flat plate material, m; and τ refers to the outgassing duration, s.

In cases where the material outgassing rate is high, the analysis results of the DLM are precise. However, in cases where the material outgassing rate is slow, analysis errors are significant. The DLM neglects the influences of material surface conditions on the outgassing process; the gas atom concentration of the material surface is normally assumed to be zero. The material outgassing rate is directly equivalent to the rate of gas diffusing from interior material to the surface. However, contrary to the situation where gas dissolves into a material, gas atoms inside the material must be recombined into molecules on the material surface before entering into the vacuum environment. Gas atoms recombined into molecules are a secondary dynamic process [40,41]. Under low pressures and temperatures, the recombined power of gas shall be determined as the squared value of the concentration of gas atoms under the material surface, and the recombination rate of gas atoms shall be lower than the diffusion rate. According to this, some scholars have proposed the recombination limited model (RLM) [42] for calculating the material outgassing rate; that is, the outgassing rate of the material can be determined by the recombination rate of gas atoms under the material surface:

$$q = K_L C^2(x,t)|_{x\to l} \tag{5.7}$$

in which K_L refers to the recombination coefficient of gas atoms, $cm^4/(mol \cdot s)$.

Some scholars have also carried out studies on the dynamic process of hydrogen on the metal surface [41,43,44]. Coverage of hydrogen atoms on a metal surface can be determined by four hydrogen flows as shown in Fig. 5.18.

1. The flow of hydrogen molecules on a metal surface caused by detachment and absorption is:

$$f_1 = 2pvs(\theta) \tag{5.8}$$

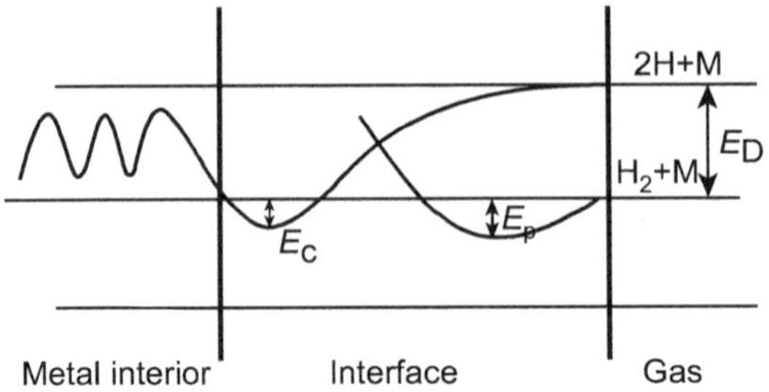

FIGURE 5.18 Potential energy of hydrogen atom and molecule on the metal surface. E_D, Thermal decomposition energy of H_2, $E_D = 218$ kJ/(mol·K) = 4.746 eV; E_P, Physical absorption heat, $E_P = 10$ kJ/(mol·K) = 0.2 eV; E_C, chemical absorption heat (chemical absorption energy), $E_C \approx 50$ kJ/(mol K).

in which p refers to the gas pressure; v refers to the speed of hydrogen molecule moving toward a unit metal area at a unit pressure; and $s(\theta)$ refers to the absorption probability (which is a function of the surface coverage θ).

2. The flow of hydrogen molecules caused by desorption is:

$$f_2 = -K\theta^2 \tag{5.9}$$

3. The flow of hydrogen atoms transiting from the surface to interior material is:

$$f_3 = -\alpha\theta(1 - x) \tag{5.10}$$

in which α refers to the hopping frequency of hydrogen atoms from the material surface to the interior; and x refers to the fraction of hydrogen atoms inside the material.

4. The flow of hydrogen atoms transiting from interior materials to the surface is:

$$f_4 = \beta(1 - \theta)x \tag{5.11}$$

in which β refers to the hopping frequency of hydrogen atoms transiting from interior materials to the surface.

When the metal surface is within a high-vacuum environment, the rate of transition of hydrogen atoms from the material surface to the interior and its reverse process exceed the rate of hydrogen molecules transiting from a vacuum environment to the metal surface by several orders of magnitude. Thus, before a balance between the gas and metal surface can be achieved in the vacuum environment, there must first be a balance between the metal surface and the interior. In the case where two hydrogen atoms are recombined into hydrogen molecules and leave the metal surface, the respective vacancies will be immediately occupied by hydrogen atoms from interior materials. The f_2 process rate, namely the outgassing rate of metal materials, including the recombination rate of hydrogen atoms, becomes a key process restricting the desorption of metal materials.

Malev combined the diffusion, desorption, and absorption processes of gases and proposed the absorption–diffusion model. This model considers the balance of gas motion on the material surface and vacuum space as shown in Fig. 5.19. The process can be expressed as:

$$kT\frac{dN_{\text{ads}}}{dt} = I_{\text{diff}} + I_{\text{ads}} - I_{\text{des}} \tag{5.12}$$

in which k is the Boltzmann constant; T refers to the thermodynamic temperature; N_{ads} refers to the quantity of absorption gas particles on material surface; I_{diff} refers to the diffusion rate of gas inside the material; I_{ads} refers to the absorption rate of gas on material surface; and I_{des} refers to the desorption rate of gas on material surface.

The material outgassing rate is the difference in gas quantities between material surface desorption and absorption within a unit time interval:

$$q = I_{\text{des}} - I_{\text{ads}} \tag{5.13}$$

All of the above theoretical models, including the DLM and RLM, assume that gas atoms are subject to only one energy level state inside the material with a unique activation energy. However, researchers have discovered through experiments [45] that gas atoms correspond to multiple diffusion energy level states inside most metal materials. In the case where a metal material has a certain constant temperature, most outgassing particles from interior materials derive from gas atoms subject to the lowest energy

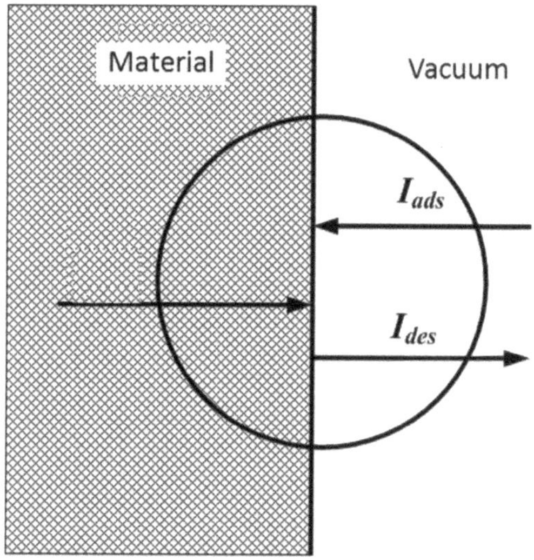

FIGURE 5.19 Balance process of gas on the material surface.

level state; thus the unique energy level state theory still applies. However, in cases where the temperature of the metal material remains changeable, the unique energy level state theory no longer applies, and the influences of gas atoms on material outgassing under different energy level states must be considered.

Based on the above analysis, no unified theory currently exists for material outgassing rules. Generally speaking, when analyzing outgassing rules for materials in low- and high-vacuum environments, the DLM is applied; when analyzing outgassing rules for materials in a super and extremely high-vacuum environments, the RLM is applied. Because many factors may influence material outgassing performance, it cannot be studied and analyzed only through theoretical studies; instead, experimental tests must be conducted on outgassing performance in parabolic trough receiver tubes, from which rational theoretical models shall be utilized to conduct studies and analysis on experimental test results.

5.7.4 Parabolic Trough Receiver Tube Gas Release Performance Test

In order to test the outgassing performance of full heat-absorbing tube in a parabolic trough, a set of outgassing performance testing methods for the parabolic trough receiver tube are here established along with a testing platform by IEECAS (Institute of Electrical Engineering, Chinese Academy of Sciences) as shown in Fig. 5.20. First, the parabolic trough

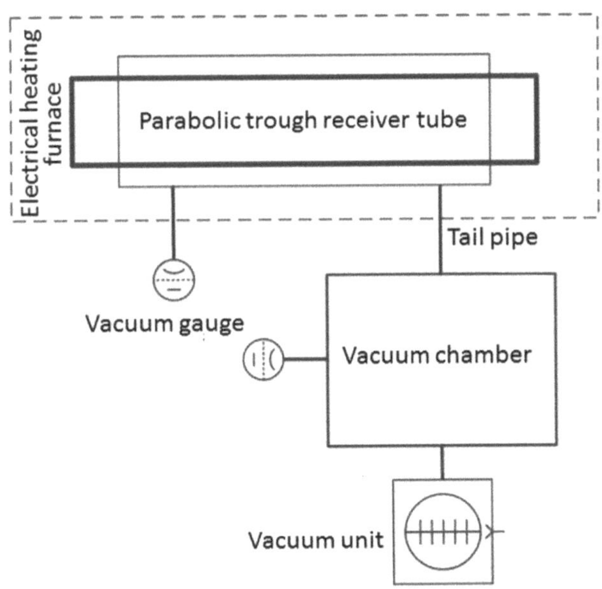

FIGURE 5.20 Principles for outgassing performance test of the parabolic trough receiver tube.

receiver tube shall be vacuumized under normal temperature; after meeting that requirement, the receiver tube shall be placed in an electrical heating furnace to be heated and outgassed; the heat collection tube shall be connected to the vacuum chamber of the vacuum system through an extraction tail pipe. When the receiver tube is being heated and outgassed, one part of the gas released from the receiver tube is extracted by the vacuum unit through the tail pipe for flow restriction while the other part of the gas results in pressure variation inside the receiver tube (in the case where the gas quantity released momentarily from the receiver tube is greater than that extracted through the tail pipe, the pressure inside the receiver tube increases; otherwise, the pressure inside the receiver tube decreases). By separately and simultaneously measuring the pressures of the parabolic trough receiver tube and the vacuum tube and calculating the gas increment and extracted amounts inside the receiver tube, the outgassing rates and quantities can be obtained for the parabolic trough receiver tube.

Devices for the full-tube outgassing performance test of a parabolic trough receiver tube mainly include the receiver tube heating system (Fig. 5.21), vacuumizing system, and vacuum measurement system.

A parabolic trough receiver tube is connected to the extraction tail pipe close to the exterior glass pipe on one side and the heated cathode-ionized gauge on the other. Except for the ability to conduct exhaust treatment, this sample tube can be manufactured completely using standard manufacturing processes for parabolic trough receiver tubes as shown in Fig. 5.22.

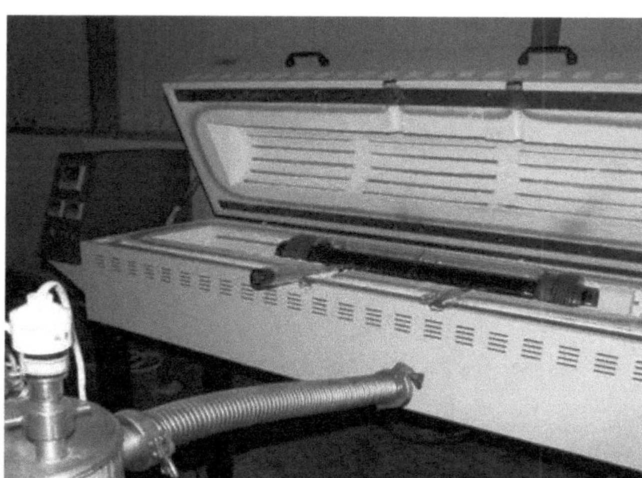

FIGURE 5.21 Parabolic trough receiver tube heating system.

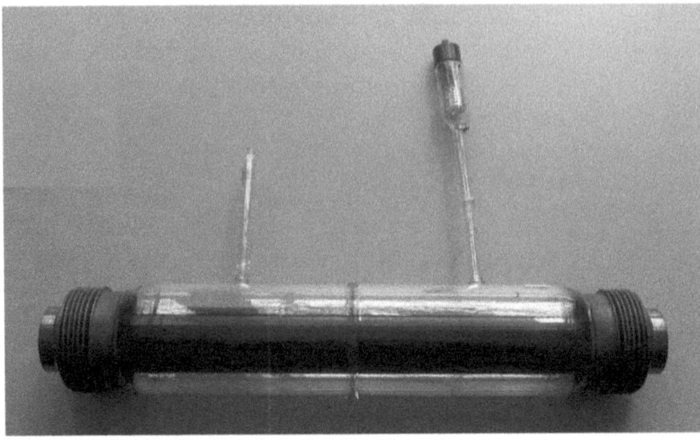

FIGURE 5.22 Sample tube of full-tube outgassing performance test.

5.7.4.1 *Outgassing Rate*

Using the outgassing rate calculation of formula (5.13), variation curves for the outgassing rate, temperature, and time corresponding to various experimental sample tubes (unit areas) can be obtained as shown in Figs. 5.23–5.33 (temperature indicated in the chart is subject to an experimental set value) [39].

According to test results from Figs. 5.23 to 5.33, the outgassing rate of the sample tube rises sharply along with temperature increases. When the temperature is constant, the outgassing rate presents an exponential decay; but when temperature rises, the outgassing rate increases quickly. It is difficult to explain these phenomena with the DLM and RLM.

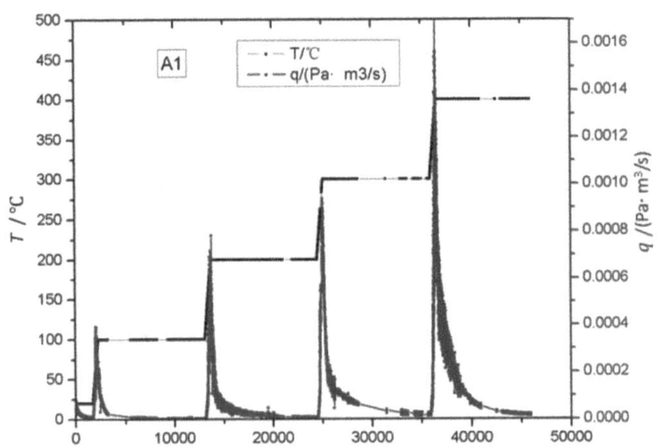

FIGURE 5.23 Outgassing rate variation curve for sample tube A1.

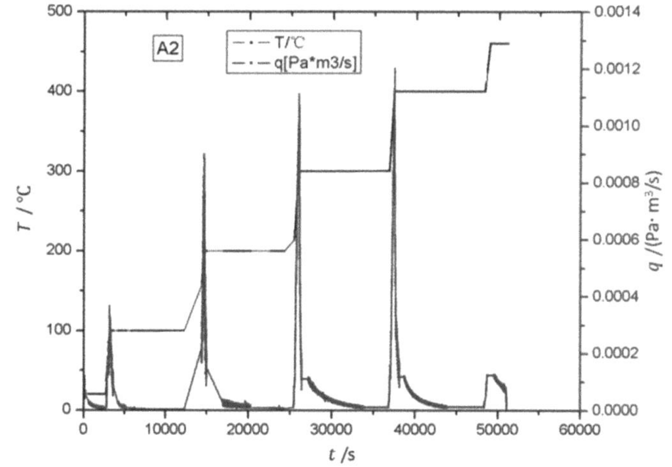

FIGURE 5.24 Outgassing rate variation curve for sample tube A2.

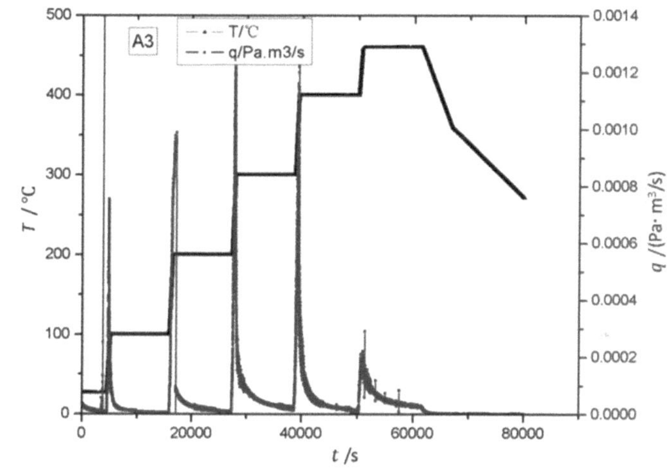

FIGURE 5.25 Outgassing rate variation curve for sample tube A3.

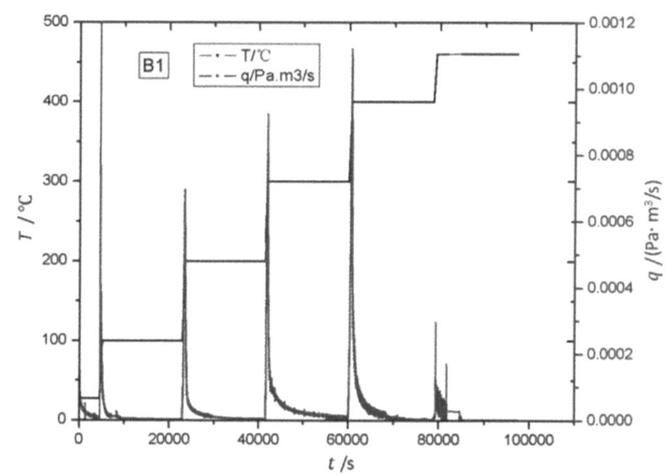

FIGURE 5.26 Outgassing rate variation curve for sample tube B1.

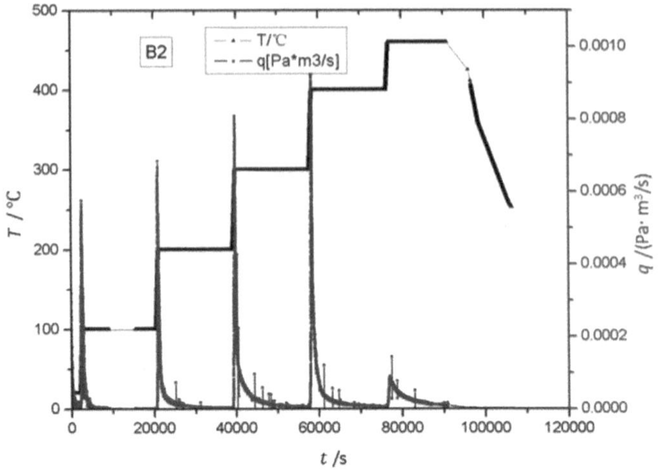

FIGURE 5.27 Outgassing rate variation curve for sample tube B2.

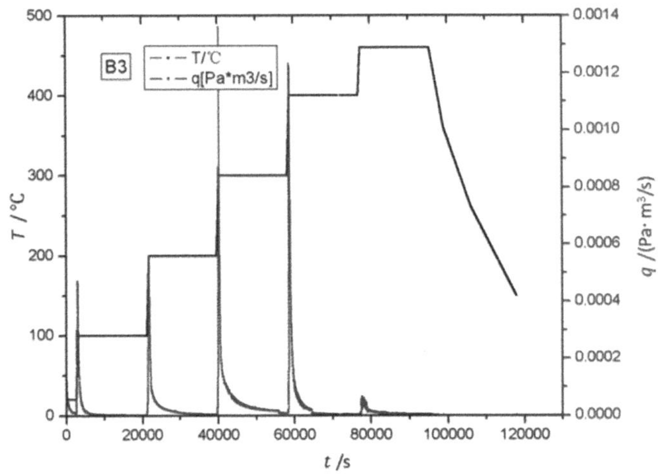

FIGURE 5.28 Outgassing rate variation curve for sample tube B3.

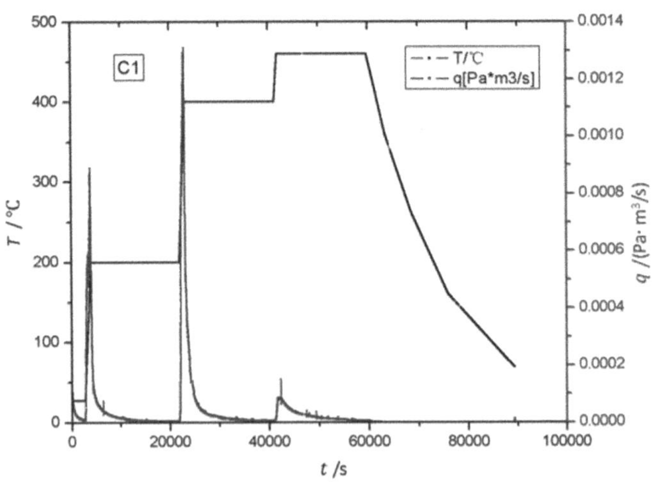

FIGURE 5.29 Outgassing rate variation curve for sample tube C1.

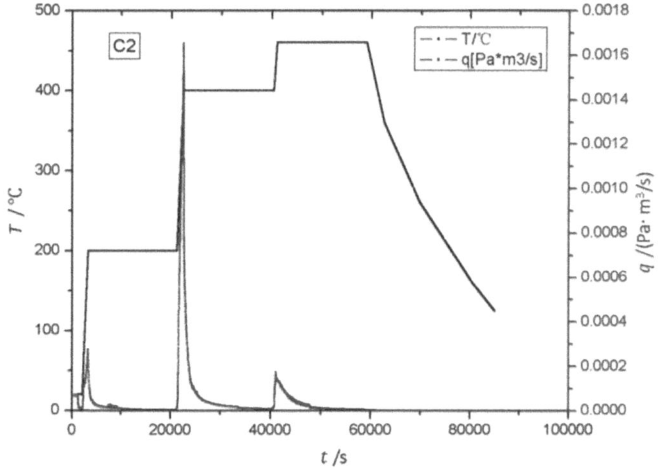

FIGURE 5.30 Outgassing rate variation curve for sample tube C2.

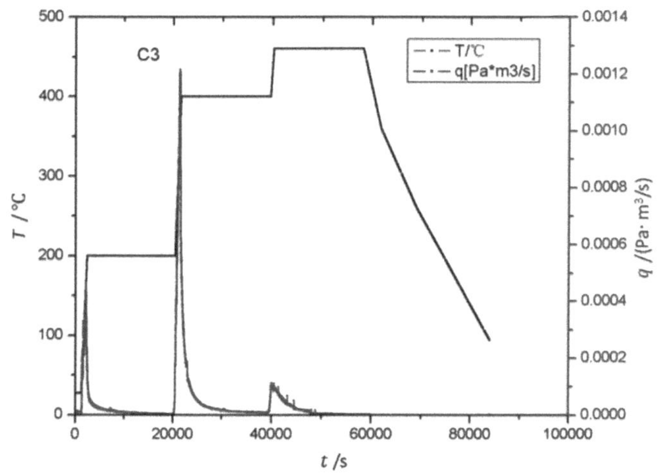

FIGURE 5.31 Outgassing rate variation curve for sample tube C3.

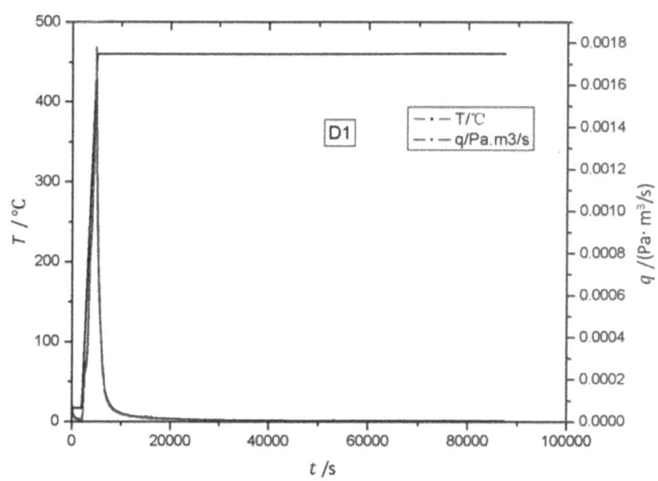

FIGURE 5.32 Outgassing rate variation curve for sample tube D1.

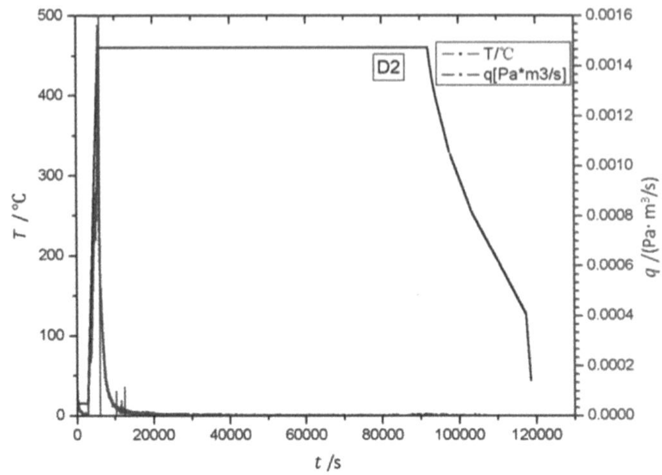

FIGURE 5.33 Outgassing rate variation curve for sample tube D2.

This is because when gas atoms stay at one energy level inside a material with small adhesive force at a given temperature, the material outgassing rate starts high but eventually reduces to the measurement limit at which most gas atoms inside the material are eliminated. By increasing the temperature, the outgassing rate increases, but only a small portion of the residual gases are released, and the respective outgassing rate drops more quickly. It has been discovered through experiments that a large amount of gas is still being outgassed after increases in temperature. In the case where gas atoms have greater adhesive force inside the material, the transition and diffusion rates of gas atoms are comparatively low, although the outgassing rate will still rise along with increases in temperature, but when the temperature is constant, the outgassing rate may approach the constant because a large number of gas atoms inside the material have yet to be outgassed. It is therefore impossible to explain such experimental phenomena with a single energy level theory.

These experimental phenomena can be explained, however, with the application of a multienergy level theory. Gases are subject to different energy levels inside a parabolic trough receiver tube. Under low temperatures, only gas atoms subject to the lower energy level can be released from interior materials through diffusion and dissolution, and the outgassing rate will also gradually decrease over time; with more activation energy, transition and diffusion of gas atoms subject to higher energy levels will not occur under low temperatures; when the temperature increases, after reaching the activation energy necessary for gas transition, gas atoms subject to higher energy levels from interior materials will be recombined into gas molecules through diffusion and dissolution and released to the vacuum environment, which increases the outgassing rate of receiver tube. Therefore, long-time baking emission at low temperatures cannot effectively reduce the material outgassing rate at high temperatures.

According to experimental results, the outgassing rate of a parabolic trough receiver tube at a certain temperature is subject to exponential decay, and from this the article proposes an outgassing rate formula for a parabolic trough receiver tube at constant temperature as shown in Eq. (5.14):

$$q(t) = q_0 + q_1 \exp(-\tau_1 t) + q_2 \exp(-\tau_2 t) \tag{5.14}$$

In which, $q_1 = \dfrac{4C_1D_1}{d_1}$, $q_2 = \dfrac{4C_2D_2}{d_2}$, $\tau_1 = \dfrac{\pi^2 D_1}{d_1^2}$, $\tau_2 = \dfrac{\pi^2 D_2}{d_2^2}$

in which q_0 refers to the outgassing rate constant of receiver tube; D_1 and D_2 respectively refer to gas diffusion coefficients of gas in glass and stainless-steel tubes; d_1 and d_2 respectively refer to thicknesses of glass and stainless-steel tubes; and C_1 and C_2 respectively refer to initial concentrations of gas in glass and stainless-steel tubes.

By applying Eq. (5.14), fitting analysis for the unit-length outgassing rates of the sample tube at various temperatures is conducted, and the fitting results for the outgassing rates of sample tube B3 are shown in Figs. 5.34–5.38 [39].

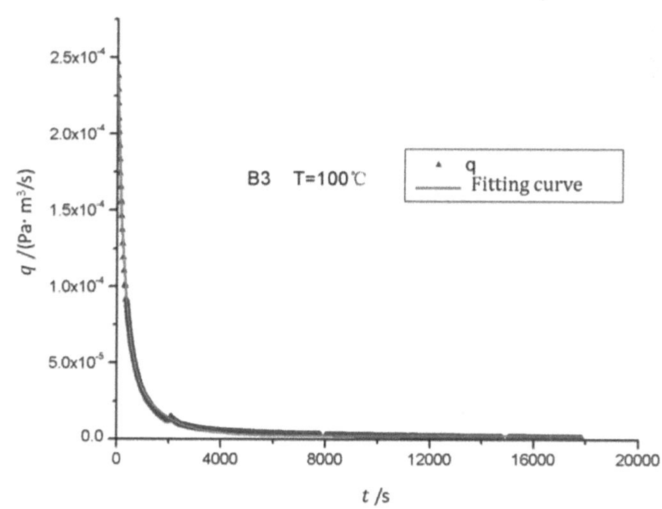

FIGURE 5.34 Outgassing rate fitting curve of sample tube B3 at 100°C.

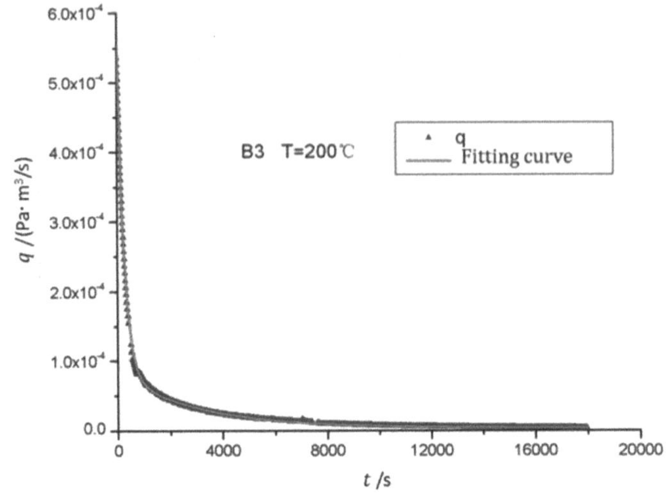

FIGURE 5.35 Outgassing rate fitting curve of sample tube B3 at 200°C.

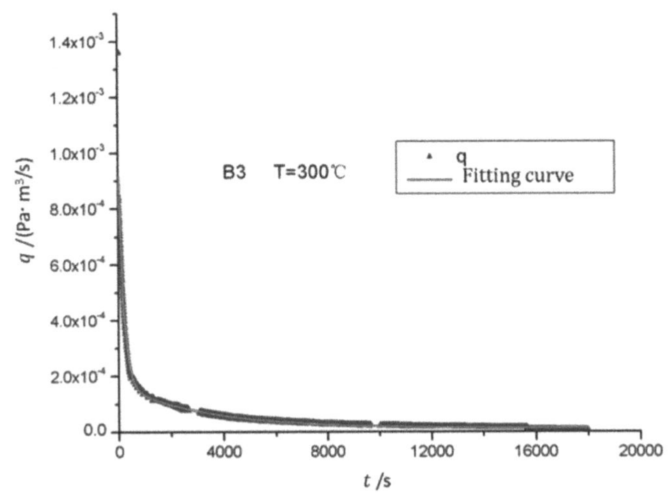

FIGURE 5.36 Outgassing rate fitting curve of sample tube B3 at 300°C.

5.7.4.2 Total Outgassing Quantity

According to outgassing rate calculation Eq. (5.14), the total outgassing quantity of each sample tube during heating can be calculated.

Heating Group D sample tubes from room temperature to 460°C takes only 50 min, but the outgassing quantity accounts for more than 55% of total outgassing quantity, which is equivalent to the outgassing quantity from Group B sample tubes heated for 10 h from room temperature to

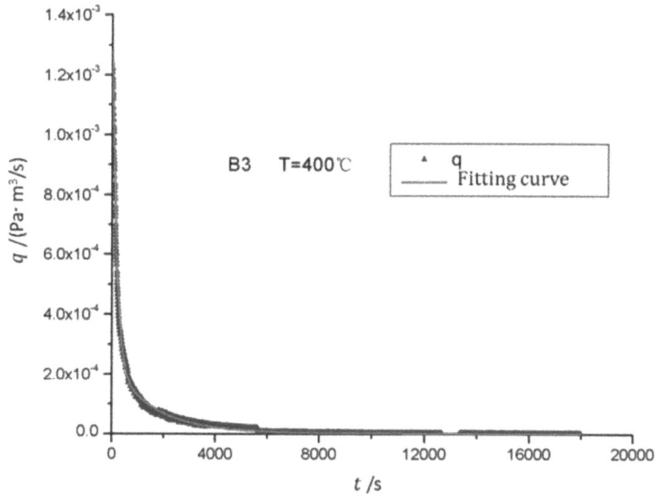

FIGURE 5.37 Outgassing rate fitting curve of sample tube B3 at 400°C.

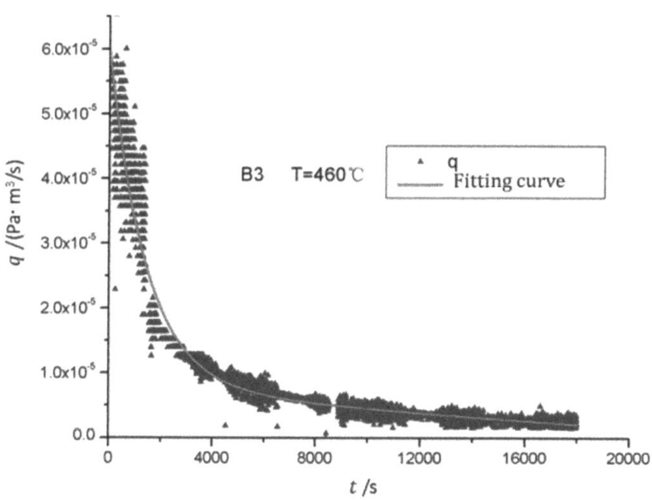

FIGURE 5.38 Outgassing rate fitting curve of sample tube B3 at 460°C.

300°C. This fully demonstrates that temperature increases can raise the material outgassing rate; the higher of the temperature, the greater the outgassing rate.

Fig. 5.39 compares the total outgassing quantities of various experimental sample tubes.

According to Fig. 5.39, although the heating parameters of sample tubes from different groups are inconsistent with each other, the total

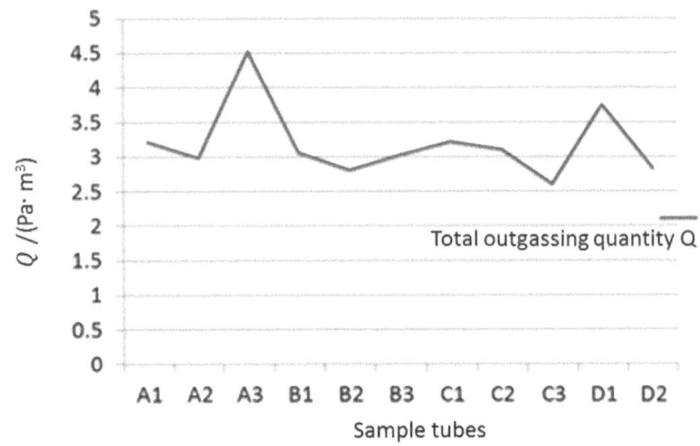

FIGURE 5.39 Total outgassing quantities for various experimental sample tubes.

outgassing quantity is around 3 Pa m^3. Based on this, the total outgassing quantity of the 4-m long parabolic trough receiver tube at 460°C while being insulated for 24 h approximates 20 Pa m^3 (20°C).

According to the stainless-steel element content measurement report prepared by the National Analysis Center for Iron & Steel for SS304 stainless steel, the mass contents of C, N, and H are 0.04%, 0.0039%, 0.074%, and 0.00028% respectively. In a stainless-steel absorber tube with a length of 4.06 m and wall thickness of 3 mm, mass is about 20 kg, and the masses of C, O, N, and H are 8 g (0.67 mol), 0.78 g (0.049 mol), 14.8 g (1.06 mol), and 0.056 g (0.056 mol) respectively. While diffusing outward, C can bond with O on the metal surface and create CO or CO_2 to be released into the vacuum chamber. According to the ideal gas state formula, N and H in stainless steel can be completely diffused and dissolved into vacuum before transforming into N_2 and H_2, with quantities of 1291 Pa m^3 (20°C) and 68 Pa m^3 (20°C) respectively. Thus it can be seen that the gas quantity contained in stainless steel far exceeds the outgassing quantity while being insulated for 24 h at 460°C. Thus in the heating process, outgassing of the receiver tube mainly comes from absorbed gases on the material surface and partial gases dissolved on the material surface layer; gases inside receiver tube material have not been completely released. In case where further reduction in the receiver tube's outgassing rate is intended, the baking duration must be extended or the baking temperature must be raised.

To sum up, through outgassing performance experiments on parabolic trough receiver tubes, it can be concluded that although experimental parameters of the four groups differed from one another, the total outgassing quantity during experiments was basically the same, demonstrating that these four groups of experiments exhibit the same degassing effects on the receiver tube. It also can be seen that gas atoms inside the receiver tube stay at diversified energy level states. Under low temperatures, degassing cannot reduce the outgassing rate as with a high-temperature state; in addition, at a constant temperature, the outgassing rate is subject to exponential decay. In the initial stage, the outgassing rate drops quickly, and thus when baking and degassing a parabolic trough receiver tube, the baking temperature must be larger than the receiver tube's applied temperature; the higher of the temperature, the better the degassing efficiency.

Material outgassing performance currently corresponds to DLM, RLM, and absorption-DLM models, multienergy level theory, and many other models and theories. Yet due to the complexity of the outgassing process, no unified theoretical model has been developed that can explain all outgassing rules, which can only be studied and analyzed according to specific circumstances on the basis of experiments utilizing theoretical models.

According to experimental results, gas atoms inside receiver tube material are subject to diversified-energy-level states. Under low temperatures, only gases subject to low energy levels can be outgassed, whereas gases in high-energy-level states can only diffuse from interior materials to the surface at higher temperatures for desorption; and at a constant temperature, the material outgassing rate is subject to exponential decay. In the initial stage it decays quickly. The total outgassing quantity of a 4-m long parabolic trough receiver tube while insulated for 24 h at 460°C is about 20 Pa m^3 (20°C). Thus when preparing the baking emission scheme for a parabolic trough receiver tube, the baking temperature shall be greater than the receiver tube's applied temperature; in order to improve emission efficiency, efforts shall be made to strive to increase the baking temperature. In addition, degassing efficiency is highest within the first few hours of baking.

By testing the outgassing composition of coated stainless steel and comparing it with that of pure stainless steel, the discovery has been made that a major outgassing component for both materials is H_2, with the content for both being above 90%. Influenced by coating film, the outgassing components of coated stainless steel also include N_2 and Ar, and the absorption effects of the getter on these two gases are insignificant, which are unfavorable results for vacuum degree maintenance inside the absorbing tube.

5.7.5 Parabolic Trough Receiver Tube Gas Permeation Performance

Outgassing of the wall material is also subject to a key factor that influences the vacuum reliability and service life of a parabolic trough receiver tube. Glass facilitates outgassing, and stainless steel facilitates hydrogen permeation; thus, helium and hydrogen are both main components to be outgassed from a parabolic trough receiver tube. Because the partial pressure of helium in the atmosphere is quite low at merely 5.3×10^{-1} Pa, even with a balance between helium and air in the vacuum jacket of a receiver tube, the influence on the heat loss of the tube is insignificant, and the thermal conductivity of helium is weaker than that of hydrogen, so the respective influence on receiver tube heat loss is quite small. Therefore, to study the gas permeability of a parabolic trough receiver tube is mainly to study the hydrogen-permeation performance of stainless steel. According to research, the hydrogen produced by organic synthetic oil in a parabolic trough solar power plant through high-temperature decomposition may permeate into the parabolic trough receiver tube via a stainless-steel absorber tube, for which in-depth analysis has yet to be carried out. Thus on the basis of tests about the hydrogen-permeation performance of coated stainless steel, by analyzing hydrogen generation and permeation processes in a parabolic trough solar power plant, a hydrogen-permeation model for the parabolic trough receiver tube has been established and used to carry out analyses and studies about various factors that influence the hydrogen-permeation process.

5.7.5.1 Gas Permeation Theoretical Research

Gas permeation is a complex physical and chemical process that includes absorption, disassociation, diffusion, recombination, and desorption of gas molecules, which is similar to the material outgassing process. Thus there has been no unified theory of gas permeation. Most scholars believe that diffusion is the key process that determines the outgassing rate. Like the DLM in outgassing theory, such a diffusion process follows Fick's law. Outgassing of internal material gas has not been considered in the outgassing process, and there is a gas concentration difference between the two ends of the metal material. When utilizing the diffusion theory to calculate the outgassing rate, because a vacuum receiver tube's stainless-steel absorber tube has a thickness that is far less than its internal diameter, it can be deemed an approximately one-dimensional flat plate. When researching outgassing, in the case that the internal gases of the material have already been completely removed and one side vacuumized, the outgassing rate of gas passing through metal material is:

$$J(x=l,t) = -D\left(\frac{\partial C}{\partial x}\right)_{x=d} = \frac{DSP_0^{0.5}}{l}\left[1+2\sum_{n=1}^{\infty}(-1)^n \exp\left(-\frac{Dn^2\pi^2 t}{l^2}\right)\right] \tag{5.15}$$

Once the outgassing process has reached stability, the rate becomes:

$$J(x = l,\ t \to \infty) = \frac{DSP_0^{0.5}}{l} = \frac{\Phi P_0^{0.5}}{l} \tag{5.16}$$

in which l refers to the thickness of vacuum absorber thickness; S refers to the solubility of gas inside the material; P_0 refers to the gas pressure of metal material on the high-pressure side; Φn, which is known as permeability coefficient, is a basic parameter for describing the permeation performance of gas-solid matched group; and under circumstances of molecular permeation, Φ has the same unit with the diffusion coefficient D, namely m^2/s.

5.7.5.2 Hydrogen Permeation Performance Test for Selective Absorbing Film

In order to research the hydrogen-permeation process of the parabolic trough receiver tube, the hydrogen-permeation coefficient of the stainless-steel absorber tube must first be acquired. Many scholars have already carried out experimental studies on various hydrogen-permeation coefficients for pure stainless steel, but the hydrogen-permeation performance of selective-absorbing film on the exterior surface of a stainless-steel tube is still unknown. Therefore, this experiment aims to acquire a hydrogen-permeation coefficient for a stainless-steel absorber tube by testing the hydrogen-permeation performance of the selective-absorbing film in order to offer major parameter data for studies on hydrogen permeation of parabolic trough receiver tubes.

The hydrogen diffusion coefficient can be measured by applying the electrochemical or vapor permeation method. The electrochemical method uses double electrolytic cells, a constant voltage (current) source, a reference electrode, and other facilities, the principle of which is to generate a high concentration of hydrogen ions on one side of the thin section sample through the electrochemical method, and to measure the seepage current on the other side. The vapor permeation testing method fills pure hydrogen at a certain pressure into one side of the sample and measures the seepage or seepage flow of hydrogen on the other side; once the seepage flow of hydrogen has stabilized, the permeability coefficient of hydrogen can be obtained through calculation.

Because the vapor permeation test method can be used within a broad temperature range and the working temperature of the parabolic trough receiver tube is high, thus, this test uses the vapor permeation method for testing the hydrogen-permeation performance of coated stainless steel.

Based on different measuring modes, measurement of the hydrogen-permeability coefficient using the vapor permeation method can be divided into dynamic and static methods. For the dynamic method, the seepage or inspection end of hydrogen is subject to a dynamic vacuum environment, and it is inspected using a quadrupole mass spectrometer while keeping records on the hydrogen-permeation flow–time curve. With the static method, the hydrogen seepage end is connected to the closed collection chamber; after beginning the test, due to the permeation and accumulation of hydrogen, the pressure of collection chamber gradually increases, and a pressure sensor is used for keeping records on the pressure–time curve of hydrogen inside the collection chamber. Therefore, the static method is also called the pressure increase method.

In this experiment, the dynamic method is applied for the test. Fig. 5.40 shows the test schematic diagram, and Fig. 5.41 displays a testing device.

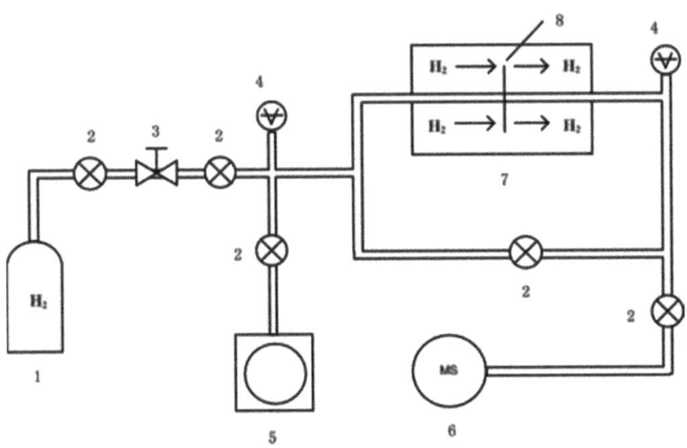

FIGURE 5.40 Principle of hydrogen permeation test.
1-hydrogen tank; 2-valve; 3-pressure control valve; 4-manometer; 5-vacuum pump; 6-quadrupole mass spectrometer; 7-heating furnace; 8-sample plate.

FIGURE 5.41 Devices for hydrogen permeation test. *Provided by the Institute of Electrical Engineering, Chinese Academy of Sciences.*

5.7.5.2.1 Data Processing and Experimental Results

Because the permeation area of the sample plate is far greater than its thickness, the hydrogen-permeation process can be deemed a one-dimensional hydrogen-permeation process. According to Fick's law and Sievert's law, the relationship between steady-state hydrogen-permeation flux and pressure is:

$$J_\infty = \frac{A\Phi\left(P_1^{0.5} - P_2^{0.5}\right)}{l} \tag{5.17}$$

in which J_∞ refers to the steady-state permeation flux of hydrogen, mol/s; A refers to the permeation area of hydrogen, m^2; P_1 and P_2 refer to gas pressures on both sides of the sample plate to be tested, MPa; Φ refers to the permeability coefficient (namely hydrogen-permeation coefficient) of hydrogen, mol/(m·s·MPa$^{0.5}$); and l refers to the thickness of the sample plate to be tested, m.

Hydrogen permeation is a thermodynamic process that is in line with the Arrhenius expression together with temperature:

$$\Phi = \Phi_0 \exp\left(-\frac{E_p}{RT}\right) \tag{5.18}$$

According to test results based on the fitting of Eq. (5.18), the permeation constant and permeation activation energy can be obtained respectively as $\Phi_0 = 5.9 \times 10^{-6}$ mol/(m s MPa$^{0.5}$) and $E_p = 57.5$ kJ/mol, namely:

$$\Phi = 5.9 \times 10^{-6} \exp\left(-\frac{57500}{RT}\right) \tag{5.19}$$

According to Eq. (5.19), variation of the permeability coefficient of hydrogen along with temperature (T) is shown in Fig. 5.42.

5.7.6 Parabolic Trough Receiver Tube Hydrogen Permeation Volume Prediction

In a parabolic trough solar power plant, hydrogen-permeation is subject to a complex process that includes hydrogen generation, stainless-steel absorber tube permeation, corrugated tube permeation, and hydrogen absorption of the getter.

5.7.6.1 Hydrogen Generation Rate in Heat-Transfer Oil

According to the research, hydrogen generation in a parabolic trough receiver tube is mainly caused by decomposition of organic synthetic oil at high temperatures. Such organic synthetic oil consists of 73.5% $C_{12}H_{10}O$ and 26.5% $C_{12}H_{10}$. Here, a certain synthetic oil product is used as an

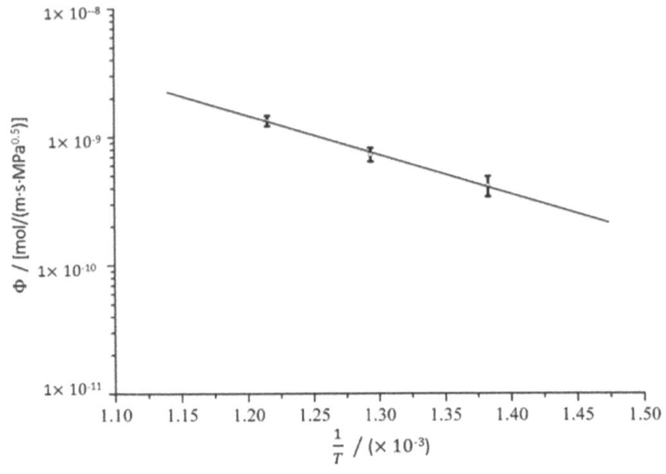

FIGURE 5.42 Variation of the hydrogen permeation coefficient along with temperature (T).

example for analysis. The condensation point of the synthetic oil is 12°C, and the decomposition rate rises along with increases in temperature; however, below 400°C, the decomposition rate is very low, with the thermal decomposition rate shown in Fig. 5.43.

Thermal decomposition expression of a certain type of organic synthetic oil can be obtained:

$$\ln k_1 = 54.4 - \frac{40208}{T} \tag{5.20}$$

in which k_f refers to the thermal decomposition rate of heat-transfer oil, %/h; and T refers to the temperature, K.

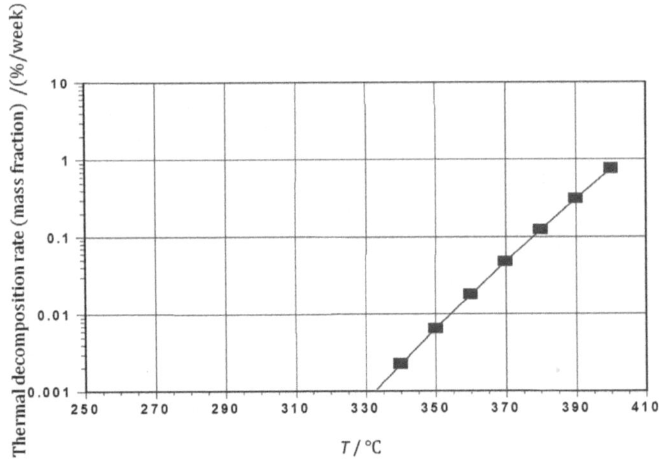

FIGURE 5.43 Thermal decomposition rate of a certain type of organic heat-transfer oil.

Primary reaction

Ph-O-Ph $\longrightarrow$ Ph* + Ph*

Ph-Ph $\longrightarrow$ 2 Ph*

Intermediate process

Ph* + Ph-O-Ph $\longrightarrow$ PhH + Ph-O-$C_6H_4^*$

Ph* + Ph-Ph $\longrightarrow$ PhH + Ph-$C_6H_4^*$

Ph* + Ph-O-Ph $\longrightarrow$ Ph-C_6H_4-O-Ph + H*

PhO* + Ph-O-Ph $\longrightarrow$ PhOH + Ph-O-$C_6H_4^*$

H* + Ph-O-Ph $\longrightarrow$ PhH + PhO*

H* + Ph-Ph $\longrightarrow$ PhH + Ph*

Final product

2 Ph-O-$C_6H_4^*$ $\longrightarrow$ Ph-O-C_6H_4-C_6H_4-O-Ph

2 $C_6H_5^*$ $\longrightarrow$ Ph-Ph

PH* + H* $\longrightarrow$ PhH

2 H* $\longrightarrow$ H_2

FIGURE 5.44 Organic synthetic oil decomposition chemical formulas. Ph—O—Ph = DPO (diphenyl ether); Ph—Ph = diphenyl; Ph* = $C_6H_5^*$ = phenyl.

According to the synthetic oil decomposition chemical formulas (refer to Fig. 5.44) proposed by Arnold and Moens based on the law of conservation of mass and considering various chemical formulas, it can be concluded that about 16 mol synthetic oil can be decomposed into 1 mol H_2. Based on this, the hydrogen-generation rate can be obtained:

$$k_H = \frac{n_f k_f}{16} \tag{5.21}$$

$$\ln k_H = \ln n_f + 47 - \frac{40208}{T} \tag{5.22}$$

in which k_H refers to the hydrogen-generation rate, mol/mol; and n_1 refers to the amount of heat-transfer oil, mol.

For a mean molecular weight of synthetic oil of 166, the amount of synthetic oil can be obtained through the following equation:

$$n_f = \frac{V_f \rho_f}{166} \tag{5.23}$$

in which V_f refers to the volume of heat-transfer oil, m^3; and ρ_f refers to the density of heat-transfer oil, kg/m^3.

5.7.6.2 Hydrogen Dissolution

When the decomposition rate of synthetic oil is low, the quantity of hydrogen generated during the initial stage is insignificant, and this part of hydrogen will be first dissolved in heat-transfer oil. Currently, there are no data on the solubility of hydrogen in such an organic heat-transfer oil, and benzoyl and organic synthetic oil are both subject to aromatic hydrocarbons; therefore, data on the solubility of hydrogen in benzoyl are used in this section instead.

In a parabolic trough receiver tube, the internal pressure of the tube's vacuum jacket must exceed the vapor pressure of heat-transfer oil, and the pressure difference of synthetic oil within the receiver tube circulation line of the collector unit approximates 0.9 MPa. Based on the physical performance parameters of heat-transfer oil, the vapor pressure at 390°C approximates 1 MPa. Therefore, the inlet pressure of synthetic oil inside the receiver tube circulation line is set at 2 MPa (290°C) in this section, and the outlet end pressure is set at 1.1 MPa (390°C). Based on data for the solubility of hydrogen inside benzoyl, the variation of hydrogen solubility data within the respective temperature and pressure variation ranges is insignificant. Therefore, the solubility data of hydrogen are deemed a constant in this section, and 0.00325 mol/mL is selected as the value for solubility of hydrogen inside heat-transfer oil.

5.7.6.3 Hydrogen Permeation Area

Once saturated, hydrogen dissolving in synthetic oil will escape and attach on the receiver tube's interior wall. In order to explore the rules and respective pattern of hydrogen escaping from the interior wall and forming on the metal wall surface, an electrolytic water experiment is conducted. Through observation, it is discovered that during the electrolytic water experiment, gas generated through electrolysis attaches to the interior wall of the electric tank in the form of hemispherical bubbles, from which it can be inferred that escaped saturated hydrogen attaches to the receiver tube's interior wall as hemispherical bubbles (refer to Fig. 5.45).

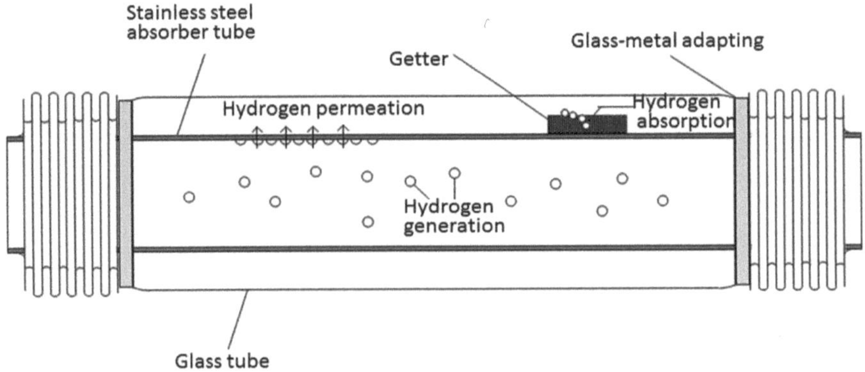

FIGURE 5.45 Schematic diagram of hydrogen permeation.

In a parabolic trough receiver tube, hydrogen flows inside a closed circulation line along with heat-transfer oil. When hydrogen flows from the receiver tube, hydrogen is flowing into another receiver tube at the same time. Thus, it is assumed in this section that hydrogen does not flow inside a receiver tube with heat-transfer oil, all of which will attach to the receiver tube's interior wall as hemispherical bubbles. According to mechanical balance, the pressure of hydrogen inside bubbles shall be equivalent to the pressure of heat-transfer oil. Therefore, the permeation area of hydrogen passing through a stainless-steel absorber tube is equivalent to the contact area of all bubbles and the interior wall of the receiver tube. The quantity of bubbles attaching to the interior wall of the receiver tube is:

$$N_{\mathrm{bub}} = \frac{n_H}{n_{\mathrm{bub}-H}} = \frac{3n_H RT}{2\pi P_{\mathrm{bub}} r_{\mathrm{bub}}^3} \tag{5.24}$$

$$n_H = k_H - 3600 \times J_{f-\mathrm{ann}} \tag{5.25}$$

in which N_{bub} refers to the quantity of hydrogen-absorbing bubbles; n_H refers to the amount of hydrogen inside the pipeline, mol; $n_{\mathrm{bub}\text{-}H}$ refers to the amount of hydrogen inside bubbles, mol; P_{bub} refers to the pressure of hydrogen inside bubbles, Pa; r_{bub} refers to the radius of the bubble, m; R is an ideal gas constant; T refers to the temperature, K; and $J_{f\text{-ann}}$ refers to the permeation rate of hydrogen from the synthetic oil pipeline to the receiver tube, mol/s.

Thus the area of hydrogen-permeation through a stainless-steel absorber tube is:

$$A_{f-\mathrm{ann}} = N_{\mathrm{bub}} \times \pi r_{\mathrm{bub}}^2 \tag{5.26}$$

5.7.6.4 Hydrogen Permeation in a Stainless-Steel Absorber Tube

When bubbles attach to the interior wall of the absorber tube, hydrogen permeates the vacuum jacket of the receiver tube through the stainless-steel absorber tube, and the permeation rate is:

$$J_{f-\mathrm{ann}} = A_{f-\mathrm{ann}} \Phi_{\mathrm{ab}} \left(P_{\mathrm{bub}}^{0.5} - P_{\mathrm{ann}}^{0.5}\right) l_{\mathrm{ab}} \tag{5.27}$$

in which $A_{f\text{-ann}}$ refers to the permeation area of hydrogen passing through the stainless-steel absorber tube, m^2; Φ_{ab} refers to the hydrogen-permeability coefficient of the absorber tube, $\mathrm{mol/(m\cdot s\cdot MPa^{0.5})}$; P_{bub} refers to the partial pressure of hydrogen inside the heat-transfer oil, MPa; P_{ann} refers to the partial pressure of hydrogen inside the jacket of tube, and MPa; l_{ab} refers to the steel wall thickness of absorber tube, m.

A stainless-steel absorber tube of a parabolic trough receiver tube can be deemed a pure stainless-steel layer and a stainless-steel layer coated with selective absorbing film (such as a hydrogen-permeation testing sample plate). The expression for the permeability coefficient of multi-layered materials is:

$$\frac{1}{\Phi} = \frac{1}{\Phi_1} + \frac{1}{\Phi_2} + \cdots + \frac{1}{\Phi_n} \tag{5.28}$$

The hydrogen-permeability coefficient of pure stainless steel SS304 is:

$$\Phi_s = 2.85 \times 10^{-4} \exp\left(-\frac{62430}{RT}\right) \tag{5.29}$$

Once the getter within the parabolic trough receiver tube is activated, hydrogen permeating into the receiver tube is absorbed by the getter. Thus the partial pressure of hydrogen inside the receiver tube is $P_{\text{ann}} = 0$; when the getter inside the receiver tube is saturated, P_{ann} changes along with the increased permeation amount of hydrogen.

Within time interval τ, the amount of hydrogen permeating into the vacuum layer of receiver tube is:

$$Q = \int J_{f-\text{ann}} d\tau \tag{5.30}$$

5.7.6.5 Hydrogen Permeation in the Corrugated Tube

Once the getter inside the parabolic trough receiver tube is saturated, hydrogen permeating into the receiver tube will accumulate there, and partial hydrogen escapes to the atmosphere through the corrugated tube. Because the partial pressure of hydrogen in the atmosphere is quite small, the partial pressure inside the atmosphere is nearly zero in this section. The permeation rate of hydrogen passing through the corrugated tube is:

$$J_{\text{ann-air}} = A_{\text{bel}} \Phi_{\text{bel}} P_{\text{ann}}^{0.5} / l_{\text{bel}} \tag{5.31}$$

in which $J_{\text{ann-air}}$ refers to the rate of hydrogen escaping from the corrugated tube into the atmosphere, mol/s; A_{bel} refers to the permeation area of hydrogen passing through the corrugated tube, m^2; Φ_{bel} refers to the hydrogen-permeation coefficient of the corrugated tube, $\text{mol}/(\text{m} \cdot \text{s} \cdot \text{MPa}^{0.5})$; and l_{bel} refers to the thickness of the corrugated tube, m.

The internal diameter of the receiver tube's corrugated tube used in this section is about 80 mm, and the external diameter is 120 mm; each corrugated tube has five waves, and the wall thickness of the corrugated tube is 0.2 mm. Thus, total permeation area of corrugated tubes of the parabolic trough receiver tube is 0.1256 m^2. In this section, the hydrogen-permeation coefficient of the corrugated tube is the same as that of pure stainless steel SS304, namely $\Phi_{\text{bel}} = \Phi_s$.

Once the getter inside the receiver tube is saturated, the amount of accumulated hydrogen inside the receiver tube is:

$$Q = \int (J_{f-\mathrm{ann}} - J_{\mathrm{ann-air}})dt \tag{5.32}$$

5.7.6.6 Model Calculation

A parabolic trough solar power plant is assumed to operate 2000 h per year. Because the temperature of synthetic oil is very low at night, the hydrogen-generation rate and permeation rate are also quite low, so power plant operation at night is not considered in this section. Calculations of this model were conducted using MATLAB, and the logic relationship for the calculation is shown in Fig. 5.46.

5.7.6.7 Results and Analysis

1. In the case of an unsaturated getter. Once the getter inside the parabolic trough receiver tube is activated and not yet saturated, hydrogen permeating the receiver tube will be absorbed by the

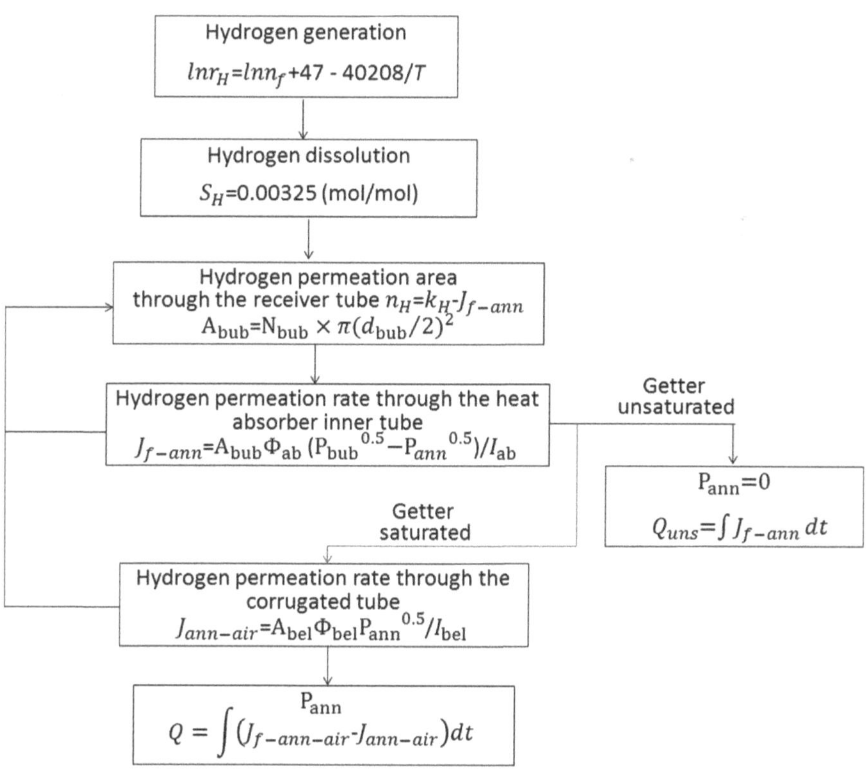

FIGURE 5.46 Schematic diagram of the calculation logic relationship of the hydrogen-permeation model.

getter, and the partial pressure of hydrogen inside the receiver tube becomes zero. Calculation results of the hydrogen-permeation model for a 25-year receiver tube service life, indicating the total hydrogen permeating the receiver tube, are shown in Fig. 5.47, showing that below 355°C, hydrogen permeation is zero; then, along with increases in the temperature of heat-transfer oil, hydrogen permeation reaches 4.4 mol at 390°C. This is because both the hydrogen generation and permeation rates are temperature-based functions that rise with increases in temperature increase. The decomposition rate of synthetic oil is subject to the Arrhenius equation; when the oil temperature is less than 355°C, hydrogen generated through decomposition completely dissolves into the heat-transfer oil with no hydrogen passing the stainless-steel absorber tube and the receiver tube; when the oil temperature is higher than 355°C, the hydrogen-generation rate increases, and when the hydrogen concentration exceeds the solubility of the heat-transfer oil, escaped hydrogen begins to attach to the receiver tube's interior wall as hemispherical bubbles and permeate the receiver tube; in addition, along with increases in temperature, permeation also increases. Therefore a parabolic trough receiver tube located at the low-temperature end of the circulation line requires fewer getters than one located at the high-temperature end, and receiver tubes located on different positions of the circulation line require different numbers of getters. In order to ensure the vacuum life of the parabolic trough receiver tube, the number of installed getters inside the receiver tube should be changed based on position.

The absorbing areas of hydrogen at different temperatures for a parabolic trough receiver with a 25-year service life are shown in

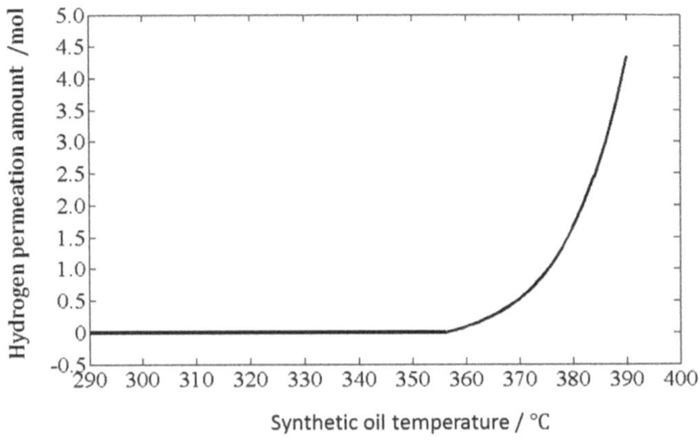

FIGURE 5.47 Variation curve of hydrogen permeation capacity along with temperature of receiver tube.

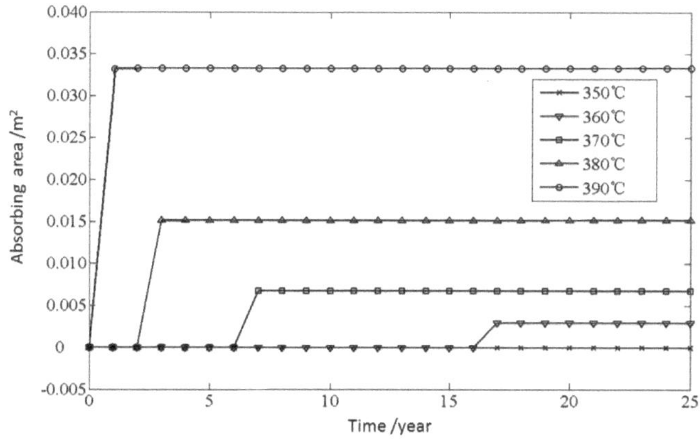

FIGURE 5.48 Variation curve of the hydrogen-absorbing area at different temperatures.

Fig. 5.48. The absorbing area of hydrogen at the initial state is zero and continuously increases thereafter until it reaches steady state within a short period. This is because at the beginning, all hydrogen is dissolved in the heat-transfer fluid; when saturated, the area of attached bubbles on the receiver tube's interior wall starts to grow so that the permeation area increases continuously, as does the permeation rate. When the rate of hydrogen permeating the receiver tube is equivalent to the hydrogen-generation rate, or the receiver tube's interior wall has been completely covered by hydrogen bubbles, the absorbing area of hydrogen no longer grows, and steady state is reached. Along with increase in temperature, the hydrogen-generation rate increases. The time required for hydrogen to dissolve into the synthetic oil and achieve a saturated state is also shortened, and more hydrogen is generated; more bubbles attach to the interior wall of receiver tube as well until the rate of hydrogen permeating the receiver tube is equivalent to the hydrogen-generation rate, and the absorbing area also reaches steady state without any increase.

In the hydrogen-permeation model, hydrogen is assumed to attach to the interior wall of absorber tube as hemispherical bubbles. Thus the radius of absorbing bubbles is a key factor that influences the hydrogen-permeation model's precision. Figs. 5.49 and 5.50 indicate variation of the hydrogen absorbing area at 290 and 390°C corresponding to different radii (r) of absorbing bubbles. Although the radii of absorbing bubbles differ from one another, all absorbing areas reach the same steady state value, and the smaller of the radius, the shorter the duration for reaching the steady state value, especially at high temperatures. Under low temperatures, the

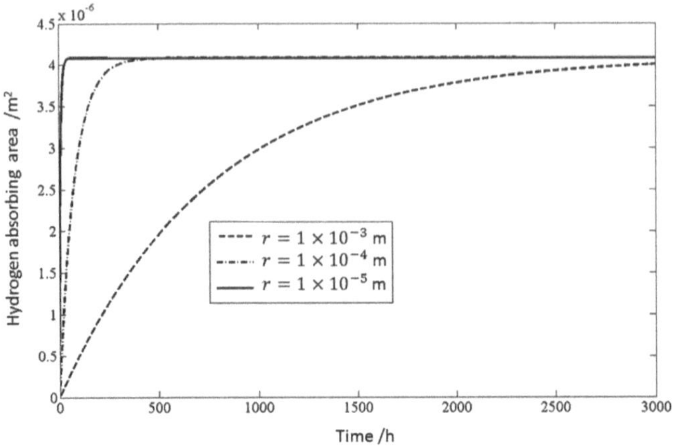

FIGURE 5.49 Variation curve of the hydrogen-absorbing area at 290°C corresponding to different bubble radii.

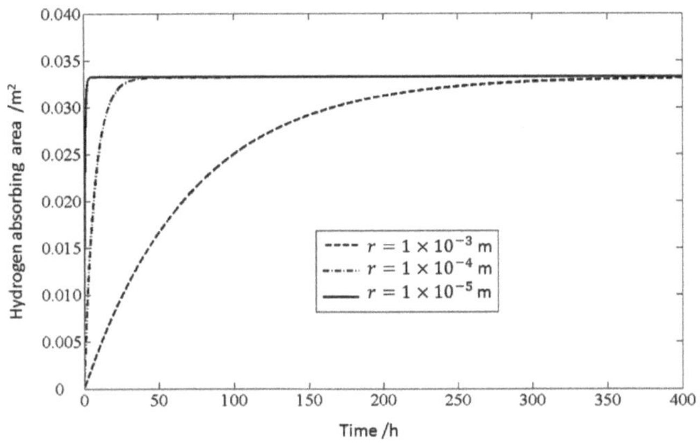

FIGURE 5.50 Variation curve of the hydrogen-absorbing area at 390°C corresponding to different bubble radii.

influences of bubble radii on the transient state process of permeation are great, but the transient state duration is far shorter than the 25-year service life of the parabolic trough receiver tube. Thus the influences of bubble radii on the precision of hydrogen-permeation model are insignificant, and the assumption of hemispherical bubbles being subject to the absorbing form is reasonable. However, according to practical engineering experiences, hydrogen bubbles inside synthetic oil are supposed to be small, and the recommended radius of hydrogen-absorbing bubbles in this calculation model is assumed to be 0.1 mm.

In a parabolic trough solar power plant, the decomposition products of synthetic oil include not only hydrogen, but also CO and hydrocarbon gases [46]. For example, when Dowtherm A is heated to 425°C and insulated for 120 h, this synthetic oil decomposes into a compound of partial gases in which hydrogen accounts for 44% of total gas volume [47]. Thus together with other decomposed gases, the hydrogen in synthetic oil will generate bubbles on the receiver tube's interior wall, with total pressure inside the bubbles equivalent to the pressure of synthetic oil inside the pipeline; the partial pressure of hydrogen in bubbles is merely one part of synthetic oil pressure. The partial pressure of hydrogen (P_H) in bubbles equal to 10%, 50%, and 100% of the pressure (P_f) of heat-transfer oil, the absorbing area and permeation rate variation curves at 390°C are shown in Figs. 5.51 and 5.52. When the partial pressure of hydrogen decreases, the absorbing area of hydrogen grows, but the permeation rate of hydrogen under steady state remains the same. This is because at the beginning, the hydrogen-generation rate is greater than the hydrogen-permeation rate, so the amount of bubbles attaching to the interior wall of receiver tube continuously increases and so does the permeation area of hydrogen until the hydrogen-permeation and hydrogen-generation rates become equal. The hydrogen-permeation rate is determined by pressure and the hydrogen-permeation as shown by Eq. (5.27). When the partial pressure of hydrogen inside bubbles decreases, the permeation area of hydrogen increases in order to make the hydrogen-permeation rate equal the hydrogen-generation rate. Therefore, hydrogen partial

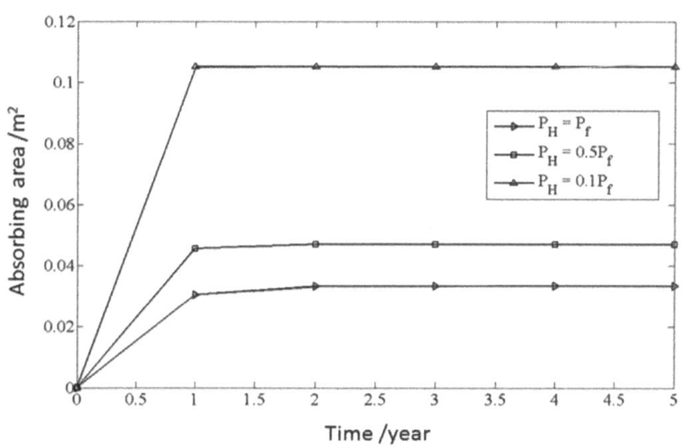

FIGURE 5.51 Variation curve of the hydrogen-absorbing area at 390°C corresponding to different hydrogen partial pressures.

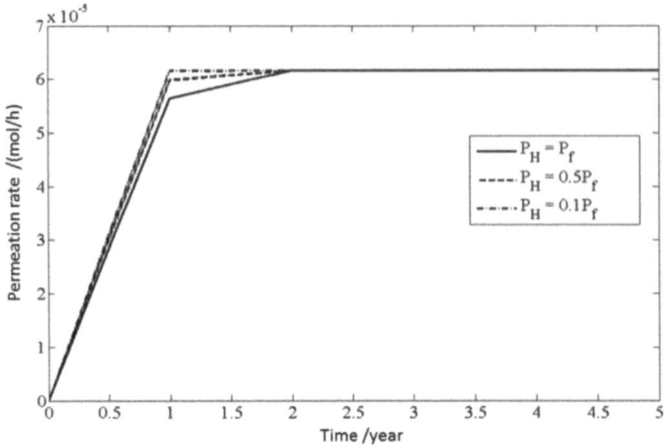

FIGURE 5.52 Variation curve of the hydrogen-permeation rate at 390°C corresponding to different hydrogen partial pressures.

pressure variation inside the circulation line only influences the hydrogen-permeation amount in the transient state process and does not influence the amount subject to steady state. The influence of hydrogen partial pressure variation on the total hydrogen-permeation amount inside the receiver tube is quite small.

Organic impurities in synthetic oil and the oxide layer on the surface of a stainless-steel absorber tube may catalyze decomposition of synthetic oil and accelerate hydrogen generation [46]. Fig. 5.53 indicates the variation curve for the hydrogen-absorbing area at

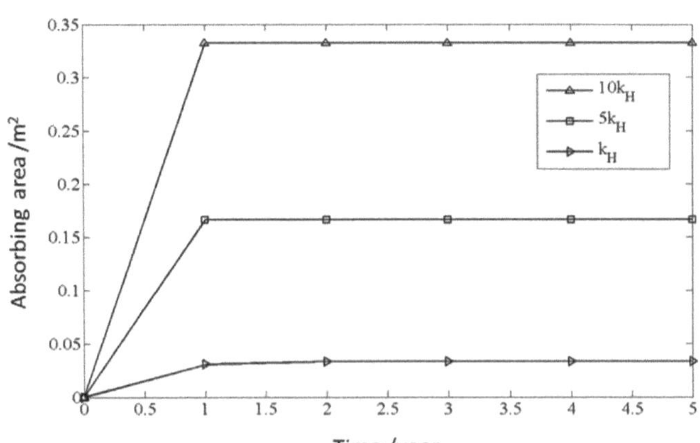

FIGURE 5.53 Variation curve of the hydrogen-absorbing area at 390°C corresponding to different hydrogen-generation rates.

390°C corresponding to different hydrogen-generation rates (k_H is the rate constant). In the case of increases in the hydrogen-generation rate, the absorbing area and permeation rate both increase until the hydrogen-permeation rate equals the hydrogen-generation rate or the interior surface of receiver tube has been completely covered by hydrogen bubbles.

A hydrogen barrier is often used to restrict hydrogen permeation and thereby prolong the vacuum life of the parabolic trough receiver tube. In this section, the hydrogen-permeation model is utilized to carry out studies on the influences of three hydrogen barriers having different coefficients for hydrogen permeation as shown in Figs. 5.54 and 5.55. When the hydrogen-permeability coefficient of the parabolic trough receiver tube is reduced by the hydrogen barrier to 10% or 5% of its original value, the hydrogen-absorbing area increases, while under steady state, the hydrogen-permeation rate remains the same. When the hydrogen-permeability coefficient is reduced to 1% of its original value, the hydrogen-absorbing area reaches the maximum level and is equivalent to the receiver tube's interior surface area, while under steady state, the hydrogen-permeation rate decreases to 25% of its original value. This is because the hydrogen-permeation rate always approaches or equals the hydrogen-generation rate; when the hydrogen-permeability coefficient is reduced, the hydrogen-absorbing area continuously increases until it reaches the maximum surface area. When the absorbing area is equivalent to the interior surface area of the receiver tube, the hydrogen-permeation rate decreases along with reductions in the hydrogen-permeation coefficient.

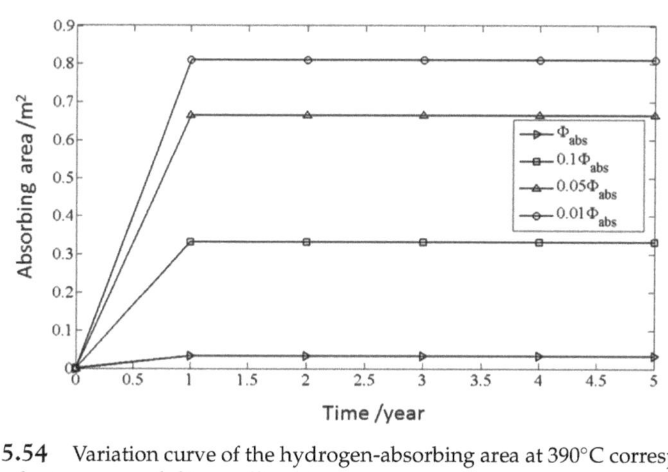

FIGURE 5.54 Variation curve of the hydrogen-absorbing area at 390°C corresponding to different hydrogen-permeability coefficients.

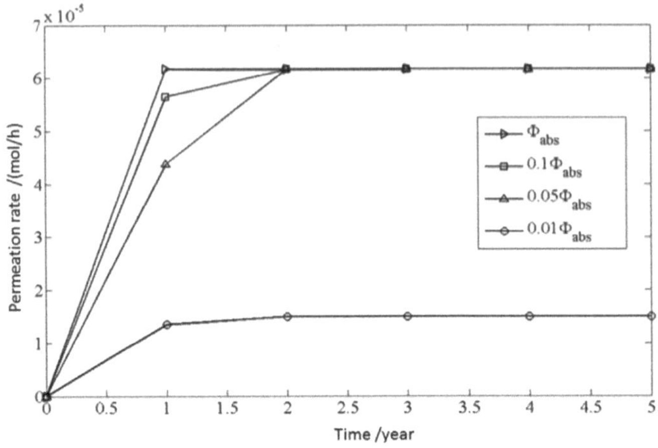

FIGURE 5.55 Variation curve of the hydrogen-permeation rate at 390°C corresponding to different hydrogen-permeability coefficients.

To sum up, the hydrogen-generation rate has a major effect on the permeation of hydrogen into the parabolic receiver tube's vacuum jacket. The partial pressure of hydrogen, permeability coefficient, and absorbing area interact so that the hydrogen-permeation rate equals the hydrogen-generation rate. By reducing hydrogen partial pressure or expanding the hydrogen barrier, the hydrogen-absorbing area is increased until the interior surface of receiver tube is completely covered by hydrogen before the hydrogen-permeation rate starts to decrease.

2. In the case of a saturated getter. When the getter inside the parabolic trough receiver tube's vacuum jacket becomes saturated, dissolved hydrogen in the synthetic oil has already been supersaturated. Thus in the case of utilizing the hydrogen-permeation model to calculate the hydrogen-permeation performance of a receiver tube with saturated getters, the influence of hydrogen solubility in synthetic oil is not considered.

When the getter is saturated, hydrogen cannot be further absorbed by the getter, and the pressure of hydrogen inside the receiver tube will continuously climb. Fig. 5.56 indicates the variation curve for hydrogen pressure inside the vacuum layer along with the temperature of synthetic oil after the vacuum tube's getter been saturated for 36 months. As both the hydrogen-generation and permeation rate rises with increases in temperature, once the temperature increases, the pressure of hydrogen inside the receiver tube climbs quickly, and 350°C is a turning point. Once the getter is saturated, the vacuum duration of the parabolic trough

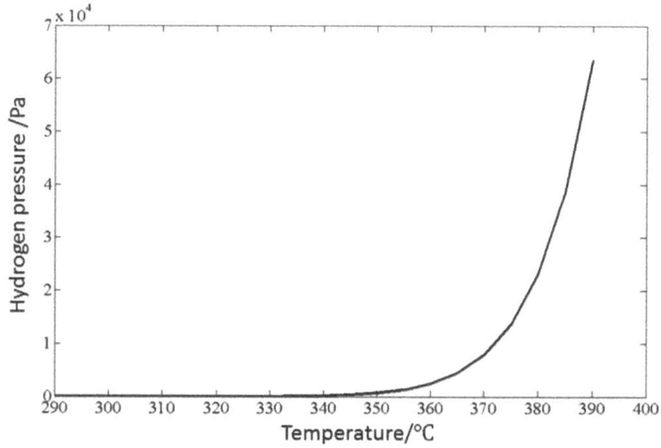

FIGURE 5.56 Variation curve of hydrogen pressure in receiver tube along with synthetic oil temperature after getter saturation for 36 months.

receiver tube mounted on the low-temperature end of the circulation line is longer than it is on the high-temperature end.

Figs. 5.57 and 5.58 indicate the variation curve for hydrogen pressure inside the receiver tube along with the temperature of synthetic oil at different hydrogen partial pressures and permeability coefficients after the getter has been saturated for 36 months. When reducing the hydrogen partial pressure or permeability coefficient inside the pipeline at low temperatures, the cumulative influence on hydrogen in the receiver tube is

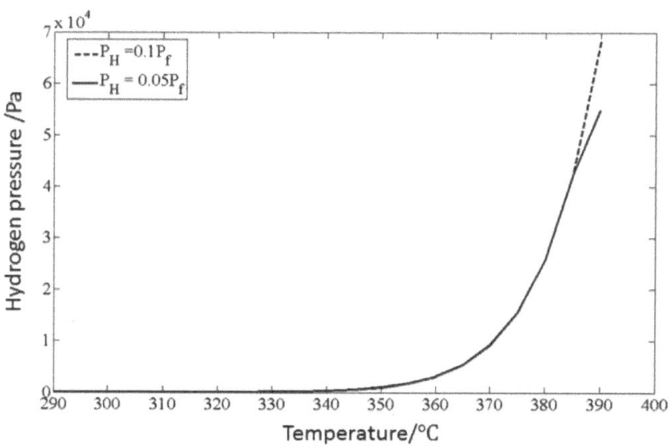

FIGURE 5.57 Variation curve of hydrogen pressure in receiver tube under different partial pressures along with synthetic oil temperature after getter saturation for 36 months.

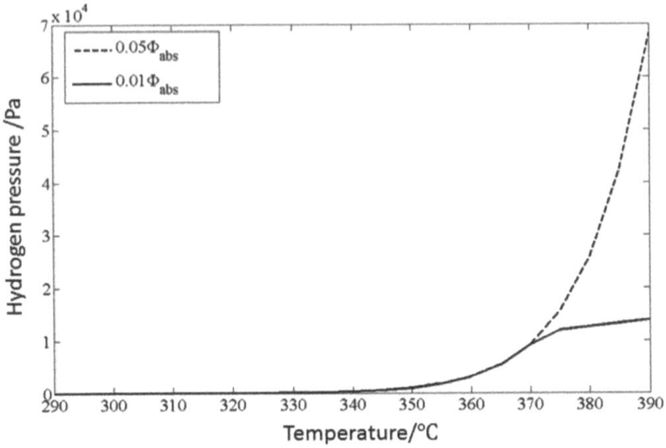

FIGURE 5.58 Variation curve of hydrogen pressure in receiver tube under different permeability coefficients along with synthetic oil temperature after getter saturation for 36 months.

insignificant. This is because both the hydrogen-generation and hydrogen-permeation rate are quite small at low temperatures and so is the hydrogen-absorbing area; when reducing the hydrogen partial pressure or permeability coefficient inside the pipeline, the hydrogen-absorbing area increases; while under steady state, the hydrogen-permeation rate remains the same and is equivalent to the hydrogen-generation rate. Under high temperatures, the hydrogen-absorbing area on the interior surface of absorber tube is large; once the hydrogen partial pressure or permeability coefficient inside the pipeline is reduced, the absorbing area will easily reach the maximum value, after which the hydrogen-permeation rate will start to decrease. Therefore, at high temperatures, the reduction of the hydrogen partial pressure or permeability coefficient inside the pipeline has obvious effects on the alleviation of hydrogen accumulation in the receiver tube.

CHAPTER

6

Thermal Storage Systems

Sensible, latent, and composite thermal storage are the three common thermal storage methods.

In light of current studies, research on sensible thermal storage is comparatively mature and has been developed to a commercially exploitative level; however, as the density of sensible thermal storage is low, sensible thermal devices typically have certain limitations due to their large sizes. Although chemical reaction thermal storage has multiple advantages, the chemical reaction process is complex. It sometimes requires catalyzers and has certain safety requirements, and there are other difficulties such as a huge one-time investment and low overall efficiency. Thus it currently remains in the small-scale experimental stage with plenty of problems yet to be solved before any large-scale application. As a superior system, phase-change thermal storage attracts people to carry out extensive studies and enjoys strong development momentum. However, regular phase-change materials (PCMs) used in actual applications are accompanied by various problems, such as inorganic PCM supercooling and phase separation, low-thermal-conductivity organic PCM, and the like, that have severely restricted the application of phase-change technology in solar thermal storage. Furthermore, the reduction of application costs for phase-change thermal storage is a practical problem that must be solved before its large-scale application in solar thermal storage. With the appearance of composite phase-change thermal storage materials, shaped PCMs, functional thermal fluid, and other new types of PCMs in recent years, the foregoing problems are expected to be solved. Research on these new types of PCMs will greatly propel the application of phase-change technology in solar thermal storage.

Design of Solar Thermal Power Plants
https://doi.org/10.1016/B978-0-12-815613-1.00006-7

6.1 GENERAL SYSTEM DESCRIPTION

6.1.1 Functionality of the Thermal Storage System

Thermal storage system functions include a cushion for the thermal power plant in variable weather conditions, the transfer of generating hours, an improved annual utilization rate, and the even distribution of generating capacity:

1. Cushion for variable weather. Cloud cover over the solar power plant results in transient changes in solar radiation input to the system. Such transient changes may severely influence the functioning of generating equipment. This is because, along with the variation of sunshine, the steam turbine generator unit will be frequently adjusted between its half-load and transient modes; in this case, the generating efficiency of the system will be greatly reduced, which can lead to a forced shutdown. A thermal storage system can eliminate such transient changes by offering a cushion to the generation system. For a thermal storage system that serves as a cushion, the respective thermal storage capacity is typically small; it is capable of satisfying a request for the full-load operation of the steam turbine unit for 1 h.
2. Transfer of generating hours. A thermal storage system can store partial solar energy collected during daytime and release that energy during subsequent peak periods for power generation. This type of thermal storage system normally does not need extra solar concentration areas because its thermal capacity is typically large enough to satisfy a request for full-load operation of the steam turbine for 3–6 h.
3. Improved annual utilization rate. The thermal capacity of a thermal storage system used for improving a power plant's annual utilization rate can satisfy a request for full-load operation of the steam turbine unit for 5–16 h. Such a thermal storage system is mainly used to prolong the power plant's generating hours through solar energy and improve the utilization rate of solar energy. However, with the introduction of a thermal storage system, the power plant must in turn have larger concentration areas.

The development tendency of solar high-temperature thermal storage is as follows:

1. Materials used during the life cycle are environmentally friendly;
2. Thermal storage materials have stable thermal properties and low corrosivity;
3. Liquid or solid inorganic nonmetallic materials, sensible thermal storage, and direct-method chemical thermal storage technology have good prospects;

4. Polymer thermal storage and PCMs feature great difficulties and still remain in the basic research stage.

6.1.2 Thermal Storage Costs for Solar Thermal Power Generation

Thermal storage systems currently applied in commercial concentrating solar power (CSP) plants include the dual-tank thermal storage system, which uses molten salt as the thermal storage medium, and steam thermal storage, which directly stores and extracts steam. Expanding the thermal storage capacity of steam thermal storage may result in a sharp rise in the cost of thermal storage tanks. Therefore, steam thermal storage is not applicable for long-term, large-capacity, High working parameters and low-cost thermal storage.

The dual-tank thermal storage system uses molten salt as the thermal storage medium and is currently the most widely applied thermal storage method and the most mature technology. Right now, it is mainly applied as an indirect thermal storage system for parabolic trough power plants that use synthetic oil as the thermal-absorbing and heat transfer fluid, and as a direct thermal storage system for tower power plants that use molten salt, which is the thermal storage medium for both systems. In 2018, the cost of a molten-salt dual-tank indirect thermal storage system was 50–80 US dollars/kWh (thermal), while the cost of a dual-tank direct thermal storage system was 30–50 US dollars/kWh (thermal). Along with continuous improvements in technology, thermal storage costs are expected to decline greatly in the near future. It is predicted that by 2020, the system cost will be reduced to 20–25 US dollars/kWh. The target of the SunShot Initiative of the US Department of Energy is to reduce the system cost to 15 US dollars/kWh by the year 2020.

A thermal storage system mainly consists of materials, tanks, tank foundations, pumps, and pipelines as well as thermal insulation, antifreezing, control, and electrical equipment. The total cost for a 50-MW parabolic trough indirect thermal storage system (with a thermal storage period of 7.5 h) in Spain, for example, is estimated at around 38.4 million US dollars, with the cost of thermal storage material (molten salt) accounting for about 50% of the total. The cost for a thermal storage tank mainly includes the costs of tank steel and thermal insulation material, with the materials cost accounting for about 75% of the total tank cost.

6.1.3 Categories of Thermal Storage Systems

Based on different materials, thermal storage systems can be categorized as sensible, latent, composite, or chemical:

1. Sensible thermal storage system. Materials can be inorganic nonmetallic materials, oil and other liquids, and thermal storage

materials made by mixing oil and inorganic nonmetallic materials. The sensible thermal storage system stores thermal energy with increases in the thermal storage medium's temperature. The principle of the sensible thermal storage system is simple; it has been widely applied as well. When utilizing a sensible thermal storage system, during energy storage and release by the thermal storage material, the material itself undergoes temperature changes but no other variations. Such a thermal storage method is simple and low-cost; however, the material has low thermal conductivity, so when energy is released, the thermal release capacity is low. In addition, such materials have low energy-storage density, leading to large sizes of the respective devices. Thus its industrial application value is yet to be improved. Common sensible storage materials include water, water steam, synthetic oil, molten salt, and gravel. Sensible thermal storage is mainly used to store thermal energy with a low temperature, for which liquid, rocks, etc. are often used as storage material. In order to facilitate thermal storage with a high volumetric thermal storage density, the thermal storage medium must have high specific thermal capacity and density. Currently, water and gravel have been most widely used as thermal storage media. The specific heat capacity of water is about 4.8 times of that of gravel, whereas the density of gravel is merely 2.5–3.5 times that of water. Therefore, water has a larger volumetric thermal storage density than that of gravel. Gravel enjoys the advantage of not having leakage loss and corrosion like water does, but its thermal conductivity is low, so the thermal charging and discharging system is more complex. Normally, a stone bed is used together with the solar air heater system, serving not only as thermal storage, but also as a heat exchanger. When high-temperature thermal storage is necessary, it is not suitable to use water as the thermal storage medium because the high-pressure tank corresponds to high expenses.

2. Latent thermal storage system. For PCM thermal storage, water, salts, and metal alloys can be used. Latent thermal storage means that one material in the system is heated until it melts down, evaporates, or results in other types of state changes under certain constant temperature conditions. Such a material not only has a high energy density, but also corresponds to devices that are easily structured, small-sized, and flexibly designed for convenient usage and easy management. Furthermore, it also has the huge advantage that its temperature during the phase-change thermal storage process can be deemed an approximate constant; it can be used to control the system temperature. The thermal storage medium of a latent thermal storage system that uses solid-liquid phase-change is normally called PCM. The latent thermal storage system takes advantage of the features of high-temperature phase change. Once the temperature of the storage medium reaches the melting point, a phase change that absorbs the melting potentials of the material will appear. Nevertheless, when thermal energy is absorbed through the storage system, it can be released through phase inversion. Compared with the sensible thermal system, a prominent advantage of this method is the ability to obtain thermal energy under necessary constant-temperature conditions. High energy and great latent thermal energy are potential advantages of the latent thermal storage system. Raw materials that can be used in the solid–liquid latent thermal storage system include fluoride, chloride, phosphate, sulfate, nitrate, Al–Si and Pb–Bi system alloys, and the eutectic mixture of hydroxide. Characteristics of these materials include fusion heat, heat quantity thermal conductivity, and thermal decomposition rate. Basically each material experimented with to date has a certain corrosiveness, and most tend to decompose under high temperatures.

3. Composite thermal storage system. Composite materials refer to materials consisting of two or more components with different chemical properties, which can be composites of PCM and inorganic nonmetallic materials, such as liquid salt and ceramics, liquid metal and ceramics, and various kinds of nitrates. A composite of thermal storage materials is made in order to fully utilize the advantages of various storage materials and overcome their individual shortcomings. For example, by applying a certain compounding process, molten salt can be compounded with proper substrate materials; molten salt has great phase-change latent thermal energy, chemical stability, and other advantages, while substrate materials are able to intensify heat transfer during the thermal storage and release process without the liquid phase leakage and corrosion of some other thermal storage materials.

6.1.4 Selection of Thermal Storage Modes

1. When the thermal storage temperature is less than 500°C, numerous thermal storage methods can be applied; when the temperature exceeds 500°C, carbonic acid, inorganic salt, ceramics, and metal storage methods, as well as chemical thermal storage methods, can be applied.
2. During the selection of different kinds of thermal storage materials and methods, compatibility of materials and tanks under high temperatures must be fully considered.

3. For a ceramic high-temperature thermal storage system, control means shall be taken to reduce thermal shock inside the thermal storage tank during the thermal charging process.
4. When a chemical thermal storage system is applied, the leakage and emission of toxic gas and liquid pollutants shall be considered.
5. Local raw materials can also be selected as thermal storage materials, such as gravel for a power plant located in a desert.

6.1.5 Storage of Thermal Storage Materials

When a molten-salt thermal storage system is used, molten-salt raw materials should be piled up indoors and sealed in order to prevent raw materials dust from being blown away by the wind and polluting the air and other metal and glass mirrors on-site. Nitrate chemicals must be piled up while adopting explosion-proof measures and storing them far away from the main powerhouse and solar tower.

When oil is used as the thermal storage working medium, treatment methods and pollution prevention measures for accidents resulting from various degrees of leakage by oil tanks, pipelines, and valves must be considered. If the equivalent length of the high-temperature oil transmission pipeline is shorter than 200 m, a negative-pressure pneumatic cleaning system can be applied; if the length is equivalent to or longer than 200 m, a positive-pressure pneumatic cleaning system shall be applied.

The thermal storage system shall be equipped with high-pressure air or water cleaning equipment for the cleaning of tanks and pipelines; the high-pressure system can also be used for firefighting. The diameter and length of the pipeline shall be designed while considering the relative convenience of cleaning the tube interior of corrosion or fouling.

6.2 TECHNICAL REQUIREMENTS OF THERMAL STORAGE SYSTEMS

During selection of a proper solid thermal storage system, it is necessary to comprehensively consider the cost-effectiveness and technical standards of the system.

1. The cost of a thermal storage system is mainly determined by the following.
   a. cost of thermal storage materials
   b. cost of heat exchanger for thermal charging and discharging
   c. land expense and the cost of other auxiliary equipment
   d. operational and maintenance costs; for the molten-salt system, the respective cost is high

2. In light of technology, the thermal storage system shall satisfy the following standards.
   a. The thermal capacity of the thermal storage material shall be as high as possible to reduce the size of the thermal storage system.
   b. Good heat exchange shall be guaranteed between the heat transfer fluid and thermal storage medium to improve the heat exchange efficiency of the system.
   c. Thermal storage materials must have good mechanical and chemical stabilities to ensure that the thermal storage system still has complete reversibility after many thermal charging and discharging cycles, indicating that the system will have a long service life.
   d. Thermal storage materials shall have good thermal conductivities to improve the dynamic properties of the system.
   e. The thermal expansion coefficient of the thermal storage material shall match the thermal expansion coefficient of the metal heat exchanger embedded in the thermal storage medium in order to always guarantee good heat exchange features between the heat transfer fluid and thermal storage medium.

6.3 THERMAL STORAGE MATERIALS AND MODES

6.3.1 Molten-Salt Thermal Storage and Room-Temperature Ionic Liquid Material

Molten-salt thermal storage systems have been experimented on or commercially operated in various power plants, including Solar Two in the United States, THEMIS in France, Gemasolar in Spain, Asola in Germany, and ENEL in Italy. Currently, the molten-salt thermal storage system typically applies dual-tank techniques.

For a tower power plant, molten salt normally serves as the heat transfer fluid as well. In this case, cold salts enter the hot salt tank after being heated by the receiver. After coming out of the hot tank, salts enter an evaporator for heat exchange and become cold salts before entering the cold tank; a complete cycle finishes.

For a parabolic trough power plant, in consideration of safety and reliability, synthetic oil is normally used as the thermal storage fluid. When salt is used as the thermal storage material, an oil/salt heat exchanger shall be added between oil and salt for transferring thermal energy.

Newly developed techniques mainly include molten-salt single-tank thermocline thermal storage techniques. Pacheco et al. once carried out a theoretical and experimental analysis on thermocline molten-salt thermal storage techniques applied in the parabolic trough system with a certain type of filler, the general idea of which was to substitute expensive salts with cheaper materials to reduce costs. Compared with conventional dual-tank molten-salt thermal storage, such techniques reduced costs by around one-third. Among single-tank thermocline thermal storage techniques, the selection of durable fillers and optimization of thermal charging and discharging methods and equipment are all main research objects.

New thermal storage materials can also be used, namely room-temperature ionic liquid (RTIL), which overcomes a shortcoming of molten salt by remaining liquified even under extremely low temperatures. As an organic salt, RTIL's steam pressure within the relevant temperature range can be neglected; its melting point is below 25°C. Currently, the stability and cost of this material after use under high temperatures remain uncertain.

6.3.2 Concrete Thermal Storage Material

Among solid sensible thermal storage materials, both pourable ceramics and high-temperature concrete have good application prospects. Deutsches Zentrum für Luft-und Raumfahrt (DLR) analyzed the physical properties of the two materials at 350°C. The basic parameters include a thermal conductivity coefficient of 1.2 W/(m K), density of 2250 kg/m^3, and specific thermal capacity of 1100 J/(kg K).

Fig. 6.1 shows sectional views of these two materials after being embedded with heat exchange steel pipes. According to the shear stress analysis, at the ambient air temperature of 350°C, the heat exchange tube contacts well with materials.

To sum up, high-temperature concrete seems to be a better material because it has lower cost and higher material strength; it also is a premixed material that can be more easily controlled. However, pourable ceramics have a thermal storage capacity 20% higher than that of the high-temperature concrete as well as 35% higher thermal conductivity; it also has potential for further cost reductions. Fig. 6.2 indicates thermal cycle and strength testing experiments on high-temperature concrete, which are mandatory tests for thermal storage concrete.

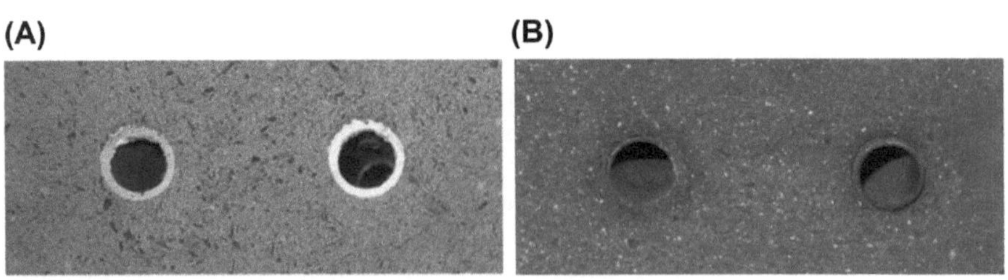

FIGURE 6.1 Sectional views of high-temperature concrete (A) and pourable ceramics (B) of embedded heat exchange steel pipe.

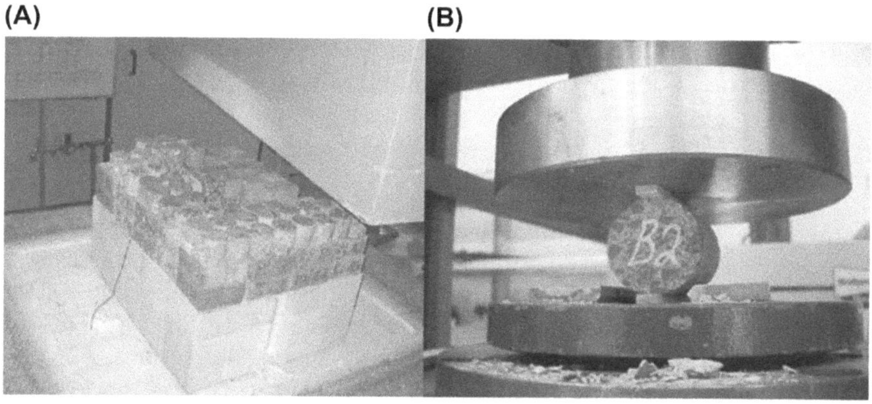

FIGURE 6.2 Thermal circulation and strength tests on high-temperature concrete. (A) Thermal circulation test. (B) Strength test.

Methods to improve the thermal conductivity of high-temperature concrete include adding metal or graphite shreds with high thermal conductivities to concrete to improve its thermal conductivity. Expansion graphite has an extremely high thermal conductivity of up to 150 W/(m K). However, due to the restriction of concrete while being cast, the maximum additive volume of graphite shreds is 10%. By adding graphite shreds, the thermal conductivity of concrete can be improved by about 15%. Some researchers have also carried out research on the influence of graphite shreds on the thermal conductivity of concrete, the results of which are shown in Fig. 6.3. According to the figure, with an increase in graphite content, the thermal conductivity of concrete rose significantly.

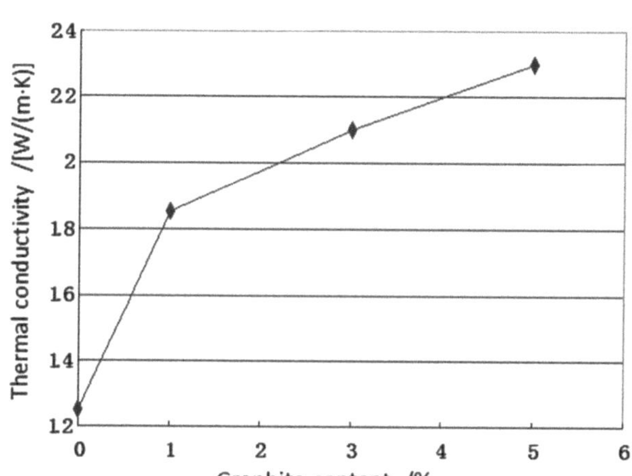

FIGURE 6.3 Relationship between thermal conductivity of concrete and content of graphite.

When the content of graphite was 5%, the thermal conductivity was as high as 2.34 W/(m K), which was higher than the thermal conductivities of concrete mentioned in other relevant literature.

6.3.3 Concrete Thermal Storage Heat Exchange Design

1. Pipeless once-through heat exchanger. In this design, precast concrete thermal storage blocks directly contact with heat transfer fluid. This design is low-cost and can directly transfer thermal energy. However, as concrete has a certain permeability, leakage may occur when synthetic oil flows inside pipelines. Furthermore, bonding the interfaces of pipelines and thermal storage blocks has certain technical difficulties, and the cost is high. For these reasons, design of a pipeless once-through heat exchanger shall be further studied.
2. Optimum tubular heat exchanger. Thermal conductivities of thermal storage materials have major influences on the tubular heat exchanger design. Along with improved thermal conductivity, spacing between heat exchange tubes increases while the quantity of heat exchange tubes decreases accordingly. For a concrete thermal storage unit with a capacity of 950 MWh, the inlet temperature of synthetic oil in the thermal absorbing process is assumed to be 390°C, and the outlet temperature of synthetic oil in the thermal release process is assumed to be 290°C. When the thermal conductivity is increased from 1 to 1.8 W/(m K), the length of the total heat exchange tube is reduced by 46% (Fig. 6.4). Therefore, improvement in the thermal conductivity of the material is crucial to the design of the entire thermal storage unit.

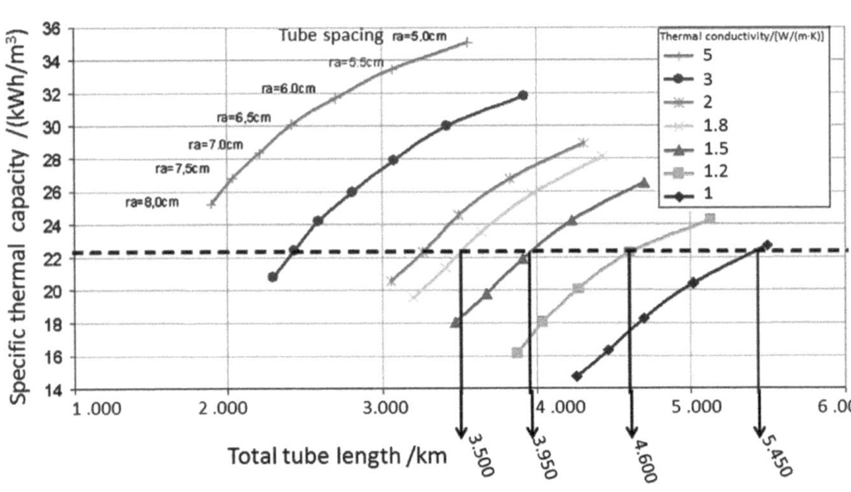

FIGURE 6.4 Relationship of total tube length of heat exchange tube and specific thermal capacity under different thermal conductivities and tube spacing.

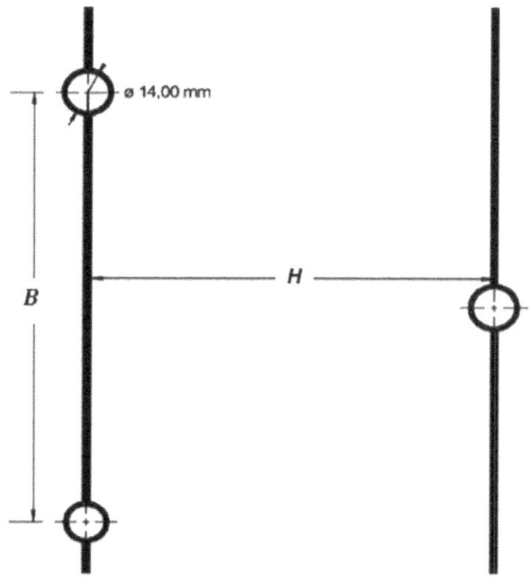

FIGURE 6.5 Axial thermal radiation fins.

Another way to improve heat exchange with thermal storage materials is to add a structure with a high thermal conductivity to the heat exchange tube, namely to mount axial thermal radiation fins on the heat exchange tube as shown in Fig. 6.5.

For two tubular heat exchangers with the same tube spacing, finite element analysis results of the respective temperature distribution are shown in Fig. 6.6, in which the left one is a heat exchanger without thermal radiation fins and the right one is a heat exchanger with axial

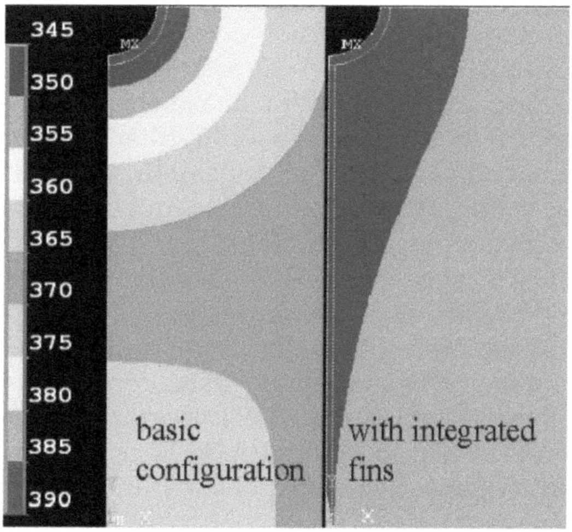

FIGURE 6.6 Finite element analysis of tubular heat exchanger (unit: °C).

thermal radiation fins. Fig. 6.6 displays the temperature distribution of thermal storage blocks after being connected with synthetic oil at 390°C for 1 h. The initial temperature for both structures is 350°C. According to Fig. 6.6, fins have effectively improved the temperature distribution of thermal storage blocks and will increase the cost of the material; in addition, the manufacture of heat exchangers is more difficult. Thus adding a heat exchange structure has no obvious advantages; it is suggested that conventional heat exchange pipelines be used [48].

6.3.4 Phase-Change Material Thermal Storage

Utilization of thermal charging and discharging features of PCMs in the phase-change process for thermal storage/release and temperature control has been an active research trend in recent years for solar thermal utilization and materials. Ambient air temperature is controlled by PCMs in the phase-change process through heat exchange with the environment. PCMs that have been the most studied include polybasic alcohol, alkane, ester, fatty acid, and other organics; crystalline hydrated salt, molten salt, and metal alloy inorganics; and organic and inorganic eutectic mixtures. PCM has the advantages of high thermal absorbing density, constant temperature control, small size, obvious energy-saving effects, a broad phase-change temperature selection range (−20 to 1000°C), and a simple and reliable structure. According to the phase-change temperature, PCM can be divided into either low-temperature PCM or medium- and high-temperature PCM. Normally, PCM with a phase-change temperature of less than 100°C is referred to as low-temperature PCM and is used in energy-saving buildings, electronic device encapsulation and radiation, aerospace system constant-temperature control, constant-temperature packaging of temperature-sensitive drugs, constant-temperature sportswear, military engineering, etc. PCM with a phase-change temperature above 100°C is referred to as a medium and high-temperature PCM and is used for industrial surplus thermal utilization, CSP generation, power peak regulation, and other uses that require medium and high-temperature thermal storage systems.

Studies on phase-change thermal storage materials originated from the construction field. Dr. Telkes of the Massachusetts Institute of Technology started studies in the 1950s on the application of PCMs in solar energy buildings. During the 1970s the energy crisis propelled studies on PCMs. In 1982 the solar energy sector of the US Department of Energy started to sponsor studies on PCMs; in 1988 the US Office of Energy Storage and Distribution further propelled these studies. In 1998 the International

Energy Agency launched the 3-year "Phase-change and Chemical Reaction Energy Storage" Multinational Joint Research Plan (Annex 10), which considered the application of phase-change thermal storage systems for energy savings in buildings as the major research orientation. In 2001, on the basis of Annex 10, "Technical Research Plan for Advanced Phase-change and Chemical Energy Storage Materials" (Annex 17) was launched. Countries participating in the plan included the United States, Canada, Japan, European countries, and other developed countries and regions. In 2004, Europe launched "Advanced Energy Storage and Transportation Technical R&D Framework," with the aim of pushing forward the applications of phase-change energy storage materials in solar energy utilization, energy-saving buildings, and so on, involving the six European countries of Germany, Spain, France, Sweden, Denmark, and Holland. This framework included studies in universities as well as the application and promotion of PCMs. Before this, Germany had independently organized a research development program, Innovative PCM-Technology, that involved universities (such as Universität Stuttgart), large enterprises (such as BASF and some construction and materials firms), and research institutions (such as DLR, the Fraunhofer Institute for Solar Energy Systems, and the Bavarian Center for Applied Energy Research). Some Chinese colleges, universities, and scientific research institutions also carried out many studies of PCM application in high-temperature and energy-saving buildings with support from the 863 Program, National Natural Science Fund, and some local scientific research programs.

PCM is the potential candidate material for latent thermal storage techniques. It is especially important for systems with a large proportion of latent thermal energy, such as direct steam generation systems. PCM thermal storage is not restricted to solid–liquid conversion; solid–solid and liquid–gas conversion can also be applied. Yet compared with other phase-change patterns, solid–liquid conversion has certain advantages. Two theoretical methods currently under research are:

1. small amount of PCM encapsulation technique (sealing); and
2. PCM embedded in a matrix consisting of other solid materials with high thermal conductivities.

The first method has been considered for reduced PCM interior distance, whereas the second method improves thermal conductivity through other materials. PCM techniques are currently in the initial R&D stage. Many recommended systems are still theoretical or laboratory-scale experimental works. Thus it is very difficult to predict the cost, but it should be less than 20 euros/kWh.

6.3.5 Solid Material Thermal Storage for Solar Air Receiver Systems

A sensible thermal storage system made of solid materials is normally used for a volumetric air or compressed-air system in which thermal energy is transferred to another medium that can be any solid material with high density and high thermal capacity. Other parameters of a solid material thermal storage system include dimension and shape, with an aim at minimizing pressure loss (the greater the pressure loss, the higher the energy self-consumption).

A new concept was developed by DLR from the use of fixed solid materials as thermal storage medium namely using borax as an intermediate heat transfer medium to avoid the adverse factors caused by completely filling the thermal storage tank field with fixed solid materials during application of an open volumetric air receiver technology solar-tower system.

Thermal storage techniques that use fixed solid materials can be realized within 5 years, but mobile solid material thermal storage technology cannot be achieved within such a short period. Furthermore, solid material medium are subject to intermediate-level risk and uncertainty, whereas mobile solid material medium have high-level risk and uncertainty.

Another technical innovation is the research and development of thermal storage tanks for a compressed closed-loop air receiver system. This type of thermal storage tank must be able to withstand pressure of 1.6–2.0 MPa, with the specific value depending on the pressure ratio of the gas turbine. In a system such as this, the receiver and concentration field must produce more energy than the amount required for the gas turbine under good solar irradiation. The extra energy is used to charge the thermal storage system through an external air blower. In the thermal discharging mode with no sunshine, the receiver is under bypass effect, and the flow passing through the system is reversed. Furthermore, in order to utilize thermal energy from the receiver and the thermal storage system in the event of poor solar irradiation, it is feasible to split the compressor airflow. The R&D and realization period of this technology is about 5–10 years, with intermediate-level uncertainty and risk.

6.3.6 Saturated Water/Steam Thermal Storage

Theoretically, a drum is also a type of thermal storage system because it contains a certain amount of pressurized water. By reducing pressure, steam can be produced. This kind of thermal storage method has been widely applied in the industrial field, and thus it is more mature. The main problems right now are the cost for steam tanks with large thermal storage capacities and the deterioration of steam in the thermal release process. This kind of thermal storage features a simple process and a high thermal release speed, and it is especially ideal for short-term buffer-type thermal storage that can be used to compensate concentration field shade losses caused by drifting small clouds.

In terms of R&D and commercial applications within a short period, the potential for cost reduction falls in the range of 30%–60%, and the respective uncertainty and risk are both quite low.

6.3.7 Alloy Phase-Change Thermal Storage Material

PCMs that can be used right now in CSP generation high-temperature thermal storage mainly include dualistic or polybasic alloys rich in Al, Si, Cu, Mg, and Zn. All of these alloy elements are regular light metal elements and even trace elements in the human body. Therefore, compared with other thermal storage materials, the direct adverse influences of phase-change thermal storage alloys on the environment may be insignificant.

Various techniques are compared in Table 6.1.

6.4 CATEGORIES AND CONSTITUTIONS OF THERMAL STORAGE SYSTEMS

The type of thermal storage system in an CSP plant can be either active or passive (refer to Fig. 6.7).

The thermal storage medium used in an active-type thermal storage system is normally fluid, which is used for forced convection and heat exchange in solar receivers steam generators, and other heat exchange equipment. According to the different thermal storage medium that participate in the heat exchange process, the active-type thermal storage system can be further divided into active-type direct and active-type indirect systems. The thermal storage medium of the former is also the heat transfer fluid of the power plant, whereas that of the latter is only used for thermal storage and release without functioning as the heat transfer fluid of the receiver. A passive-type thermal storage system is normally a double-medium system. The thermal storage medium itself has not been used in heat exchange equipment for forced convection and heat exchange; instead, thermal charging and discharging can be realized through the functioning of the heat transfer fluid.

A thermal storage unit consists of thermal storage tanks, thermal storage materials, thermal exchangers, and the respective control system. The evaluation index for a thermal storage unit is to achieve a low cost while satisfying performance conditions (Fig. 6.8). For a thermal storage system required to offer energy for power generation, thermal release is crucial. The thermal discharge power and temperature of the solid thermal storage system decrease over time.

TABLE 6.1 Thermal Storage System Technical Innovation [32]

Category	Tank Type	Power Plant Type	Duration	Cost Reduction Amplitude	Uncertainty
Molten Salt	Dual	Tower, parabolic trough	Commercialized, mainly use within 5 years	Within 30%	L
	Single thermocline	Parabolic trough	5–10 years	30%–60%	M
	Dual	Parabolic trough	Within 5 years	Within 30%	M
Room-Temperature Ionic Liquid	Tubular	Parabolic trough	Over 10 years	30%–60%	H
Concrete	Pipeless	Parabolic trough	5–10 years	Above 60%	M
	Advanced thermal charging and discharging	Parabolic trough	5–10 years	30%–60%	M
PCM	All	Tower, parabolic trough	Over 10 years	30%–60%	M
Solid Material	Fixed	Tower	Within 5 years	30%–60%	L
	Mobile	Tower	5–10 years	30%–60%	H
	Fixed solid and pressurized	Tower	Over 10 years	30%–60%	H
Drum	Saturated water	Tower, parabolic trough	Within 5 years	30%–60%	L

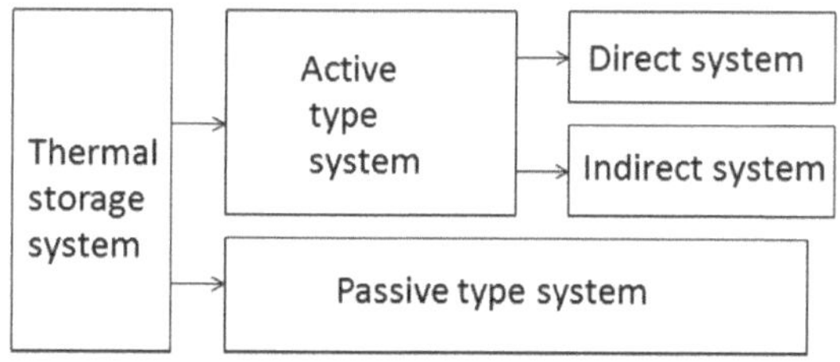

FIGURE 6.7 Classification of thermal storage systems in CSP plants.

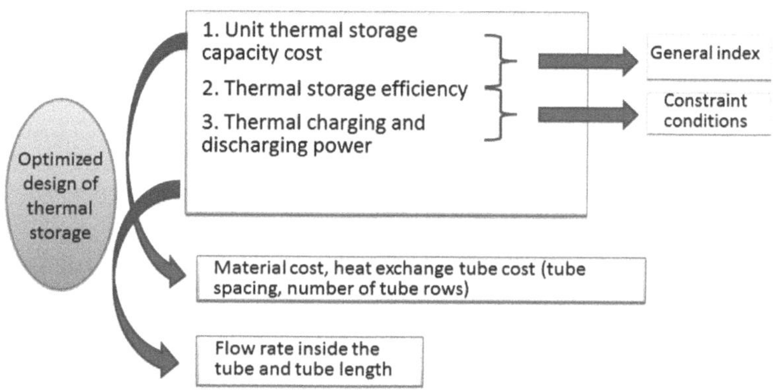

FIGURE 6.8 Schematic diagram of evaluation indices for a thermal storage system in an CSP plant.

6.4.1 Active Direct Thermal Storage System

The active direct thermal storage system can be divided into direct steam and dual-tank direct systems.

In a direct steam thermal storage system, steam not only serves as the heat transfer fluid, but also functions as a thermal storage medium. The working principles of the entire system are shown in Fig. 6.9. Water passes through the solar concentration field while being heated to become superheated steam; partial surplus superheated steam turns into liquid water after pressurization and is then stored in the steam thermal storage. If necessary, high-temperature liquid water in the steam thermal storage can be depressurized to become saturated steam, which can be used to propel the steam turbine to generate power. The thermal storage system of the PS10 power plant in Spain (Fig. 6.9) is a direct steam system. The power generation capacity of the power plant is 11 MW, and its thermal storage capacity is 20 MWh and can be used to propel 50% steam turbines to work for 50 min. Saturated steam meeting specific parameters (250°C, 4 MPa) is stored in the steam thermal storage.

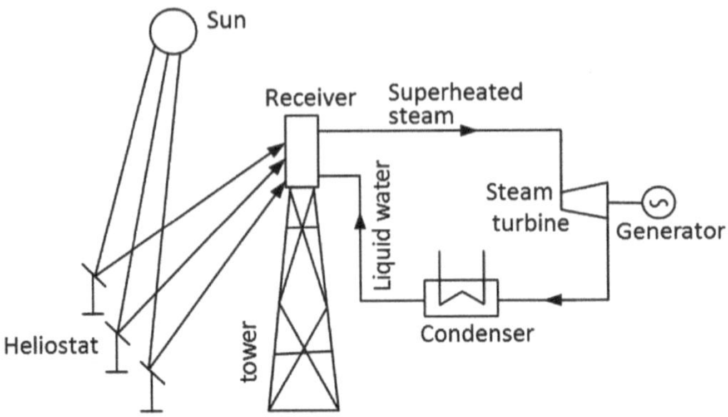

FIGURE 6.9 Working principles of the direct steam thermal storage system (PS10 power plant).

The efficiency of the thermal storage system is 92.4%. The advantage of this system is that intermediate heat transfer fluid and heat exchange equipment are not necessary.

The dual-tank direct thermal storage system is a typical active-type direct system. Among existing power plants, it is the thermal storage pattern that has been most widely applied. It stores high-temperature heat transfer fluid in a high-temperature thermal storage unit (hot tank) to be used in case of night or shading clouds; it also stores low-temperature heat transfer fluid in a low-temperature thermal storage unit (cold tank). The working principles are shown in Fig. 6.10.

The dual-tank direct thermal storage system has been applied in the SEGS I and Solar Two power plants in the United States. Mineral oil is both the thermal storage medium and the heat transfer fluid for the SEGS

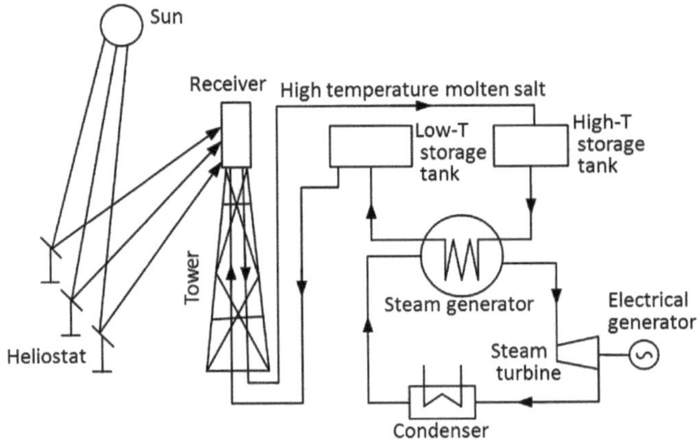

FIGURE 6.10 Working principles of the dual-tank direct thermal storage system (Gemasolar power plant).

I power plant. The temperatures of mineral oil in the cold tank and hot tank are 240 and 307°C respectively. The power generation capacity of the SEGS I power plant is 14 MW, and its thermal storage capacity is 120 MWh, which can be used to propel the steam turbine to work under full load conditions for 3 h. The greatest disadvantage of this thermal storage system is its high cost; the cost of mineral oil alone accounts for 42% of the total investment. Molten salt is both the heat transfer fluid and the thermal storage medium for the Solar Two power plant. The temperatures of molten salt in the cold tank and hot tank are 290 and 565°C respectively. Molten salt is a mixture of 60% sodium nitrate and 40% potassium nitrate with a melting point of 207°C and good thermal stability even under 600°C. The power generation capacity of the Solar Two power plant is 10 MW, and its thermal storage capacity is 105 MWh, which can be used to propel the steam turbine to work under full load conditions for 3 h. The thermal storage efficiency is 97%. The dual-tank direct thermal storage system has also been applied in the Gemasolar power plant in Spain (Fig. 6.10), and molten salt is also its heat transfer fluid and thermal storage medium. The Gemasolar power plant is the first commercial tower power plant to utilize a molten-salt thermal storage system. Its power generation capacity is 19.9 MW, and its thermal storage capacity is 600 MWh, which can be used to propel the steam turbine to work for 15 h. The energy storage utilization factor of the system is 74%. The major advantage of the dual-tank direct thermal storage system is the separated storage of cold and hot thermal storage medium for better control. Compared with the single-tank thermal storage system, its disadvantages are a higher cost and greater power plant operational and maintenance expenses.

6.4.2 Active Indirect Thermal Storage System

The active-type indirect thermal storage system can be divided into dual-tank and single-tank indirect systems (the single-tank system is also known as the thermocline system).

In the dual-tank indirect thermal storage system, energy is not directly stored in heat transfer fluid; instead, another kind of fluid is used as the thermal storage medium. Energy in the thermal storage system is transferred to the thermal storage medium by relying on the heat transfer fluid through a heat exchanger. Fig. 6.11 indicates the working principles of the dual-tank indirect thermal storage system in a parabolic trough power plant in which synthetic oil is used as the heat transfer fluid with molten salt as the thermal storage medium. In the thermal charging process, partial synthetic oil from the collector concentration field enters the synthetic oil–molten salt heat exchanger. Meanwhile, molten salt in the cold tank enters the heat exchanger from the opposite direction. After this process, synthetic oil is cooled and the temperature is reduced; meanwhile,

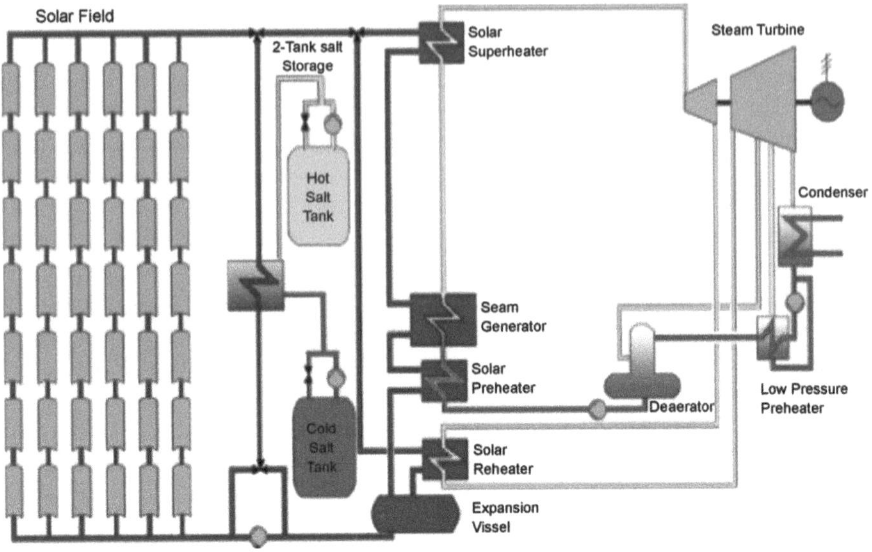

FIGURE 6.11 Working principles of the dual-tank indirect thermal storage system.

the molten salt is heated to a higher temperature before being stored in the hot tank. In the thermal discharging process, the directions of the synthetic oil and molten salt entering the heat exchanger are the opposite of those during the thermal charging process. In this case, thermal energy is transferred from molten salt to synthetic oil in order to offer thermal energy to the steam turbine.

The Andasol 1 power plant in Spain is an CSP plant with parabolic trough concentration techniques that applies the dual-tank indirect thermal storage system, molten salt as the thermal storage medium, and synthetic oil as the heat transfer fluid. In the thermal storage system, the temperature of the hot tank is 384°C and the temperature of the cold tank is 291°C; molten salt is a mixture of 60% potassium nitrate and 40% sodium nitrate with a melting point of 221°C. The thermal storage capacity of the Andasol 1 power plant is 1010 MWh, which can be used to propel the steam turbine with a rated power generation capacity of 50 MW to work under full load conditions for 7 h. The mean annual efficiency of the power plant is 14.7%. Compared with the direct thermal storage system, the advantage of the dual-tank indirect system is that cold and hot heat transfer fluids are separately stored; the thermal storage medium flows only between the cold tank and hot tank without passing through any collector. Its disadvantages are higher costs and greater power plant operational and maintenance expenses.

In the single-tank indirect thermal storage system, cold and hot fluids are stored in the same thermal storage tank. When hot heat transfer fluid

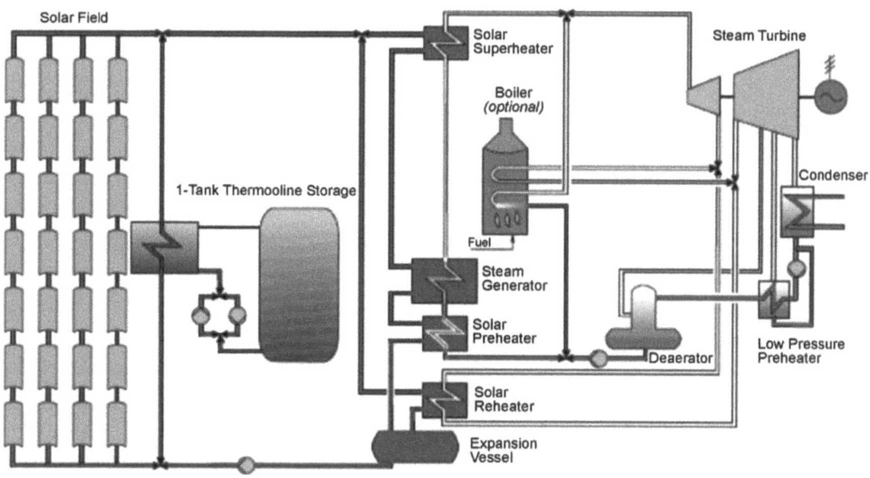

FIGURE 6.12 Working principles of the single-tank indirect thermal storage system based on the parabolic trough power plant.

flows through the heat exchanger, the thermal storage fluid medium (Fig. 6.12) in the single tank is heated. Due to separated temperature layers, cold and hot fluids can be separated, namely the hot fluid is at the upper level of the thermal storage tank while cold fluid is at the lower level. Separated layers of cold and hot fluids are referred to as the thermocline, which normally requires fillers to facilitate formation of the thermocline. Experimental research indicates that fillers are subject to the main thermal storage medium in a single-tank thermal storage system. Thus fillers with a low cost (such as quartzite and gravel) can be used as substitutes for most of the thermal storage medium.

Single-tank indirect thermal storage (thermocline) system was applied in the Solar One power plant in the United States. Solar One was a power plant functioning from 1982–88 with rocks and gravel as the thermal storage medium and mineral oil as the heat transfer fluid. The power generation capacity of Solar One was 10 MW. After the introduction of a thermal storage system, the power plant was able to operate for 8 h during summer and 4 h during winter. The main advantages of this system are the reduced cost of a single thermal storage tank and the use of low-cost fillers (rocks and gravel) as thermal storage medium. The resulting cost of the thermocline system was 35% lower than that of the dual-tank system. The main disadvantage of the thermocline thermal storage system is the comparative difficulty of separating cold and hot fluids; in order to maintain separated temperature layers inside the thermal storage tank, strict thermal absorption and release procedures, and appropriate methods or equipment, must be used to prevent cold fluid from being mixed with hot fluid. Thus, the design of this thermal storage system is quite complex.

In the Badaling tower-power-generation experimental plant located in Beijing, the power generation capacity of the steam turbine generator is 1 MW. A thermal storage system that integrates the active-type direct steam thermal storage and dual-tank indirect thermal storage is applied with synthetic oil and high-temperature steam as the thermal storage medium. The working principles of this system are shown in Fig. 6.13.

Thermal charging process: Transferred by oil pump, synthetic oil in the cold tank (low temperature oil tank) exchanges heat in the left thermal charging exchanger with high-temperature superheated steam from the receiver or boiler; after absorbing thermal energy, high-temperature synthetic oil is stored in the hot tank (high-temperature oil tank) so that it can be used during cloud overcast, nights or rainy days. After discharging thermal energy in the thermal charging exchanger, superheated steam turns into saturated steam before being stored in the low-temperature steam thermal storage.

Thermal discharge process: After flash distillation from the low-temperature steam thermal storage, saturated steam enters the right thermal discharge exchanger, absorbs thermal energy, and turns into

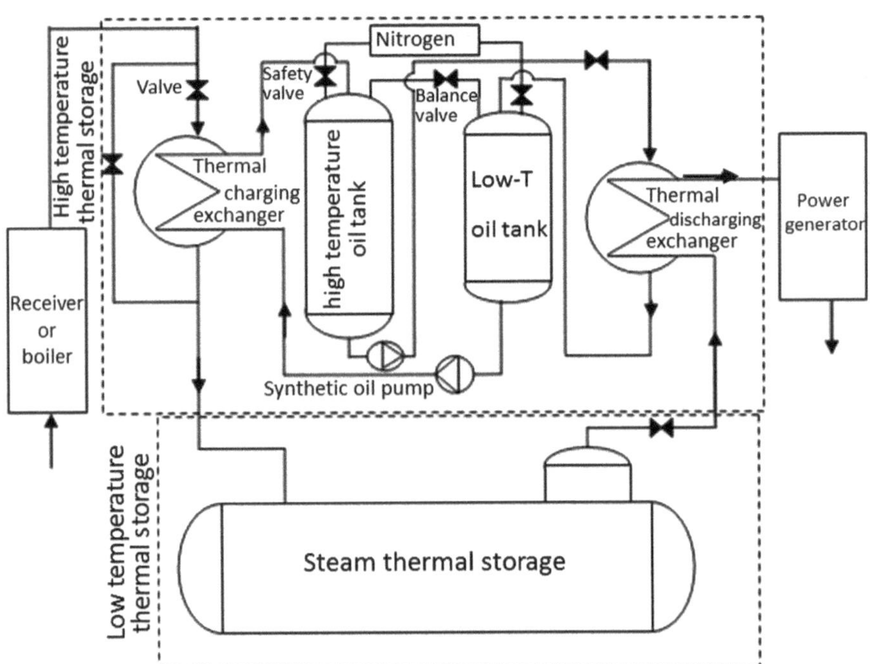

FIGURE 6.13 Working principles of a thermal storage system integrating direct steam and dual-tank indirect thermal storage (Badaling tower power plant).

high-temperature superheated steam before being supplied to the steam turbine for power generation. Meanwhile, high-temperature synthetic oil is pumped by the oil pump from the high-temperature oil tank before entering the thermal discharging exchanger to heat the steam; after discharging thermal energy, the temperature is reduced and the synthetic oil is stored in the low-temperature oil tank.

6.4.3 Passive Thermal Storage System

In the passive-type thermal storage system, thermal storage material itself doesn't circulate, and thermal charging and discharging to the system are mainly accomplished by the circulation of heat transfer fluid. A passive-type thermal storage system is mainly a solid thermal storage system, typically with concrete, pourable material, and PCM as the thermal storage media. Fig. 6.14 is the schematic diagram of working principles for a passive-type thermal storage system with concrete as the thermal storage medium and molten salt as the heat transfer fluid.

Heat transfer fluid transfers thermal energy to the solid thermal storage material through a tubular heat exchanger. This tubular heat exchanger is integrated with the thermal storage material, and the exchanger cost constitutes a large proportion of the system's total cost. The design of the geometric parameters of the heat exchanger (such as diameter and quantity of pipelines) is crucial to the performance of the heat exchanger. The advantages of the indirect thermal storage system using solid thermal storage materials include an extremely low cost

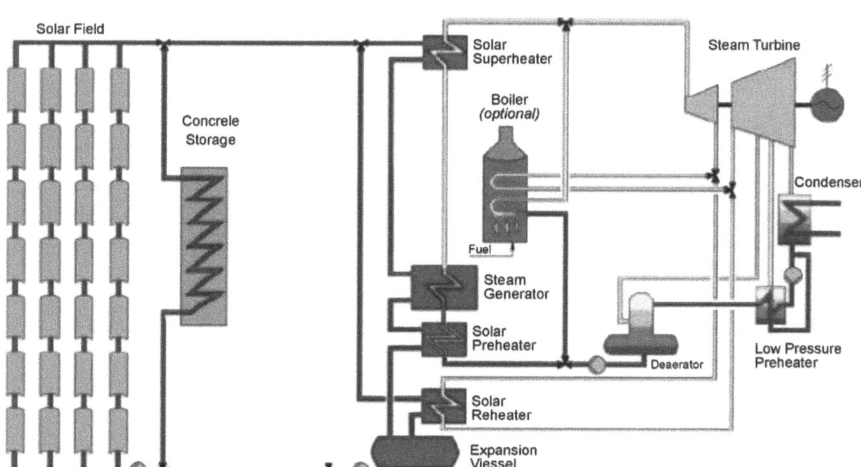

FIGURE 6.14 Working principles of the passive-type thermal storage system on the basis of concrete thermal storage medium.

for thermal storage materials, a high heat exchange rate due to good contact between the solid thermal storage material and the heat exchange pipeline, and a low heat exchange gradient between the thermal storage material and the heat exchanger. The disadvantages include the additional cost of the heat exchanger and possible system instability during long operational periods.

The following factors should be considered when selecting solid thermal storage materials: cost shall be low in order to reduce total investment in the thermal storage system; unit volume thermal capacity shall be as high as possible to reduce system size; good heat exchange between the heat transfer fluid and the thermal storage medium to improve the system's heat exchange efficiency; good mechanical and chemical stability to ensure the system still has complete reversibility after many thermal charging and discharging cycles and a long service life; and good thermal conductivity to improve the dynamic properties of the system. The thermal expansion coefficient of the material shall match the thermal expansion coefficient of the metal heat exchanger embedded in the thermal storage medium to always ensure good heat exchange features between the heat transfer fluid and thermal storage medium. Due to low materials costs, higher-volume thermal capacity, acceptable thermal conductivity, and stable mechanical and chemical properties, concrete is a solid thermal storage material with extensive application prospects.

6.4.4 Constitution of the Thermal Storage System

1. The system should have a high-temperature thermal storage tank including tank body, tank support and protection system, pressure release, temperature detection, material overheating protection, low-temperature protection, leakage detection system, internal fuel or electric heater, fluid mixer, and filling materials inside the thermal storage tank and discharge system. Fig. 6.15 shows the thermal storage system in the Beijing Badaling tower power plant, which uses high-temperature synthetic oil as the thermal storage medium. The temperature of the high-temperature thermal storage tank can be as high as 385°C.
2. The low-temperature thermal storage tank should include tank body, tank support and protection system, pressure release, temperature detection, material overheating protection, low-temperature protection, leakage detection system, internal fuel or electric heater, and filling materials inside the thermal storage tank and discharge system;

FIGURE 6.15 A thermal storage system that uses high-temperature synthetic oil (Beijing Badaling tower power plant).

3. The thermal storage tank connecting pipeline should include pipeline, valve, pump, pressure-release device, overheating protection, material leakage detection system, alarm system, pipeline electrical tracing, pipeline thermal insulation, and pipeline preheating and temperature control systems.
4. A tank fouling discharge and cleaning system should be included. Foulings in the tank mainly include various chemicals after high-temperature chemical decomposition of fluid and rust inside the pipeline. It also includes a compressed-air blowing system; for oil and the like, nitrogen and other inert gases shall be blown in instead of air. Molten salt can be blown away by air. The cleaned tank must be sealed in accordance with equipment requirements as soon as possible.
5. The thermal energy charging and discharge unit and control system includes a thermal charging exchanger side that is supplied with high-temperature fluid while the other side is supplied with low-temperature fluid. For the non-phase-change thermal charging exchanger, a tubular heat exchanger can be used. The heat exchanger tube can be filled with high-pressure steam, while a high-temperature fluid passageway, such as molten salt, can be mounted on the tube casing. A heat exchanger that uses PCM to charge heat can be integrated with the thermal discharging exchanger. Both thermal energy charging and thermal energy discharging pipelines are located within the heat exchanger.

6.5 SELECTION OF THERMAL STORAGE MATERIALS AND TANKS

1. Thermal storage material selection principles:
   a. Thermal storage materials normally include synthetic oil, molten salt, saturated water steam, inorganic nonmetallic materials, alloy PCMs, and chemical thermal storage materials. It should be noted that these materials shall be nontoxic, nonexplosive, and nonflammable industrial products.
   b. It should have a high boiling point, low freezing point, high flash point, and no coke charring under high temperatures.
   c. The material shall have good thermal physical properties and high thermal conductivity. The regular solid material is required to exceed 2 W/(m K). It should also exhibit high specific thermal capacity and high density, as well as low fluid viscosity under high temperatures.
   d. The thermal storage material and tank, as well as the pipeline and valve, shall be compatible with each other.
2. Thermal storage tank selection principles
   a. The design table of technical features for a regular tank normally includes tank type, designed pressure, designed temperature, medium, geometric volume, corrosion allowance, welding seam coefficient, and materials of major pressurized elements.
   b. The stress yield point of the tank material exceeds the thermal storage working temperature of 100°C.
   c. The pressure tank shall be designed in accordance with the regulations of "Pressure Tank Safety and Technical Supervision Regulation" issued by the General Administration of Quality Supervision, Inspection and Quarantine.
   d. Drilling on the tank shall be in line with the regulations of Section 8.2 of Chinese State Standard GB150, the reinforcement calculation of which is normally required, unless the conditions in Section 8.3 of Chinese State Standard GB 150 are satisfied. When selecting connecting pipes, the conditions in Section 8.3 of Chinese State Standard GB 150 shall be satisfied, and optimum safety and economy shall be achieved as much as possible while avoiding extra reinforcement rings.
3. Thermal storage tanks shall be arranged for facilitating waste discharge while satisfying the following requirements:
   a. The thermal storage tank shall be designed with waste liquid discharge holes at the bottom and a slope with a grade of not less than 1%.

b. A waste ditch shall be arranged while striving for shortness and straightness; its direction and elevation shall not influence the expansion. This ditch mainly handles problems such as the treatment of charred heat transfer fluid and thermal storage fluid and the discharge of aged wastes. Measures shall be taken to handle wastes with oil contamination. The volatile toxicity of high-temperature oil or molten salt when discharging wastes under high temperatures is significant, and thus personal protection must be considered.
c. No sewage, wastewater within the power plant, or plant area rainwater shall be discharged into the thermal storage waste ditch so as to avoid the occurrence of any chemical reactions and the release of harmful gases.

6.6 CHARGING AND DISCHARGE EQUIPMENT OF THE THERMAL STORAGE TANK AND RESPECTIVE PROCESS DESIGN

1. Equipment selection principles. The charging and discharge of the thermal storage tank are conducted through the heat exchanger, which can be selected according to the following principles:
   a. thermal load and flow rate, fluid properties, temperature, allowable range of pressure and pressure drop, requirements for cleaning and servicing, equipment structure, dimension, weight, price, operational safety, and service life;
   b. heat exchanger properties that usually include shell-and-tube-type pressure from fine vacuum to 41.5 MPa and temperature from −100 to 1000°C. The shell-and-tube-type heat exchanger shall be designed according to Chinese State Standard GB151-2011; other heat exchanger types mainly include plate, air-cooling, spiral-plate, multitube, baffle-plate, plate-fin, spiral-tube, and thermal pipe.
2. Solid material heat exchanger. For solid thermal storage materials, a heat exchanger can be mounted inside the thermal storage. For example, for material used in ceramic and concrete heat exchange, the temperature difference at the low temperature end shall be not less than 20°C.

   Due to the discontinuity of solar irradiation, the heat exchanger inside the thermal storage material shall be designed while fully considering problems such as inconsistency of the thermal expansion coefficient and separation of the heat exchanger from solid material caused by multiple times of thermal shock.

3. Evaporator. The evaporator shall be designed with a focus on the designed pressure, temperature difference, fouling coefficient, and boiling point range of the thermal charging and discharging fluid. For the high-pressure evaporator, it is better to select the tank type or built-in type. The oil/water heat exchanger shall be designed while fully considering the pressure difference of fluids under thermal state.
4. Dry-cooling heat exchanger. The dry-cooling heat exchanger shall be designed by following standard GB/T 15386-1994.

6.7 THERMAL STORAGE SYSTEM CONTROL

6.7.1 Constitution of the Control System

The control system's purpose is to charge and discharge thermal energy toward the thermal storage. The thermal storage control system consists of:

1. temperature measurement including thermal storage material temperature, tank wall surface temperature, and heat exchanger fluid inlet and outlet temperatures;
2. flow rate measurement and heat transfer fluid flow rate;
3. pressure measurement and heat transfer fluid resistance;
4. electric motor of pump or fan;
5. heat exchanger valve;
6. thermal storage system DCS.

6.7.2 Control Logic of the Thermal Storage System

The thermal storage control mainly includes thermal storage charging and discharging processes:

1. Heat charging process. Firstly, the temperatures of the pipeline, pump, valve, heat exchanger, tank, etc. shall be checked to determine whether they have reached the set value above the fluid freezing point. If not, they shall be preheated in advance. When fluid flows into the thermal charging exchanger or inside the tank for charging heat, the temperature and liquid level shall be monitored to ensure that they are below the set value. After the thermal charging process, the pump or valve shall be closed. In case of unsteady state variation of heat charging fluid during the respective process due to irradiation and changes in meteorological conditions, the flow rate control device for fluid on the other side of the heat

exchanger shall be interlocked at the same time to ensure a safe and normal heat exchange process.

2. Heat discharging process. Firstly, the temperatures of the pipeline, pump, valve, heat exchanger, tank, etc. shall be checked to determine whether they have reached the set value above the fluid freezing point. If not, they shall be preheated in advance. Heat discharging is normally a water evaporation process, so the thermal discharging exchanger is also referred to as the evaporator. Pumping of the thermal storage thermal discharging exchanger shall be initiated to heat or evaporate water. The residual thermal storage capacity of the thermal storage tank shall be monitored so that the working mode of the fluid pump and other elements of the thermal discharging exchanger can be adjusted in a timely manner. To change the steam from subcooled to superheated, it is necessary to control different flow rates of thermal storage media of the heater/evaporator so that the evaporator can steadily produce superheated steam.

6.8 FACILITIES FOR THERMAL STORAGE SYSTEM INSPECTION

1. Cleaning. The heat exchange pipeline in the thermal storage tank shall be cleaned on a regular basis. A pipeline cleaning system shall be designed; in addition, when pipelines are designed in the thermal storage tank, the inclination angle shall be considered so that it can facilitate cleaning and discharge.
2. Transportation. A thermal storage heat exchanger overhaul yard and hoisting facilities shall be arranged and shall be equipped with overhauling tools and spare parts.
3. Temporary discharge point. For a large-scale thermal storage tank that uses a liquid thermal storage medium, temporary discharge points for thermal storage fluid during overhaul shall be designed. For a facility that uses inflammable fluid as the thermal storage medium, during overhaul of the tank, attention shall be paid to fire hazards caused by welding and other high-temperature operations.
4. Material replacement. In a solid thermal storage system such as a concrete or ceramic thermal storage system, for the heat transfer pipeline embedded in the thermal storage material, the replacement of metal pipelines shall be considered when experiencing corrosion and the like.

CHAPTER

7

Site Selection, Power Load, and Power Generation Procedures

7.1 SITE SELECTION

Compared with a photovoltaic system as a technical form of solar power generation, solar thermal power (also known as concentrating solar power, CSP) generation features steady power output, which is an important factor for the electricity grid. Furthermore, large-scale CSP plants have the potential to supply power for the basic-load power market. Thus CSP technology will occupy an important position in China's future energy strategy. Site selection of CSP plants has a direct influence on generating costs, and therefore site selection is quite important. Siting of CSP plants shall consider various factors, including solar direct normal irradiation (DNI) resources, land and topography, local water resource conditions, and traffic and power grid coverage, for which solar DNI serves as the most basic and important reference for CSP plant site selection. The precision and reliability of data directly influences the generating costs of CSP generation. Earthquake probability, meteorology, topography, water sources, traffic and transportation, outgoing lines, the thermal supply pipeline, geology, hydrology, environmental protection, and comprehensive utilization are also site selection factors. Under the premise of considering solar irradiation as proposed in this section as the primary reference factor, areas in China currently appropriate for the construction of CSP plants have been screened.

7.1.1 Principles of Site Selection

Sites for solar power plants shall be selected by integrating the national and local solar energy utilization scheme, land use plans, thermodynamic

Design of Solar Thermal Power Plants
https://doi.org/10.1016/B978-0-12-815613-1.00007-9

and power system plans, regional construction and planning, combined thermodynamic and power loads, and solar resource planning.

7.1.2 Macro Site Selection

In terms of macro site selection, quicksand, swamps, forests, salt lake basins, and other inappropriate areas with a slope above 1% can be removed as possible sites by using a geographic information system while considering the distance between the power plant and power grid. A map of locations appropriate for CSP plant construction can then be obtained.

7.1.3 Ecological Protection

In areas under land use control, both urban and rural, arrangement of the power plant's concentration field shall consider plant growth and the migration paths of birds.

Water used to clean heliostats shall be circulated or integrated with the spray or drip irrigation of ground-plant growth. Cleaning agents containing chemical detergents shall not be allowed.

7.1.4 Placing of Thermal Storage Tank

The determination of high-temperature, high-pressure thermal-storage tank locations shall consider (1) the achievable minimum heat transfer distance and (2) their influence on plant area safety. Because thermal storage is a high-temperature process, long-distance thermal transmission features great thermal losses. Thus thermal storage tanks shall be located around the turbo-generator unit to the greatest extent possible.

7.1.5 Solar Resource and Site Selection

The range of cumulative annual solar DNI during site selection is shown in Table 7.1.

TABLE 7.1 Annual Cumulative Solar Direct Normal Irradiance and Site Selection [49]

Not Recommended	DNI < 1600 kWh/(m^2·a)
Recommended	DNI $= 1600–2000$ kWh/(m^2·a)
Good	DNI > 2000 kWh/(m^2·a)

- During site selection, the concentration field shall be designed to avoid any light pollution toward the surrounding ground and atmosphere at any time.
- During site selection, in order to determine the feedwater source as well as the water usage, consumption, and source for the design, the following requirements shall be satisfied. Unrecoverable water usage includes cooling tower evaporation loss, cooling tower waste discharge, domestic and firefighting water discharge, and mirror surface cleaning water. Cooling methods for CSP plants normally include two technical forms, namely water cooling and air cooling. According to data from the US Department of Energy (2007), when applying water-cooling techniques besides the disc-type Stirling power system [0.0757 $m^3/(MWh)$], water usage for other technical forms is normally within 2.27–3.02 $m^3/(MWh)$, with water consumption by a tower power plant of about 2.27 $m^3/(MWh)$ and that of a parabolic trough power plant of about 3.02 $m^3/(MWh)$. When air cooling techniques are applied, water consumption by an CSP plant is greatly reduced and is about 0.299 $m^3/(MWh)$; meanwhile, this also results in increased investment cost and decreased generating capacity, with the former accounting for about 7%–9% and the latter accounting for about 5%.
- The feedwater source must be reliable. When determining the feedwater capability of the source, local agricultural, industrial, and domestic water usage conditions as well as the influences of water conservancy planning and climate on water source variations must be determined. Potentially frozen water during winter in northwest areas shall be considered, and attention shall be paid to corrosion by the chemical components of water on absorber metals.
- When using underground water sources, existing underground prospection data shall be fully utilized; when current data are insufficient, hydrogeological prospection shall be carried out while providing a hydrogeological prospection evaluation report in accordance with relevant regulatory requirements for hydrogeological prospection.

7.1.6 Selection of Land

The following requirements shall be satisfied:

- In terms of selecting land to construct CSP generation projects, under the premise of being in line with the overall land utilization plan, priority shall be given to the usage of barren mountains, wastelands, deserts, and other land that is difficult to utilize and is inappropriate for agricultural, ecological, and industrial development; efforts shall be made to not occupy farmland, or if that is not possible, to occupy less farmland. CSP generation companies are encouraged to utilize roofs or land with underlying mineral reserves to construct photovoltaic power generation projects. The relationship between CSP generation project construction and the natural environment, ecological protection, military facilities, mineral resource exploitation, and land for the construction of other industrial projects shall be coordinated on a reasonable basis. Instead of occupying fertile farmland, the project shall save land or occupy less fertile farmland as well as striving to utilize wasteland and scabland. In terms of the tower power plant, mountain slopes also can be utilized. High mountains or slopes to the south and north of a solar tower, when present, may facilitate a reduction in system costs.

- The land-use range of the power plant shall be determined according to planned capacity while offering a map of phased land expropriation or leasing based on the requests of housing and construction. Land for constructing CSP generation projects includes land for the solar collector field, the power generation production area, the domestic area, and permanent roads.
- The nature of land use for CSP generation, because CSP plants typically have large land coverage, is an important element in government approval. Currently, the nature of land use for CSP generation has not been specified in any government files, especially the nature of and calculation methods for land coverage of the collector field. Referring to land use calculation principles used photovoltaic power generation projects, some local regulations have proposed that "as for the photovoltaic power generation project being supplied with land in the form of allocation, land between its solar panel arrays shall maintain the original type without any change, and shall not be transformed into construction land" [50].

7.1.7 Determination of Site Elevation

The following requirements shall be satisfied:

- The site elevation shall exceed the level of the hundred-year flood. If lower than this, the plant area shall be equipped with reliable flood-control facilities and be fully completed during the initial stages of construction. When implementing flood-control measures in the concentration field, attention shall be paid to ground surface protection. Because the concentration field area is large and the ground ecology in northwest areas is fragile, recovery costs that result from significant ground surface damage are also large.

- The outdoor floor elevation design around the main powerhouse shall exceed the level of the hundred-year flood by 0.5 m, and the concentration field area shall be large. The concentration field floor elevation for a power plant built in South China shall satisfy the foregoing requirements, whereas for a power plant built in the arid areas of the northwest, waterproof drainage ditches shall be designed.
- For a power plant located on a riverside, next to a riverbank, or next to a lakeside, the crown elevation of the flood protection embankment shall exceed the level of the hundred-year flood by 0.5 m. When operating an CSP plant in combination with a seawater desalination system, if the power plant is located next to the ocean, the stability of the designed foundation of the solar concentration field must be considered. In the event of foundation shaking or sinking, the entire concentration field would be damaged.
- For a power plant located on a riverside, the crown elevation of the flood protection embankment shall be determined as the sum of the 50-year high-water or sea level, the wave run-up corresponding to 1% of the cumulative frequency of the 50-year return period, and the safety elevation of 0.5 m.
- When constructing a power plant in an area where waterlogging is dominant, the crown elevation of the waterlogging harnessing embankment shall be determined as the sum of the maximum waterlogging level in history and the safety elevation of 0.5 m. If waterlogging harnessing facilities are present, it can be determined as the sum of the designed waterlogging level and the safety elevation of 0.5 m. An embankment shall be fully completed during the initial stage. As the concentration field has large land coverage, the economy of the embankment shall be considered; adding a protecting embankment for each foundation shall also be considered.
- Flood control standards for the company's self-owned power plant shall be consistent with those of the company.
- During site selection, geological engineering data and regional geological site conditions must be determined. Where local geological conditions are appropriate, it is suggested that buildings and structures be built on natural foundations; considering the stability requirement, the foundation of the concentrator must be made of reinforced concrete.

7.1.8 Seismic Intensity at the Power Plant Site

Seismic intensity shall be determined according to the Earthquake Intensity Zoning Map of China released by the China Earthquake Administration. Considering that the concentrators cannot be used after an earthquake due to displacement and minor deformation, the CSP plant shall not be built on an active seismic zone.

7.1.9 Determination of Site Location

The following requirements shall be satisfied [51]:

- The power plant site shall be selected according to the range of annual cumulative solar DNI recommended in Table 7.1.
- The power plant site shall not be located on dangerous rocks, landslide sections, karst development areas, mudslide sections, seismic fracture zones; not be located in sections with landslide, avalanche, and subsidence during earthquakes; and shall avoid quicksand, swamps, forests, salt lake basins, and other inappropriate areas.
- The power plant site shall avoid cultural relics and scenic spots under special protection. It is suggested that it not be located in densely populated residential areas or on promising deposits; and shall avoid areas with military purposes and those that require the demolition of numerous buildings. The heliostat concentration field of a high solar tower shall avoid mutual interference with aircraft routes.
- A power plant site selected in a mountainous area on a hillside or hilly land shall not ruin natural terrain. In addition, the site will require a large area of flat ground for installing collectors, mirrors, and the like. Such a plant requires sufficiently broad ground for the layout of all equipment and shall be very flat with few slopes; the allowable slope technical standard is 3% or less so that both the oblique sunlight loss and the ground leveling workload can be reduced. In the Northern Hemisphere, it is hoped that the sloped site faces south so that losses related to incident angle can be reduced. Low latitudes shall be selected to reduce incident angle losses as much as possible. It is recommended that the latitude does not exceed 42 degrees.
- The power plant shall be located in an area with low wind speeds so as to reduce concentrator costs.
- Efforts shall be made to avoid locating a power plant in an area with frequent hailstorms or sandstorms. Hailstorm impacts may damage mirrors, and sandstorm sands cover mirror surfaces, thus making sunlight concentration on the collection tubes more difficult. In such a case, the cleaning frequency of mirror surfaces will rise, and maintenance costs will increase accordingly.
- During site selection, construction and installation fields shall be planned, including concentrators, thermal storage facilities, and concentration field lightning protection.

- In addition to power-generating equipment, the power plant shall be equipped with a large-span high-rise assembly workshop. Dimensions of the building shall be determined in accordance with the dimensions of the concentrator, whereas gate dimensions shall satisfy the traffic requirements of vehicles that transport concentrators.
- During site selection, according to meteorological and topographic factors, the influences of wastewater, waste oil, and waste thermal storage materials discharged by the power plant into the ambient environment shall be in line with the relevant regulations of existing national environmental protection standards.

7.1.10 Location Selection of Power Plant Residential Area

The following requirements shall be satisfied:

- The location of the power plant residential area shall be determined in accordance with the facilitation of production and life while being in line with the relevant regulations of existing national hygienic standards.
- The residential area is best located on the windward side of the plant in the area that experiences the minimum wind direction frequency throughout the year.
- The residential area of the company's self-owned power plant shall be planned in a unified manner with the residential area of the company.
- When planning a residential area, it must avoid the influences of harmful substances discharged by surrounding industrial companies.

7.2 POWER LOAD AND POWER GENERATION PROCEDURES

7.2.1 Power Load Data

The construction unit shall offer the design unit the short-term and long-term annual power load data for the construction area.

- Power load data shall include the following:
  - Major power users' existing and new production scales, main products and output, power consumption, power load constitution and properties, maximum power load and respective utilization hours, first-level power load proportion, and other details;

  - The annual power load of industrial production and development in the area to be supplied with power;
  - The annual power load of agricultural production, irrigation, and water conservancy construction and development in the area to be supplied with power;
  - The annual power load of municipal life development in the area to be supplied with power.
- Power load data shall specify load distribution.
- Power load data shall be reviewed; users with large power loads shall be analyzed and verified.

7.2.2 Power Load Plan

According to the power supply development scheme and power load data of the construction area, short-term and long-term power for the area shall be balanced; if necessary, power capacity shall be balanced as well.

7.2.3 Time Selection of Power Output

Overall procedures for CSP generation shall be designed by following technical and economic principles. For a peak-regulation power plant, the layout of the concentration field shall satisfy the peak power load requirement, which might vary by season. For a basic-load power plant, the layout of the concentration field shall satisfy the requirement of achieving maximum mean annual efficiency. The range of steam turbine parameters for an CSP plant shall be determined according to the concentration ratio and type of heat transfer media, while the thermal storage capacity and working temperature shall be determined based on the requirements of the steam turbine. For a hybrid power plant with both fossil fuel and CSP generation, the ratio of fossil fuels to total energy shall be in line with national laws and regulations.

CHAPTER

8

Plant Layout Planning

8.1 BASIC RULES

8.1.1 Principles of Power Plant Layout Planning

Buildings, structures, concentrators, solar towers, pipelines, and transportation routes within the plant area shall be arranged reasonably while giving overall consideration to the requirements of geographical latitude and longitude, elevation, solar radiation resources, wind speed, wind direction, production processes, transportation, fireproofing, explosion-proofing, environmental protection, hygiene, construction, and living, as well as integration with natural conditions like plan area terrain, geology, hydrology, earthquake risk, and meteorology according to planned capacity and focused on the short term so as to smooth the process flow, conveniently conduct overhaul and maintenance, facilitate construction, and assist expansion.

Plant area planning for the self-owned power plant of the company shall be consistent with the company's general layout.

8.1.2 Requirements for Power Plant Layout Planning and Design

- Area planning for the power plant shall be designed according to planned capacity. For a tower power plant, the design height of the solar tower shall fully consider future expansion of the solar power plant; for a parabolic trough power plant, the expansion of concentrator length as well as total area shall be considered. Attention shall be paid to the distance between high-temperature heat transfer fluid and generating facilities, and shall be shortened to the largest extent possible. During phased constructions of the power plant, the overall plan shall correctly handle the relationship between

Design of Solar Thermal Power Plants
https://doi.org/10.1016/B978-0-12-815613-1.00008-0

short-term and long-term targets; the short-term target is to achieve a centralized layout, whereas the long-term target is to reserve space for development. As for a hybrid power plant using both solar energy and fossil fuels, fossil fuel pipelines shall be piled up separately from solar concentration and thermal storage units.

- Area planning for an expanded power plant shall be conducted and modified in overall scale while integrating with the production system and layout characteristics of the old plant. The concentration field of the power plant shall be constructed with attention paid to soil and vegetation cover protection, and reservation of the steam-generating system header for concentration field expansion.
- The plant layout and spatial combination of buildings and structures within the plant area shall be compact and reasonable, colors shall be consistent with those of the concentrator, and the respective heights shall not block light in the concentration field.
- The solar concentration field shall be arranged while reserving routes for clean vehicles, and the respective sewage treatment pipelines shall be reserved. Parking space for overhaul vehicles, space for hoisting equipment during overhaul, and temporary storage space for items shall also be reserved.
- While completing planning in front of the plant, buildings shall not cast shadows on the concentrators. As for auxiliary workshops and accessory buildings, alliance buildings and multilayered buildings are suggested; residential areas shall apply multilayered buildings.
- The building types and layout of the company's self-owned power plant shall be consistent with the style of the company and building; a regional power plant shall be consistent with the style of buildings in the respective town.

8.1.3 Notices for Plant Layout Planning

The plant area shall be arranged by centering the main powerhouse; concentrators and the thermal storage system shall be located close to the main powerhouse to the largest extent possible.

In sections featuring complex terrains, terrain features can be considered in selecting an appropriate plant layout for buildings and structures. It is suggested that main long axes of buildings and structures be located along the natural contour line.

As for a power plant that is required to be protected based on the seismic intensity, the construction field is suggested to be located at a favorable section and buildings shall be concisely and neatly shaped. During design of concentrators, earthquake prevention measures shall be taken, especially that foundation of concentrator shall be designed while considering seismic deformation, etc.

Inlets and outlets of heat transfer medium of collectors within the power plant shall be located not far from the main power generation system or thermal storage system. As for the molten salt power plant, the chemical properties and toxicity of salt shall be fully considered, salt storage tanks shall be sufficiently protected, salt pipelines shall be designed with leakage-proof early-warning measures, and the molten salt shall be as far from living and office areas as possible.

8.1.4 Plant Landscape Layout

The following requirements shall be satisfied:

- The plant landscape layout shall be implemented according to planned capacity and production features while integrating requirements of the overall plane layout, environmental protection, plant appearance beautification, and local natural conditions.
- Major afforestation sections shall be scheduled on both sides of the main access, at the main entrance and exit of the plant area, and around the main power house, main auxiliary buildings, and coal yard.
- Afforestation in outdoor power distribution unit sections shall satisfy requirements for the safe distance of electrical equipment.
- The afforestation coefficient is suggested as 10%–15%, and the concentration field's afforestation coefficient shall be not less than 60%; as many low plants or crops shall be planted as possible at the bottoms of tower-power heliostats and parabolic trough collectors.
- Plant area afforestation of the company's self-owned power plant shall tally with company requirements for afforestation planning.

8.1.5 Directions of Main Buildings

It is suggested that building direction be determined by integrating factors such as sunshine, natural ventilation, and natural day lighting. Directions of the heliostat concentration field mainly depend on the annual efficiency of the concentration field and requirements for ground plantation.

8.1.6 Solar Tower Thermal Resistance and Fireproofing

Protective measures shall be taken for the architectural supporting structure around the aperture of the solar tower receiver so that it can withstand 200°C. Overheating protection equipment for this part shall consist of high-temperature thermal insulating materials, the temperature sensor, and the heliostat concentration field controller. When the temperature of this part exceeds the set safety temperature, heliostats will be quickly reset to initial states to prevent concentrated solar beam from burning and damaging the solar tower.

8.2 LAYOUT OF THE MAIN BUILDINGS AND CONCENTRATION FIELD

8.2.1 Determination of the Main Powerhouse's Location

The main powerhouse has been mounted with steam turbines, the water production and chemical water system, and the control room. The following requirements shall be satisfied:

- It shall satisfy the process flow, smoothen the route, be shortly and promptly connected to the exterior pipeline, and be close to the solar concentration field and thermal storage tank;
- When once-through cooling water is used, it is suggested that the main powerhouse be close to the water intake;
- The orientation of the turbine house shall smooth the high-voltage power transmission of outgoing lines; in areas with high temperatures during summer, it is suggested that the turbine house faces the air duct of the concentration field.

8.2.2 Solar Tower Layout of a Tower Power Plant

The following requirements shall be satisfied:

- The distance between the tower power plant's solar tower and the first row of heliostats shall be not less than the length of shadow to the north on winter solstice. The solar tower shall not be too wide. It is better to apply the hollowed-out structure or steel structure tower so the shadow of the tower has minimum influence on the concentration field, and the first row of heliostats is close to the tower.
- The distance between the solar tower and thermal storage tank, as well as the distance between the solar tower and evaporator, shall not be too large, so as to strive for the reduction of thermal fluid heat losses during transmission. The steam turbine system and the like can be mounted inside the solar tower in order to shorten the transmission distance of high-temperature fluid and control heat losses during transmission.
- Antiexplosion and insulation measures must be taken in the area between thermal storage tanks with high-pressure, explosive, and toxic media and the solar tower.

- It is suggested that the cooling tower and/or spray fountain be close to the turbine house while satisfying the minimum requirement for protective spacing.
- The cooling tower and/or spray fountain shall not be mounted on the windward side of the outdoor power distribution unit and concentration field under prevailing winter winds; if they were, the concentrator surface will become frozen.

8.2.3 Road Arrangement Within the Concentration Field

Roads inside the concentration field shall be designed while considering the turning radius of large maintenance vehicles and the hoisting height of concentrators; road load shall be designed while considering the weight of hoisting.

Expansion of the power plant shall be designed with specific entrances and exits for construction. As for large concentrators delivered to the plant area, the road grade within the plant area shall be designed while considering the load of large vehicles and loaded weight.

8.2.4 Enclosure Wall of the Plant

When enclosure walls are designed outside the concentration field, settlement of sands in the concentration field shall be considered. After designing enclosure walls, the wind speed is reduced, and sands can easily settle in the concentration field in the downwind direction of enclosure walls. Normally, as the concentration field is a large area, priority can be given to the application of railing pattern enclosure walls. In terms of the concentration field built in the Gobi Desert, its protective structure shall prevent large wild animals from entering so as to protect interior controllers and cables. Around outdoor power distribution unit, oil depot, oil tank farm, and other areas with burning and explosion dangers, solid enclosure walls shall be designed according to requirements regarding fire and explosion prevention.

8.3 COMMUNICATION AND TRANSPORTATION

1. Layout requirement on roads within the plant area.
   a. It shall satisfy the requirements for production and firefighting and be consistent with vertical and pipeline layouts as well as roads within the concentration field.
   b. Circumferential roads shall be designed around the main powerhouse.

c. To maintain and clean concentrators, circumferential roads are suggested to be designed within the concentration field. As for the parabolic trough power plant, roads within the concentration field and thermal collection fluid pipelines shall be designed while causing no mutual interference. In the case that heat transfer pipeline layout on the parabolic trough section lies on the same height level with the floor contour line, dead-end roads shall be designed. Distance of neighboring parabolic trough concentrators shall be designed while considering turning radius of vehicles, and the turnaround loop or a turnaround yard with the area of not less than 12 m × 12 m shall be designed.

d. The main access road to the power plant shall be connected to the existing road leading to the town. It shall be short in length; efforts shall be made to avoid the crossover between it and the railway line. In case of grade crossing, crossings and other safety facilities shall be designed.

e. Feed water and discharge buildings, water sources, and dock and residential areas within and outside of the plant area shall be connected by roads.

2. Design requirements for roads within the plant area where the main power generator has been located.
   a. It is suggested to use concrete road surface or asphalt road surface.
   b. Width of the travelling section of the main access road is suggested to be 6–7 m.
   c. As for a hybrid power plant that uses automobiles to transport fuel or natural gas, the width of the travelling section of its main access road is suggested to be 7 m.
   d. Width of other major roads shall be determined according to traffic and service conditions, in which the width of the one-way lane can be 3.5–4 m. Roads shall be designed while considering the transportation of major thermal storage tanks.
   e. Sidewalk width shall be not less than 1 m.
3. Roads within the concentration field shall be designed while mainly considering overhaul and cleaning; it is not necessary to design asphalt or concrete hardened road surfaces; instead, sands or detritus can be used to pave the roads.

8.4 VERTICAL LAYOUT

1. The pattern and designed elevation of the vertical layout of the entire plant including the concentration field. It shall be determined according to production process requirements, shading and blocking of the concentrators at work, transportation, pipeline layout,

and embedded depth of the foundation by integrating with plant area terrain, engineering geology, and specific hydrological and meteorological conditions.

2. The power plant area drainage structure shall be designed in a unified manner according to planned capacity and land coverage while preventing rainwater gathered by external roads of the plant from flowing inside. The drainage for a small-scale CSP plant (<500 kW) concentration field shall be designed together with the power plant area.
   Site drainage of the company's self-owned power plant shall be consistent with the site drainage design of the company.
   As for a large-scale CSP plant, because the concentration field's land coverage is large, there are plenty of small wild animals. An internal drainage ditch is not suggested in the concentration field during drainage design. Foundation design of the concentration field and cable design of the concentrator shall consider the functions of flood control and waterlogging prevention:
3. Site drainage modes within the power plant area:
   a. Site drainage within the power plant area should be carried out by applying the integrated method of urban-type road surface drainage groove and ditch drainage or buried drainage. If possible, free drainage shall be applied.
   b. For the tower power plant with a multistep layout, drainage measures shall be taken for each step.
   c. When the outdoor channel is higher than the designed floor elevation, discharge measures shall be taken.
   d. Drainage of the concentration field in a large-scale power plant shall be designed while taking advantage of topography and gradient drainage to the largest extent possible. As for the heliostat concentration field in a tower power plant, drainage can apply underground natural seepage. As for the concentration field in a parabolic trough or a linear Fresnel power plant, because the floor has already been handled, it is also possible to consider designing a drainage ditch when handling the floor. Considering winds and sands in the wild, it is suggested that the ditch drainage method not be applied.
4. Protective treatment shall be conducted to the drainage ditch inside the power plant. Its longitudinal grade shall be not less than 0.3%, and the start point depth shall be not less than 0.2 m. The bottom width of the trapezoidal cross-section shall be not less than 0.3 m, and the bottom width of rectangular section shall be not less than 0.4 m. The bottom width from the urban-type road surface drainage groove to the flume of the drainage ditch shall be not less than 0.3 m.

5. The level grade of the concentration field shall be subject to 0.5% –2%; in sections with difficulties, the minimum-level grade shall be not less than 0.3%; for the parabolic trough power plant, the maximum level grade in partial areas shall be determined according to the soil texture, but it shall not exceed 1%. As for the tower power plant, it shall not exceed 2%.
   During the design of the ground drainage grade, efforts shall be made to prevent groundwater from flowing into the cable trench, trench, and building.
6. Terrain layout. If the natural terrain of the plant area along with site conditions results in large engineering quantities of leveling earthwork on-site, it is suggested that it be arranged in multistep mode. According to the production process flow, transportation, buildings, structures, and the pipeline layout requirement, the power plant area shall have no more than three steps, and the height difference of the adjacent two steps shall be not more than 5 m.
   As for the solar tower plant heliostat concentration field or solar dish large-scale power plant, as they are large in land coverage, and the terrain might be nonuniform, drainage facilities can be designed in sections. Measures shall be taken to prevent rats in the drainage ditch.
7. Indoor bottom elevation of buildings and structures shall be higher than the outdoor floor elevation by 0.15–0.30 m. In terms of soft foundation, difference between indoor bottom elevation and outdoor floor elevation shall be increased according to precipitation amount.

8.5 PIPELINE LAYOUT

8.5.1 Underground Pipeline Layout of the Plant

The following requirements shall be satisfied:

- Pipeline layout shall facilitate construction and overhaul. Antifreezing measures shall be taken, especially on water pipelines for cleaning mirrors in winter. As the outdoor temperature is low in winter, non-water-cleaning methods can also be considered.
- In case of any accident to the pipeline, heat transfer fluid, especially such as synthetic oil and molten salt, shall not damage the foundations of buildings and structures, and also, sewage and toxic substances shall not permeate into the domestic feed water pipelines or flow into the cable trench.
- Pipelines inside the concentration field shall be designed with consideration given to the running load of large vehicles and the

required positions and loads of concentrators during overhaul. Lanes and service load points shall be designed outside the concentration field as much as possible.
- Embedded pipeline depth shall prevent the freezing of inside fluid, especially for pipelines inside the concentration field. In case of water, the embedded depth shall be below the local permanent frozen layer; for oil or molten salt, tracing measures shall be considered. For an oil pipeline, steam tracing shall be applied; normally, electric tracing shall not be applied.
- Main pipelines shall avoid passing through any expansion site.

8.5.2 Overhead Pipeline Layout

It shall not impede power plant traffic, as well as the natural daylighting and ventilation of buildings and concentrators, while striving for being neat and beautiful.

8.5.3 Coordination Between Pipelines and Ambient Environment

Pipelines shall be arranged to be parallel to roads and the axes of buildings, concentrators, and structures. The pipeline shall be laid outside of the road surface range; meanwhile, it shall not block sunlight to the concentrator throughout the year. Main pipes should approach buildings and the side with additional branch pipes. Intersections shall be eliminated between pipelines or between pipelines, railways, and roads; if necessary, a right-angle intersection shall be applied.

Under difficult conditions, underground pipelines can be laid within the road surface range.

When conditions for safe production and convenient overhaul are satisfied, pipelines may be arranged in the same trench or frame. Overhead pipelines should overlap underground pipelines, and partial secondary pipelines can be directly buried.

Horizontal distance between pipelines and buildings, structures, railways, roads, and other pipelines shall be determined according to various factors, including site geology, frame foundation form, manhole structure, embedded depth of the pipeline, diameter of the pipeline, and the properties of medium inside pipes.

CHAPTER

9

Main Powerhouse Layout

9.1 DIRECTION OF MAIN POWERHOUSE

Layout of the main powerhouse of power plant shall be in line with the thermal and power production process flow while achieving compact and reasonable equipment arrangement, short and neat pipeline connections, and concise and bright workshop layout.

9.2 MAIN POWERHOUSE AND THERMAL STORAGE

The main powerhouse and shall be located on the windward side of the power plant's thermal storage in order to avoid toxic gases and fire hazards in the event of accidents. The height of the main powerhouse shall not block sunlight from the concentration field at any time of the year.

9.3 SOLAR THERMAL STORAGE SYSTEM LAYOUT

As it is large in volume, the thermal storage system should not be located inside the main powerhouse when considering an expansion of thermal storage capacity.

Design of Solar Thermal Power Plants
https://doi.org/10.1016/B978-0-12-815613-1.00009-2

CHAPTER

10

Water Treatment Equipment and System

10.1 RECEIVER AND EVAPORATOR MAKEUP WATER TREATMENT

The receiver and evaporator makeup water treatment system shall be determined according to the quality of raw water, the quality standard of feed or boiler water, the makeup water rate, the blowdown rate, equipment and drug supply conditions, and effluent discharge as well as other factors after conducting comprehensive technical and economic comparisons. Meanwhile, the following requirements shall also be satisfied:

1. Treatment of receiver and evaporator makeup water shall adapt to devices inside the receiver and evaporator as well as the temperature reduction method for superheated steam.
2. If drums of the receiver and evaporator are connected through expansion joints, the selected chemical water treatment system must maintain the relative basicity of boiler water below 20%. If this requirement cannot be satisfied, a corrosion inhibitor should be added to the boiler water.

10.2 CALCULATION OF WATER TREATMENT EQUIPMENT

1. The output of water treatment equipment shall be determined by the sum of the overall normal water vapor loss, the water vapor loss caused by turbo-generator unit initiation or accidents, and water consumption for cleaning the power plant's concentrators.
2. Selection of the quantity of ion exchangers can be determined according to the following principle: (1) the quantity of each type

Design of Solar Thermal Power Plants
https://doi.org/10.1016/B978-0-12-815613-1.00010-9

of ion exchanger shall not be less than two; (2) normal regeneration per ion exchanger shall be subject to a schedule of one to two times daily; and (3) when program control is adopted, the schedule can be changed to two to three times daily.

When a piece of equipment is being overhauled, the outputs of other equipment shall satisfy the requirement of overall length normal makeup water consumption.

10.3 FEED WATER AND BOILER WATER MODIFICATION AND THERMAL SYSTEM STEAM SAMPLING

1. The receiver and evaporator shall be equipped with boiler water phosphate treatment facilities. Each receiver and evaporator shall be equipped with one dosing pump and one backup pump. In the case that several dosing pumps of the receiver and evaporator are mounted together, it is suggested that one public backup pump is added.
   The phosphate solution shall be prepared using desalinated (demineralized) or condensate water.
2. Feedwater of the receiver and evaporator shall be ammoniated. The dosing pump shall be equipped with one backup pump. In the case of several machine units using a single dosing pump, a ballast chamber shall be designed at the outlet pipeline of the dosing pump, and each dosing tube shall be mounted with a flowmeter.
   Ammonia solution shall be prepared using desalinated (demineralized) water or condensate water.
3. Correction treatment facilities for feed and boiler water shall be arranged inside the main powerhouse.
4. In configuring the water vapor sampler of the thermodynamic system, the system design, layout, and materials selection shall satisfy the following requirements:
   a. The temperature of the water vapor sample should be less than 30°C and shall not exceed 40°C.
   b. The water vapor sampling cooler of the receiver shall be mounted on the operating floor of the main powerhouse while facilitating sample-taking and access for the operator.
   c. The sampling tube shall not be too long.

10.4 ANTICORROSION

Equipment, pipelines, and valves that contact corrosive media or influence water quality shall be made of noncorrosive materials or be coated with appropriate anticorrosive coatings on the contacting surface.

CHAPTER

11

Power System

11.1 POWER GRID CONNECTION OF POWER PLANT

When a regional power grid is located near the power plant, the power plant shall be connected to the regional power grid. The voltage grade of the power plant connected to the regional power grid shall be determined according to the unit capacity and construction scale of the power plant and the specific conditions of the regional power grid after technical and economic comparisons of the access system design.

11.2 SYSTEM PROTECTION

The design of system relay protection and automatic safety devices shall be in line with the relevant regulations of the existing China state standard "Code for Design of Relay Protection and Automatic Device of Electric Power Installations."

11.3 SYSTEM COMMUNICATION

1. The design of system communication for the power plant connected to the power grid shall be determined according to the power system's communication design or the respective generator access system's design.
2. There shall be one reliable dispatching channel between the power plant and dispatching station.
3. A power-line carrier or other reliable communication method shall be applied as the system communication method.

Design of Solar Thermal Power Plants
https://doi.org/10.1016/B978-0-12-815613-1.00011-0

4. An automatic-switching double-circuit shall be used as the AC power supply for communication, whereas a floating-charge from a rectifier and storage battery shall be used as the DC power supply for communication. The capacity of the storage battery shall be selected based on 1 h discharging, or two groups of direct-supply rectifiers can be used to supply power. In the case of using a direct-supply rectifier to supply power or requiring the supply of AC power to be used in the communication device, a reliable emergency backup power supply shall be designed.
5. The system communication device of the power plant can share the machine room with the communication device in the plant area, the area of which shall be determined according to the number of communication devices in the long-range plan.

CHAPTER

12

Electrical Equipment and System

12.1 HIGH-VOLTAGE ELECTRICAL INSTALLATIONS

High-voltage electrical installations shall be designed according to the relevant rules in the existing China state standard "Code for Design of High Voltage Electrical Installations (3–110 kV)."

12.2 MAIN ELECTRICAL CONTROL ROOM

1. An independent main control room shall be designed in the power plant. The area of the main control room shall be designed according to the planned capacity and fully completed within the first phase of construction. It is best to directly observe the entire solar concentration field from the main control room; this requires that the height of the main control room exceeds the top of the solar concentrators by 2.5 m.
2. Layout of the first-phase engineering platform shall be integrated with the long-range plan in order to determine the mutual distance and channel width. It shall also satisfy requirements for phased expansion, operation, maintenance, and commissioning convenience.

12.3 DC SYSTEM

When calculating the capacity of the storage battery in the power plant connected to the power system, the outage time due to AC power accidents shall be calculated on the basis of 1 h; for the direct lubricating oil pump, the basis shall be 0.5 h; for power consumption of the solar concentrators, the basis shall be 5 min; and for the circulating pump of receiver, the basis shall be 0.5 h.

Design of Solar Thermal Power Plants
https://doi.org/10.1016/B978-0-12-815613-1.00012-2

12.4 ELECTRICAL MEASURING INSTRUMENT

Electrical measuring instruments in the power plant shall be designed according to the relevant rules in the existing China state standard "Code for Design of Electrical Measuring Device of Power System" (GB/T 50063).

12.5 RELAY PROTECTION AND AUTOMATIC SAFETY DEVICE

Relay protection and automatic safety devices in the power plant shall be designed according to the relevant rules in the existing China state standard "Code for Design of Relay protection and Automatic Device of Electric Power Installations" (GB/T 50062).

12.6 LIGHTING SYSTEM

1. The lighting system of the power plant shall use a power supply network with separate normal and emergency lighting, the voltage of which is suggested to be 220 V.
   Normal lighting shall be supplied with power by a low-voltage plant transformer that has been directly grounded at the common neutral point of the dynamic and lighting networks, whereas emergency lighting shall be supplied with power by storage batteries.
   Emergency lamps can be used as the required emergency lighting at the entrance and exit, in passageways, in the staircase of the main powerhouse, and for important workplaces far from the main powerhouse.
   Because a concentration field has a large area, night lighting is typically not designed. If necessary, mobile light sources can be used.
2. Low-voltage (36V and below) lighting is applied in the following locations:
   a. The voltage of portable working lamps for regular overhaul shall be 36 V.
   b. The voltage of the portable working lamp for the receiver, thermal storage, and metal tank overhaul shall be 12 V.
   c. The voltage of the portable working lamp for solar concentration field overhaul shall be 36 V.
   d. Requirements for obstacle lighting equipment mounted on the solar tower and other towering buildings or structures in the power plant shall be subject to the relevant rules of regional aviation and traffic authorities.

12.7 CABLE SELECTION AND LAYOUT

Cable selection and layout design of power plant shall be in line with the relevant rules of the existing China state standards and regulations.

The cable layout of the concentration field shall focus on damage caused by rats and accumulated water while considering cleaning and service vehicle access as well as the hoisting equipment for concentrators under service. As the concentration field has large land coverage, and considering economics, large-scale power plants shall directly bury armored cables.

12.8 OVERVOLTAGE PROTECTION AND GROUNDING

The generating module as well as the overvoltage protection and grounding of the concentration field shall tally with the relevant rules of the existing China state standard "Code for Design of Over-Voltage Protection of Industrial and Civil Electrical Installations" (GBJ 64) ["Code for Design of Over-Voltage Protection and Insulation Coordination of AC Electrical Installations" (GB/T 50064), Revised] and "Code for Design of AC Electrical Installations Earthing" (GB/T 50065).

12.9 ELECTRICAL INSTALLATIONS IN A DANGEROUS ENVIRONMENT WITH POTENTIAL EXPLOSIONS AND FIRE HAZARDS

The design of power plant electrical installations for a dangerous environment with the potential for explosions and fire hazards shall be in line with the relevant rules of the existing China state standard "Code for Design of Electrical Installations in a Dangerous Environment with Potential Explosions and Fire Hazards" (GB 50058).

CHAPTER

13

Thermal Automation

13.1 BASIC RULES

1. Thermal automation design shall include meteorological instruments, a solar radiometer set, a meteorological station, thermal inspection, a thermal alarm, thermal protection, thermal control, and a thermal automation laboratory. Each side of the concentration field shall be mounted with a meteorological station anemometer in order to obtain the maximum wind speed and guarantee concentrator safety.
2. The solar radiometer set shall include a pyrheliometer (measuring solar beam radiation at normal incidence), a pyranometer (measuring total hemispherical solar (beam plus diffuse) radiation, usually on a horizontal surface) and a pyranometer measuring diffuse radiation. The glass windows of the radiometer set shall be cleaned regularly.
3. During phased construction of the power plant, the control mode, equipment selection, and facilities related to thermal automation shall be comprehensively planned and reasonably arranged. The site requires the design of only one solar radiometer set and one ambient air temperature indicator.
4. The thermal automation design of the power plant shall apply mature control techniques and highly reliable, good-performing equipment, especially for the pyrheliometer. New products and techniques shall be qualified through testing and identification before their use in engineering applications.

Design of Solar Thermal Power Plants
https://doi.org/10.1016/B978-0-12-815613-1.00013-4

13.2 CONTROL MODE

1. The receiver, thermal storage, steam turbine, and deoxygenating feedwater system shall be controlled locally; control rooms shall be set up accordingly.
2. When using the receiver, thermal storage, and main machinery control room to monitor and control machine units, the following basic rules shall be satisfied:
   a. appropriate initiation and stoppage of machine units coordinated with local operators while observing the application of local control methods on thermal storage system initiation equipment;
   b. proper monitoring and control over regular operating conditions;
   c. adequate alarming for abnormal operating conditions and emergencies.

13.3 THERMAL INSPECTION

1. Thermal inspection design shall satisfy the requirements of machine unit safety and operational economy while being able to precisely measure and indicate the thermal technical parameters of various equipment in the technological system.
2. Indicating instruments shall be configured and equipped to reflect the major parameters of the receiver and concentrator with regard to safety and operational economy under various working conditions as well as general parameters that require regular monitoring.

13.4 AUTOMATIC ADJUSTMENT

1. The drum and liquid levels of the receiver and thermal storage system shall be designed to include an automatic feedwater adjustment function.
2. An receiver with mixed-type water spray desuperheater shall be included and shall provide automatic temperature adjustment of superheated steam.
3. A parallel-running steam generator (receiver or thermal storage) shall be designed with automatic output flow adjustment and shall be equipped with an automatic main steam pipe pressure adjustment valve.
4. Automatic adjustment of the steam-generating device can be realized through a microcontroller; if possible, it should be combined with a microcomputer monitoring system.

5. The pressure and temperature reducer of the receiver shall be equipped with an automatic pressure and temperature adjustment function.
6. Tanks that need to maintain certain liquid levels while functioning (such as the thermal storage oil tank and the drum) shall be equipped with an automatic liquid level adjustment function.

13.5 THERMAL PROTECTION

1. The thermal protection design shall be stable and reliable; contact signals used for protection shall be derived from specific passive primary instruments.
2. The work receiver shall be designed to include the following protection items:
   a. receiver steam overpressure protection
   b. receiver overheating protection, including for the temperatures of the various thermal surfaces of the receiver as well as temperatures at various points on the supporting structure of the solar tower

13.6 INTERLOCKING

1. Conditions for technological system interlocking shall be determined in accordance with the requirements of main and auxiliary equipment and technological system design.
2. Automatic charging and interlocking devices for the major auxiliary equipment in the solar concentration, thermodynamic, and auxiliary technological systems shall meet the following requirements:
   a. Backup auxiliary equipment, such as the feedwater or circulating pump, shall be equipped with automatic charging devices.
   b. The quick pressure reducer and temperature control for the receiver shall be equipped with automatic charging devices.
   c. Complete automatic interlocking devices shall be mounted among the auxiliary equipment in the solar concentration system.

13.7 POWER SUPPLY AND STEAM SUPPLY

1. Thermal instruments and controls shall be designed with safe and reliable AC/DC power. Computer-monitoring devices and concentrators, as well as the receiver system main pumps, shall be

designed with uninterruptible power supplies, especially for the receiver and concentrator.
2. The heliostat power distribution in a tower power plant shall be supplied with power from two routes in order to avoid concentrator outages and shutdowns. The concentrator in a parabolic trough power plant can be supplied with power in a single route; the power supply of heliostat may also rely on local photovoltaic batteries, which can greatly reduce the use of on-site cables.
3. The pneumatically driven receiver valves shall use compressed-air pumps mounted around the receiver; one pump shall be used as a backup pump.

13.8 CONTROL ROOM

1. The position and area of the control room layout shall be in line with the following requirements:
   a. The control room shall be located in an appropriate position relative to control equipment—for instance, in the middle of the solar concentration field. This position shall facilitate access to the emergency location in the concentration field.
   b. It shall facilitate the connection of conduits and cables to the control room.
2. Environmental facilities in the control room shall be in line with the following requirements:
   a. The side of the control room that faces equipment and the concentration field shall be designed with large glass windows so that the concentrator's operating conditions can be observed.
   b. Good antiglare facilities shall be mounted inside the control room in order to prevent stray heliostat and concentrator light from interfering with work.
   c. For safety, a power tower plant's control room shall not be located near the solar tower.
3. No process pipelines shall pass through the control room. No high-temperature gas, water pipelines, hot-air ducts, or oil pipelines shall pass through the cable interlayers and main channels below the control room.

13.9 CABLES, CONDUITS, AND LOCAL EQUIPMENT LAYOUT

1. Temperature-resistant wires and compensating conductors shall be used in high-temperature areas. Attention shall be

paid to preventing rats from damaging concentration field cables.

2. When digging cable trenches in the concentration field, water drainage shall be considered. Because concentration field topography is nonuniform, the drainage well and equipment shall be designed with full consideration of terrain features. Ideally, cables should be directly buried.
   When burying cables, the subgrade hardness of roads shall be considered in terms of future maintenance and measurement. Considering the high temperatures occurring with lightning protection and on the concentration light path, overhead cables are not advised in the heliostat concentration field of a power tower plant.
3. When various types of pipelines are laid parallel or intersect, the minimum clearance shall be in line with the rules of relevant existing national regulations.

13.10 BASIC RULES FOR BUILDING SPACE HEATING

The concentrated heating area is defined as an area with a stable daily mean temperature less than or equal to +5°C and with mean annual cumulative days exceeding or equal to 90 days.

When people regularly stay or work indoors, or there is a requirement for a certain room temperature, in the production workshop, auxiliary buildings, and accessory buildings located in the concentrated heating area, concentrated heating shall be designed using the thermal energy in the thermal storage tank as the thermal source.

Necessary space-heating equipment also includes thermal storage and a heat exchanger; hot water from heating can be considered for integration with antifreezing of the parabolic trough synthetic oil pipes.

13.11 SOLAR TOWER

1. In a power tower plant, the height of the solar tower shall be determined while considering both optical efficiency and height.
   In the case of a high solar tower, the intercept efficiency of the absorber may decrease due to tower swing.
2. The interior of the solar tower may not adopt heating. The thermal medium temperature shall be in line with rules in the existing China state standard "Code for Fire Protection Design of Buildings" (GB 50016).

3. The solar tower's interior elevator shall be designed and operated in line with China state standards, including fire protection inside the elevator and means of escape for personnel. For a steel tower, quick escape for personnel under high wind conditions shall be considered.
4. The same elevator shall not be used as both the manned and freight elevator inside the solar tower.
5. The designed load of the solar tower shall consider not only the dead load of the receiver, but also live load and weight during installation.

13.12 HEATING NETWORK AND HEATING STATION IN THE PLANT AREA

1. If hot water is the on-site heating medium, the heating network shall be equipped with heaters. Thermal storage can be used as the thermal source for plant area heating. To avoid continuous cloudy days, electric heating can be designed in the main buildings for auxiliary heating.
2. There shall be at least two sets of circulating water pumps in the heating network that connect plant area heating with thermal storage tanks, with one used for backup. Different water pumps shall have the same characteristics.
   The water pump flow rate shall be determined according to the designed thermal load and the heating network's designed supply–return water temperatures. The water pump head shall include the heating user's internal system resistance, outdoor pipe network resistance, heating network heating station internal equipment, and pipeline resistance.
   The water pump flow rate shall have a 10% margin; the water pump head shall have a 20% margin.
3. Deoxygenated demineralized water, continuously discharged sewage from the receiver, thermal storage, and evaporator, and condensing water from the steam heating system can be used as makeup water in the heating network system.
4. The heating pipeline valves lying inside the trench as well as accessories that need to be regularly maintained shall be designed with manholes.

CHAPTER

14

Architecture and Structure

14.1 BASIC RULES

1. The construction and structural design of a power plant must be conducted while implementing the "safe, applicable, economic, and beautiful" principles. Buildings within the plant area shall be consistent with the surrounding environment; buildings can be integrated with the solar tower; positions, heights, and shapes of buildings shall not influence the sun beam collection of the concentration field throughout the year.
2. Design and construction shall consider the possibility of using a solar tower to reduce building temperatures in summer through ventilation, and using high-temperature thermal storage to heat buildings in winter.
3. The architectural design shall allow for potential expansion; the land requirement for a multiple-energy complementary-type power plant shall also be considered.
4. The design of electrical and chemical buildings shall be in line with the following requirements:
   a. Measures shall be taken to prevent snakes, rats, and other small animals from damaging solar concentration field electrical installations, especially outdoor and network cables.
   b. The design of the retaining structure, floors, grounds, acid–base drainage ditch for the acid-storage battery chamber, acid-regulating chamber, acid–base measuring room, and dosing room, as well as drug storage, shall be in line with relevant rules of the existing China state standard "Code for anticorrosion design of industrial constructions" (GB 50046).
5. Settlement observation points shall be designed in the main powerhouse for observing steam turbine generators, thermal storage, and concentrators.

Design of Solar Thermal Power Plants
https://doi.org/10.1016/B978-0-12-815613-1.00014-6

6. If necessary, buildings and structures must be checked for explosion-proofness and vibration.
   The design of the solar tower and its floors shall be in line with the rules of the existing China state standard "Code for design of reinforced concrete silos" (GB 50077), and the solar tower and its floors must also be designed to withstand possible receiver explosions.
7. The fortification intensities of buildings and structures shall be determined in accordance with relevant national regulations. The seismic fortification requirements used for buildings and structures in the area that have an estimated China seismic intensity scale impact of six degrees and above shall be in line with the relevant rules of the existing China state standard "Code for seismic design of buildings" (GB 50011). A power tower plant's solar tower and heliostat foundations shall be subject to a fortification requirement of seismic intensity plus two degrees.

14.2 FIRE PROTECTION

1. The fire protection design of power plant buildings shall be in line with the relevant rules of the existing China state standard "Code for fire protection design of buildings" (GB 50016).
2. Vertical firewalls shall be designed among the power plant's turbine house, deaerator bay, synthetic oil storage tank, molten-salt storage tank, coal storage bunker, diesel tank, and auxiliary absorber chamber. Partition walls above the operating floor shall be constructed of noncombustible materials with a fire resistance limit of not less than 0.9 h, whereas partition walls below the operating floor shall use noncombustible materials with a fire resistance limit of not less than 4 h.
3. Fire-resistant or antiablative materials with a width of not less than 2 m shall be designed next to at all sides of the extension of the thermal-absorbing aperture on a power tower plant's solar tower while taking water-cooling protection measures. This section shall be designed while considering the structural deformation of materials after thermal expansion. Combustible materials may not be used in any section of the aperture and surrounding area of the solar tower.
4. In a workshop that uses and stores flammable and explosive liquids, underground trenches shall not be connected to the trenches of the adjacent workshop, and the sewer shall be equipped with water-seal or oil-separation facilities.

14.3 INTERIOR ENVIRONMENT

Building and window layouts shall be conducive to indoor cross-ventilation for a better view of the concentration field; they shall also avoid glare in the concentration field under various solar angles.

CHAPTER

15

Auxiliary and Affiliated Facilities

1. The power plant shall be designed according to turbo-generator unit capacity, type, quantity, equipment overhaul features, transportation, and other conditions and equipped with necessary metalworking facilities. Attention shall be paid to the processing of large parts and sophisticated pieces as well as castings. Assembly of solar concentrators can be solved by establishing temporary construction facilities on-site.
2. The structural design of thermal insulation of the solar thermal-absorbing section shall be in line with the following requirements:
   a. A good outer protective layer shall be applied to the insulating layer of the cavity receiver in the power tower plant, which shall be waterproof and flame-retardant, and its mechanical intensity shall satisfy the operational requirements of construction.
   b. When using hard insulating materials, expansion joints shall be reserved on straight pipes and elbows; if the vertical pipeline length exceeds 3 m, insulation weight holders with a clearance of 3~6 m shall be designed.
   c. Valves, flanges, and other components that must be disassembled during overhaul shall adopt a movable thermal insulation structure.
   d. If the pipeline can come in contact with concentrated solar radiation, highly reflective devices shall be mounted on the pipeline and the outside of the receiver.
3. For solar tower power generation, the surfaces of protective layers of the upper and lower pipelines on the solar tower shall be protected from concentrator radiation. For cavity receiver support, the steel structural surface near the aperture of receiver shall be made of highly reflective and fireproof materials.

Design of Solar Thermal Power Plants
https://doi.org/10.1016/B978-0-12-815613-1.00015-8

4. A synthetic oil storage tank for the solar collector shall be mounted outdoors, while the oil storage tank of the steam turbine shall be mounted outside the turbine house. Antifreezing and fireproofing measures shall be undertaken for oil storage tanks located in cold areas.

CHAPTER

16

Environmental Protection of the Concentrating Solar Power Plant

16.1 BASIC RULES

1. Power plant environmental protection design must be conducted while implementing the decrees, regulations, policies, standards, and rules on environmental protection promulgated by the national, provincial, and local governments or governments of autonomous regions and municipalities directly under the central government.
2. Power plant environmental protection design must be conducted while executing principles for harnessing pollution and the comprehensive utilization of resources, especially as they relate to ground damage during concentration field construction.
3. Possible soil pollution caused by leaking thermal storage materials must be fully considered; response, prevention, and rehabilitation treatment measures shall be appropriate.

16.2 REQUIREMENTS FOR ENVIRONMENTAL PROTECTION DESIGN

1. During design of the power plant, environmental impact statements shall be proposed during the feasibility study stage, and a special article on environmental protection shall be prepared; in the preliminary design stage, a special article on environmental protection shall be prepared, and engineering measures shall be proposed for the prevention and harnessing of pollution.
2. Except for design norms and requirements for solar thermal power generation, the special article on environmental protection of the

Design of Solar Thermal Power Plants
https://doi.org/10.1016/B978-0-12-815613-1.00016-X

power plant during the feasibility stage shall also include the following content:

a. recovery measures after land excavation in the solar concentration field
b. wastewater cleaning treatment-process flow chart, water-balance chart, and the landscape planning diagram for the solar concentration field

16.3 POLLUTION PREVENTION AND TREATMENT

1. In a concentrating solar power (CSP) plant, especially in a mixed-type power plant combined with a receiver that uses conventional fuels, emissions of atmospheric pollutants shall be in line with the existing China state standards "Emission standard of air pollutants for coal-fired power plant" and "Emission standard of air pollutants for receivers" as well as environmental quality requirements. Emissions shall also tally with the rules of emission standards promulgated by provincial and local governments of autonomous regions and municipalities directly under the central government.
2. Auxiliary boilers of the CSP plant must be equipped with dust wipers, and the respective dust-wiping efficiency shall be in line with relevant national and local emission standards and atmospheric environmental quality requirements.
3. A water-saving design shall be used for the power plant; wastewater from production and the mirror surface cleaning system shall be recycled and reused on a reasonable basis according to water quality and volume.
4. Receiver safety valve and ignition exhaust pipes that discharge waste toward the sky shall be equipped with mufflers.

16.4 ENVIRONMENTAL PROTECTION FACILITIES

The design of the power plant's environmental protection facilities shall include the following: emission treatment of wastewater from the mirror surface cleaning system or the adoption of water-saving cleaning measures; leakage and environmental protection problems for discharge and replacement of thermal storage materials and heat-transfer fluid.

References

[1] General Administration of Quality Supervision. Inspection and quarantine of the People's Republic of China, GB 50049–51994 code for small-size power plant. Beijing: Standards Press of China; 1994.

[2] Tarquini P. A new approach to concentrating solar plant (STE) by ENEA. Casaccia, Italia: ENEA; 2011.

[3] Zhifeng W, et al. "The 11th five-year plan" 863 program major project "STE generation technology and system demonstration" topic "overall design, technology and system integration of the solar tower-type thermal power generation system" (topic serial no. 2006AA050101) self-acceptance report. Beijing: Institute of Electrical Engineering Chinese Academy of Sciences; 2011.

[4] National Development and Reform Commission. Medium and long-term development plan for renewable energy, FGNY. 2007. No.2174.

[5] Qiang Y. Modeling and dynamic simulation of concentrating-thermal absorbing system of solar power tower plant. Beijing: Chinese Academy of Sciences; 2012.

[6] Duffie JA, Beckman WA. Solar engineering of thermal processes. 2nd ed. New York: A Wiley-Intersciences Publication; 1991.

[7] Chao Z. Solar power generation system design and optimization in Lhasa, Tibet. Information and Communications 2011;5:47–50.

[8] Draft for examination of solar photovoltaic power generation master plan (2010–2030) in Golmud. Water Conservancy and Hydropower Survey and Design Institute of Qinghai, Xining; Water Conservancy and Hydropower Survey and Design Institute of Qinghai; 2010.

[9] Juan Q. Research on temporal and spatial variation characteristics and formation mechanism of atmospheric boundary layer in northwest Arid areas. Beijing: Chinese Academy of Meteorological Sciences; 2009.

[10] Aini G, Aji D, Wubuli G. Analysis on influences of global climate change on eastern Xinjiang-on the example of Turpan basin. Journal of Jinggangshan University (Natural Science Edition) 2011;32(3):70–5.

[11] Junming W, Biqing T, Jing Y. On the relationship of climate and ecological environment improvement in the Karst region of Guizhou. Journal of Guizhou Meteorology 2003; 27(5):25–8.

[12] Shaoxia T, Zhizhong Z, Hua B, et al. Climatic resource characteristics and utilization of Hainan. Journal of Hainan Normal University 2008;21(3):343–6.

[13] Kunti W, Sheng W, Ming C. Comparison on climatic conditions of Taiwan and Hainan and influences on agriculture plantation. Chinese Journal of Tropical Crops 2006;27(3): 105–10.

[14] Portal website of Local Records Office of People's Government of Harbin City. December 8, 2011. http://dqw.harbin.gov.cn.

[15] Xiaowen L, Weiliang L, Xiuji Z. Research on China's solar radiation conditions in recent three decades. Journal of Applied Meteorological Science 1998;9(1):24–31.

[16] Liqin H, Changsheng L. Influences of clouds and aerosols on atmospheric absorption solar radiation. Plateau Meteorology 2001;20(3):264–70.

[17] Guangren L, Yuesi W, Fei H, Wen Z, Yuqiong H. 320m meteorological tower solar radiation gradient observation. Acta Energies Solaris Sinica 2003;24(3):295–301.

[18] Guo M, Sun F, Wang Z. The backward ray tracing with effective solar brightness used to simulate the concentrated flux map of a solar tower concentrator. In: Conference: 22nd International conference on concentrating solar power and chemical energy systems (SolarPACES). Abu Dhabi: United Arab Emirates: Masdar Inst Sci & Technol; October 11–14, 2016. Proceedings: AIP conference proceedings, vol. 1850: Article: UNSP 030023-1; 2017.

[19] The optical code of cylinder central receiver solar field layout (CCRSFL) is developed by Institute of Electrical Engineering Chinese Academy of Sciences; May 2017. http://cnste.org/html/zixun/2017/0525/1559.html.

[20] Zhifeng W, et al. Research report on badaling solar thermal experimental power plant. Institute of Electrical Engineering Chinese Academy of Sciences, Beijing; Institute of Electrical Engineering Chinese Academy of Sciences; 2012.

[21] Zhifeng W. An analysis on thermal performance of solar parabolic dish concentrator. Deajeon: Korean Institute of Energy Research; 2001.

[22] Burkholder F, Kutscher C. Heat loss testing of Schott's 2008 PTR70 parabolic trough receiver. Technical report NREL/TP-550–45633. Golden: NREL; 2009.

[23] ASHRAE. ANSI/ASHRAE standard 93-2010 methods of testing to determine the thermal performance of solar collectors. Atlanta (Geogia, USA): American Society of Heating, Refrigerating and Air-Conditioning Engineers, Inc.; 2010.

[24] CEN. EN 12975-2: 2006. Thermal solar systems and components-solar collectors-part 2: test methods. Brussels (Belgium): European Committee for Standardization Management Centre; 2006.

[25] Lippke F. Simulation of the part-load behavior of a 30 Mwe SEGS plant SAND95-1293. Albuquerque (New Mexico, USA): Sandia National Laboratories; 1995.

[26] Dudley VE, Kolb GJ, Sloan M, et al. Test results of SEGS LS-2 solar collector SAND94-1884. Albuquerque (New Mexico, USA): Sandia National Laboratories; 1994.

[27] Hosoya N, Peterka JA, Gee RC, et al. Wind tunnel tests of parabolic trough solar collectors NREL/SR-550-32282. Golden (Colorado, USA): National Renewable Energy Laboratory; 2008.

[28] ISO 9806-1. Test methods for solar collectors-part 1: thermal performance of glazed liquid heating collectors including pressure drop. Geneva (Switzerland): International Organization for Standardization; 1994.

[29] Fischer S, Lupfert E, Muller-Steinhagen H. Efficiency testing of parabolic trough collectors using the quasi-dynamic test procedure according to the European Standard EN 12975. In: Proceedings of the solar PACES 2006 conference. Seville (Spain): SolarPACES; 2006.

[30] Li X. Research on parabolic trough solar collector thermal performance dynamic test [Thesis]. Beijing: University of Chinese Academy of Sciences; 2013.

[31] Zhihao Y. Research on modeling and control logic of power tower generation system. Beijing: Chinese Academy of Sciences; 2009.

[32] Zhifeng W, et al. STE generation industry report (2011–2012). Beijing: China National Solar Thermal Energy Alliance, Industrial Securities Corp.; 2012.

[33] Minghuan G. Research on solar heliostat error analysis and concentration performance evaluation methods. Beijing: Chinese Academy of Sciences; 2011.

[34] Garcia P, Ferriere A, Bezian J-J. Codes for solar flux calculation dedicated to central receiver system applications: a comparative review. Solar Energy 2008;82(3):189–97.

[35] Ignacio Ortega J, Ignacio Burgaleta J, Tellez FM. Central receiver system (CRS) solar power plant using molten salt as heat transfer fluid [EB/OL]. December 1, 2013. http://www.sener.es/EPORTAL_DOCS/GENERAL/FILE-cwa0fcc36424ab41b7bf04/SOLARTRES.pdf.

[36] Dudley V, Kolb GJ, Mahoney AR, et al. Test results: SEGS LS-2, solar collector. SAND94–1884. 1994.
[37] Gong B. Badaling parabolic trough concentrator field test report (Part 1: surface layer wind field characteristics and wind load on mirror surface). Beijing: Joint Laboratory of the Institute of Electrical Engineering Chinese Academy of Sciences and Himin Solar Energy Group; 2011.
[38] ASCE minimum design loads for buildings and other structures. 1999. ASCE798.
[39] Jian L. Research on vacuum reliability of parabolic trough evacuated tube [Thesis]. Beijing: University of Chinese Academy of Sciences; 2012.
[40] Shipilevsky BM, Glebovsky VG. Competition of bulk and surface processes in the kinetics of hydrogen and nitrogen release from metals into vacuum. Vacuum 1990; 41(1–3):126–9.
[41] Pick MA, Sonnenberg KA. Model for atomic-hydrogen metal interactions-application to recycling, recombination and permeation. Journal of Nuclear Materials 1985; 131(2–3):208–20.
[42] Moraw M, Prasol H. An interpretation of outgassing characteristics of metals. Vacuum 1996;47(12):1431–6.
[43] Davenport JW, Estrup PJ. Hydrogen in metals. New York: Elsevier; 1981.
[44] Davenport JW, Dienes GJ, Johnson RA. Surface effects on the kinetics of hydrogen absorption by metals. Physical Review B 1982;25(4):2165–74.
[45] Zajec B, Nemanic V. Hydrogen bulk states in stainless-steel related to hydrogen release kinetics and associated redistribution phenomena. Vacuum 2001;61(2–4):447–52.
[46] Moens L, Blake DM. Mechanism of hydrogen formation in solar parabolic trough receivers. Journal of Solar Energy Engineering, Transactions of the ASME 2010;132(3).
[47] Arnold CJ. Evaluation of organic coolants for the transportation of LMFBR spent fuel rods. 1978. SAND-77-1486 United States Fri Feb 08 08:41:03 EST 2008 Dep. NTIS, PC A03/MF A01.SNL; INS-78-017378; ERA-03-057310; EDB-78-124893 (English).
[48] Yongfang J. Research on heat transfer mechanism of concrete thermal storage system. Beijing: Chinese Academy of Sciences; 2011.
[49] Capacity Development Technical Assistance. People's Republic of China: concentrating solar thermal power development. TA NO. 7402-P.R.C. Beijing: Asian Development Bank; 2011.
[50] People's Government of the Ningxia Hui autonomous region. Administration methods on construction land use of wind power and solar photovoltaic power generation projects in the Ningxia Hui autonomous region. NZF. 2011. No.103.
[51] Xiang H. Environmental adaptability of large-scale STE generation system (2010CB227106–2) 2010 interim progress report. Beijing: Institute of Electrical Engineering Chinese Academy of Sciences; 2010.
[52] Morgan VT. The overall convective heat transfer for smooth circular cylinder. Advances in Heat Transfer 1975;11:199–264. https://doi.org/10.1016/S0065-2717(08)70075-3.
[53] Peterka JA, Derickson RG. Wind Load Design Methods for Ground Based Heliostats and Parabolic Dish Collectors. Parabolic Dish Collectors. Report SAND92-7009: Sandia National Laboratories; 1992.

Index

'*Note*: Page numbers followed by "f" indicate figures and "t" indicate tables.'

A

Absorptance, 44
Absorption, 44
Acceleration sensor MSI4000, 294
Active direct thermal storage system
  direct steam thermal storage system, 403, 404f
  dual-tank direct thermal storage system, 404, 404f
Active indirect thermal storage system
  Andasol 1 power plant in Spain, 406
  Badaling power-tower-generation, 408, 408f
  dual-tank indirect thermal storage system, 405–406, 406f
  single-tank indirect thermal storage system, 406–407, 407f
  thermal charging process, 408
  thermal discharge process, 408–409
Affiliated facilities, 455–456
Air mass number (AM), 50–51
Alloy phase-change thermal storage material, 401, 402t
Angstrom-Prescott model, 84
Annual power generation calculation, 194–195
  calculation example for
    analysis and solution, 198
    daily required thermal storage, 198
    design software, 199
    energy balance, 199, 200t
    steam turbine, 197, 198t
  design point method, 196f
    irradiation and meteorological conditions, 195
    system efficiency, 195
    system photoelectric efficiency, 196
    thermal storage and concentrating area, 197, 197f
    thermal storage capacity, 196
  hourly simulation
    annual generating capacity and daily solar direct normal irradiance, 206, 206f
    Badaling power plant, 202, 203f
    concentration power projected by heliostat field, 204–205, 205f
    generating capacity of, 205, 206f
    heliostat field of Badaling power plant, 202–204, 204f, 204t
    HFLD software, 202–204
    outlet steam flow of steam thermal storage, 206–207, 207f–208f
    outlet superheated steam flux, 204–205, 205f
    TRNSYS model, 202, 203f
  parabolic trough collector efficiency
    cosine value of daily maximum incidence angle, 210, 211f
    cosine value of daily mean incidence angle, 210, 212f
    daily cosine value variation, 210, 213f
    incidence angle, 210–214
    transient efficiency of, 209
  thermal power plant capacity optimization, 201, 201f
Anticorrosion, 438
ASHRAE 93.77, 153
ASHRAE 93 steady state test method, 159
  incident angular modifier (IAM), 161
  measurement parameters, 160–161, 161t
  thermal efficiency testing, 160
Atmospheric aerosol attenuation, 133f
  absorption efficiency factor Qa, 132–135, 136f
  definition of, 132–135
  extinction efficiency factor Qe, 132–135, 134f
  Mie-Debye scattering, 132–135
  optical wave attenuation of, 132–135
  physical parameters of, 132–135, 132t
  scattering effect factor Qs, 132–135, 135f
  types of, 132–135
Atmospheric transmittance analysis
  atmospheric aerosol attenuation, 132–135, 132t
  extinction coefficient, 129, 136–137, 136f
  Lambert's law, 129
  molecule absorption, 130–131

Atmospheric transmittance analysis (*Continued*)
molecule scattering, 131–132
monochromatic radiation, 129
Atmospheric transparency, 51
Auxiliary facilities, 455–456
Azimuth-elevation tracking heliostat, 229, 230f

B

Beijing Badaling CSP testing power plant, 15–16, 17f
Boiler water modification, 438

C

Cavity receiver, 321–322, 321f
area of, 139–150, 139f
concentration fields, 138–139, 138f
conductive heat loss, 145–147
convective heat loss, 143–145, 143f
equivalent emittance of, 141–143, 142f
equivalent reflectance, 139–141, 140f–141f
heat loss, 139
Nusselt number, 143–145, 144f
overflowed energy, 143–145
radiation heat loss PRAD, 141–143, 142f
receiver heat loss, 147–150, 149f
reflective radiation loss, 139–141
thermal-insulating performance, 145–147
Climate conditions, plant area
ambient air temperature
extreme maximum temperature, 101t–102t, 102
extreme minimum temperature, 102
multiyear mean temperature, 102
typical day temperature variation, 102
designed wind speed and ambient air temperature, 106–107, 107f
disastrous weather phenomena and parameters
freeze and snow accumulation, 105
hail, 103
parabolic dish concentrator and solar furnace after Blizzard, 105, 106f
rainfall, 105
sand blow, 105
sandstorm, 103
snowfall, 105
surface of heliostat after Blizzard, 105, 106f
thunderstorms and overcast conditions, 105
precipitation parameters
annual mean evaporation, 103
annual mean precipitation, 103
wind speed, 102–103, 104t
Collector field annual efficiency, 45
Collector field efficiency, 45
Composite thermal storage system, 391
Concentrating field design
circumferential spacing, 250–252, 251f
heliostat radial spacing, 249–250, 249f
layout pattern of, 248–249, 248f
receiving tower height and collected energy, 250–251, 253f
required space for, 246, 248f
Concentrating field supervisory controller (CFSC), 226
Concentration astigmatism, 232–233
Concentration field
annual efficiency of, 41
control, 226–227
design, 7
efficiency of, 41
layout, 223–224
heliostat, 229–231
heliostat field layout, 239–243
parabolic trough concentrator, 236–239
spherical heliostat, 232–233
toroidal heliostat, 234–236
Concentration system
CFSC control room, 227
concentration field control, 226–227
concentration field layout
heliostat, 229–231
heliostat field layout, 239–243
parabolic trough concentrator, 236–239
spherical heliostat, 232–233
toroidal heliostat, 234–236
concentration field output, calculation methods for, 227–228
concentrator adjustment and control, 225–226, 226f
constitution of, 225
dust accumulation on mirror, 228–229, 229f
solar tower power plant concentrating field, 243–256
wind load characteristics of
field measurement, 292–318
wind tunnel test, 283–292
Concentrator aperture area, 30

Concentrator field control design
structure, 282
technical conditions for, 281
Concentrator performance requirements
atmospheric attenuation loss, 36–37, 38f
cosine loss, 33–34, 34f
spillage, 37, 38f–39f, 39t–40t
shadow and shade losses, 34–36, 34f–37f
specular loss, 33, 33f
Concrete thermal storage heat exchange design
axial thermal radiation fins, 397, 397f
finite element analysis of, 397–398, 397f
optimum tubular heat exchanger, 396, 396f
pipeless straight-through heat exchanger, 396
Concrete thermal storage material
embedded heat exchange steel pipe, 394, 394f
pourable ceramics, 394, 394f
thermal circulation and strength tests, 394, 395f
thermal conductivity of, 395–396, 395f
Cosine loss, 119

D
Daily mean solar direct normal irradiance, of Badaling
four seasons of 2009, 64, 66f
weather of, 66–68, 68f
and sunshine duration, 66–68, 68t
year, 66, 67f
Data quadratic fitting, 278–279, 278f–279f
Density fluctuation, 131–132
Design point power, 44
Deutsches Zentrum für Luft-und Raumfahrt (DLR), 157
Dish-Stirling solar power generation, 26–27, 26f
Distributed control system (DCS), 226, 332
Dual-tank thermal storage system, 389
Dust accumulation on mirror, 228–229, 229f

E
Electrical equipment/system
automatic safety devices, 442
cable selection and layout design, 443
dangerous environment, 443
DC system, 441
electrical measuring instruments, 442
grounding, 443
high-voltage electrical installations, 441
lighting system, 442
main electrical control room, 441
overvoltage protection, 443
relay protection, 442
Emission ratio, 44
Energy storage utilization factor, 46
EN 12975-2 quasi-dynamic test method, 162–164, 164t
Environmental protection
basic rules, 457
facilities, 458
pollution prevention and treatment, 458
requirements for, 457–458
External receiver, 321–322, 322f
Extinction coefficient
mean value, 137
variation of, 136–137, 136f

F
Feed-in tariff, 128
Feedwater, 438
Fiat Lux, 119
Fire protection, 452
Fresnel receiver, 321–322, 322f

G
Geographic latitude, 51
Geometric concentration ratio, 30
Global direct normal irradiance distributions
and district direct normal solar irradiance distributions, 112–115, 113f
satellite-derived solar radiation, 112
Global radiation, 49
Global solar irradiation meter
accessories, 107–108
inspected and serviced, 110
installation and maintenance, 109
sensors, 107–108
structure and constitution, 107–108, 108f
Gradient correction parameter identification method, 168

H
Heat absorber, 274
Heat transfer fluid, 31
of receiver, 5
Heat-transfer medium, 274

Heliostat
azimuth-elevation tracking heliostat, 229, 230f
cosine loss of, 33–34, 34f
geometrical dimensions of, 31, 32f
parts of, 229
pitch-roll tracking, 229
polar two-axis tracking, 230
shapes for, 231
spherical heliostat, 232–233
spinning-elevation tracking heliostat, 229, 230f–231f
spinning-elevation two-axis tracking, 230–231
toroidal heliostat, 234–236
Heliostat field efficiency analysis, 119–150. *See also* Power tower power plants
Heliostat field layout
azimuth, 242
Beijing Dahan power plant, 239f
Daegu power plant, 240f
Gemasolar power plant, 240f
length, 243
level-land requirement, 242
maintenance space, 242
rodent pest prevention, 243
total mirror area, 242
Heliostat field, of solar tower power plant
heliostat tracking errors
automatic deviation correction, 262–263, 262f
Badaling solar thermal power plant, 259–262, 261f
closed-loop inspection and correction, 258
data record sheet of, 259–264, 260f
equipment used, 258, 259f
insignificant variation of, 259
manual tracking deviation correction, 259–262
special circumstances, 263–264, 264f
technical conditions for, 256–258
Heliostat optical code (HOC)
algorithmic principles of
backward ray tracing (BRT), 119–121, 120f
brightness distribution, 119–120
flat-ground-plane projection model for, 121–124, 122f
geometric topological invariance, 122f, 124
mirror curvature, 119–120
shading-and-blocking computation, 121
for solar towers, 119
two specific versions of
cross-flux-density profiles, 127
cylinder central receiver solar field layout (CCRSFL) version, 124–127, 125f
DAHAN solar tower power plant, 124, 125f
flux map reconstruction, 125f, 127
Heliostat tracking errors
automatic deviation correction, 262–263, 262f
Badaling solar thermal power plant, 259–262, 261f
closed-loop inspection and correction, 258
data record sheet of, 259–264, 260f
equipment used, 258, 259f
insignificant variation of, 259
manual tracking deviation correction, 259–262
special circumstances, 263–264, 264f
HOC. *See* Heliostat optical code (HOC)
Hydrogen permeation performance test
data processing and experimental results, 371, 372f
devices for, 370, 370f
hydrogen diffusion coefficient, 369
pressure increase method, 369–370
principle of, 370, 370f
quadrupole mass spectrometer, 369–370
selective-absorbing film, 369
vapor permeation testing method, 369
Hydrogen permeation volume prediction
corrugated tube, 376–377
hydrogen dissolution, 373–374
hydrogen generation rate, heat-transfer oil, 371–373
hydrogen permeation area, 374–375
model calculation, 377, 377f
results and analysis
hydrogen-absorbing area, 377–384, 379f–384f
hydrogen barrier, 377–384
hydrogen permeation capacity, 377–384, 378f
hydrogen pressure in receiver tube, 384–386, 385f, 386f
organic impurities, 377–384

parabolic trough solar power plant, 377–384
saturated getter, 384–386
unsaturated getter, 377–384
stainless-steel absorber tube, 375–376

I
Incident angle modifier (IAM), 155–156, 238–239
Instantaneous efficiency
data quadratic fitting, 278–279, 278f–279f
linear fitting curve of efficiency, 276, 277f
parabolic trough collector field, 279–281
test loop, 276, 276f
test results, 276, 277t
Interior environment, 453
Ivanpah power tower plant, 45–46, 46f

L
Lambert's law, 129
Latent thermal storage system, 390–391
Least-square method, 168
Linear Fresnel reflector (LFR) solar power generation, 27, 27f, 28t, 31
Liquid metal receiver, 323
Lumped Capacitance Method, 169–170

M
Main powerhouse
direction of, 435
and thermal storage, 435
Mean solar time, 49
Measuring instrument
atmospheric transmittance meter, 111–112
global solar irradiation meter
accessories, 107–108
inspected and serviced, 110
installation and maintenance, 109
sensors, 107–108
structure and constitution, 107–108, 108f
solar direct normal irradiance meter, 110–111, 111f
Mie-Debye scattering, 132–135
Mineral aerosols, 132–135
Mirror cleaning time, 128, 128f
Mirror surface errors of concentrator, 39–41, 41f
Molecule absorption, 130–131
Molecule scattering, 131–132
Molten-salt power tower generation system, 9–10, 10f
Molten-salt receiver, 323
Molten-salt thermal storage, 393–394
Multiple linear regression (MLR), 163

N
Nonpressure air receiver, 323

O
Optical concentration ratio, 30

P
Parabolic trough collector thermal performance
ASHRAE 93 steady state test method, 159
incident angular modifier (IAM), 161
measurement parameters, 160–161, 161t
thermal efficiency testing, 160
current overseas research status of
ASHRAE 93.77, 153
ASHRAE 93.2003, 158–159
ASTM E905, 153
Deutsches Zentrum für Luft-und Raumfahrt (DLR), 157
EN 12975-2, 158–159
incident angular modifier (IAM), 155–156
Institute of Electrical Engineering, Chinese Academy of Sciences (IEE-CAS), 158, 158f
Kontas test platform in Spain, 157–158, 158f
sandia national laboratories of america, 153–154, 154f
SEGS parabolic trough collector, 156–157, 157f
SimulCET, 158–159
SolarPACES, 159
dynamic test model, 168–169, 169f
and parameter identification, 175–177, 177t
thermal performance prediction of, 177–179
EN 12975-2 quasi-dynamic test method, 162–164, 164t
experimental condition I
absolute error of dynamic model, 180, 180f
collector outlet temperature, predicted value of, 179–180, 179f

Parabolic trough collector thermal performance (*Continued*)
collector output power, predicted value of, 180–182, 181f
relative error of dynamic model, 180, 181f
experimental condition II
absolute error of dynamic model, 182–183, 183f
collector outlet temperature, predicted value of, 182, 182f
prediction of dynamic model, 184, 184f
relative error of dynamic model, 183–184, 183f
experimental condition III
absolute error of dynamic model, 185–186, 186f
predicted value of dynamic model, 185, 185f
prediction of dynamic model, 187, 187f
relative error of dynamic model, 186–187, 186f
experimental condition IV
absolute error of dynamic model, 188, 189f
predicted value of dynamic model, 188, 188f, 191t
prediction of dynamic model, 189–190, 190f
relative error of dynamic model, 189–190, 189f
heat-transfer process
differential equation, 171
Lumped Capacitance Method, 169–170
outlet temperature, 171
optical model
ansmittance-absorptance-product, 173
maximum optical efficiency, 173
metal evacuated tube, 174, 174f
truncation factor, 173
parabolic trough receiver tube heat loss parameters
heat loss coefficient of ptr70, 150–152, 151f
SCHOTT PTR70 evacuated tube, 150, 151t
standards, 152
temperature measuring points, 152, 152f
thermal performance prediction
ASHRAE 93 standard, 178
cosine loss, 178
specific thermal capacity of synthetic oil, 178, 178f
thermal performance test method
collector transient efficiency curve, 165, 166f
comprehensive heat loss coefficient, 165–166
gradient correction parameter identification method, 168
hourly useful thermal collection, 165–166, 167t
least-square method, 168
mechanism analysis method, 168
probability density approximation parameter identification method, 168
significance and application, 166–167
system identification method, 168
Parabolic trough concentrator
available normal irradiance (ANI), 237
axial arrangement of
annual irradiation, 271–273, 271f, 272t
autumn equinox, 267
irradiance variations for, 268, 269f
received daily irradiations, 268–269, 270t
spring equinox, 266
summer solstice, 266–267
winter solstice, 267–268
cosine effect, 237–238
different tracking modes of, 264–265, 265f
direct normal solar irradiance (DNI), 237
incidence angles, 237
incident angle modifier (IAM), 238–239
nonvertical solar beam incidence, 237–238, 238f
thermal efficiency evaluation, 273, 274f
working principle of, 236–237
Parabolic trough receiver tube
gas permeation performance
hydrogen permeation performance test, 369–371
theoretical research, 368–369
gas release performance
absorption-diffusion model, 354–356
balance process of gas, 354–356, 355f
Boltzmann constant, 354–356
diffusion limited model (DLM), 352
Fick's law, 352
hydrogen atom and molecule, 353–356, 354f
one-dimensional diffusion equation, 352

outgassing rate of flat plate, 353
recombination limited model (RLM), 353
gas release performance test
full-tube outgassing performance test, 357, 358f
heating system, 357, 357f
principles for, 356–357, 356f
sample tube A1, outgassing rate variation curve for, 358f
sample tube A2, outgassing rate variation curve for, 359f
sample tube A3, outgassing rate variation curve for, 359f
sample tube B3 at 100°C, outgassing rate fitting curve of, 363f
sample tube B3 at 200°C, outgassing rate fitting curve of, 364f
sample tube B3 at 300°C, outgassing rate fitting curve of, 364f
sample tube B3 at 400°C, outgassing rate fitting curve of, 365f
sample tube B3 at 460°C, outgassing rate fitting curve of, 365f
sample tube B1, outgassing rate variation curve for, 359f
sample tube B2, outgassing rate variation curve for, 360f
sample tube B3, outgassing rate variation curve for, 360f
sample tube C1, outgassing rate variation curve for, 360f
sample tube C2, outgassing rate variation curve for, 361f
sample tube C3, outgassing rate variation curve for, 361f
sample tube D1, outgassing rate variation curve for, 361f
sample tube D2, outgassing rate variation curve for, 362f
total outgassing quantity, 364–367, 366f
hydrogen permeation volume prediction
corrugated tube, 376–377
hydrogen dissolution, 373–374
hydrogen generation rate, heat-transfer oil, 371–373
hydrogen permeation area, 374–375
model calculation, 377, 377f
results and analysis, 377–386
stainless-steel absorber tube, 375–376
structure, 340f
corrugated tube, 342
gas getter, 342
glass cover tube, 339–340
glass-metal sealing adapting piece, 341
metal absorber tube, 340–341
vacuum reliability, 350f–351f
borosilicate glass, 349, 349f
heat conduction experiment and simulation, 347, 348f
heat loss of, 343, 344f
hydrogen, 343
hydrogen barrier development, 350, 351f
hydrogen-permeation rate, 344–345
hydrogen pump, 345
impurities, 343
installation position, 345, 345f
SCHOTT, 345–346, 345f–346f
SEGS V power plant, 347
stainless-steel outgassing conditions, 349, 350f
steady-state assumption, 343
variation curve of, 346–347, 346f
xenon capsule, 347–348, 348f
Parabolic trough solar power generation, 20f, 31
Beijing parabolic trough solar power collector, 24, 24f
equipment of, 20
getter, 23–24
glass-metal sealing transition piece, 23
glazed shield tube, 23
interior and exterior tube, 22
metal receiver tube, 23
parameters for, 24, 25t
problems, 20
receiver tube of, 22, 22f
selective absorption film, 24
structural diagram of, 21, 21f, 25f
thermal stress buffer section, 23
Parabolic trough surface, 42, 42f
Passive thermal storage system, 409–410, 409f
Phase-change material thermal storage, 398–399
Pipeless straight-through heat exchanger, 396
Plant layout planning
communication and transportation, 429–430
directions of main buildings, 427
enclosure walls, 429
fireproofing, 427–428

Plant layout planning (*Continued*)
main powerhouse's location, 428
notices for, 426–427
pipeline layout
and ambient environment, 433
overhead pipeline layout, 433
underground pipeline layout, 432–433
plant landscape layout, 427
solar tower layout of a tower power plant, 428–429
principles of, 425
requirements for, 425–426
road arrangement, 429
solar tower thermal resistance, 427–428
vertical layout
drainage structure, 431
indoor bottom elevation, 432
pattern and designed elevation, 430–431
protective treatment, 431
site drainage modes, 431
terrain layout, 432
Power generation units, pressure parameters for, 4–5
Power load data, 423–424
Power load plan, 424
Power plant capacity, 5–6
Power plant design
data required, 192
parameters and principles of, 193–194
point
calculation examples of, 117, 118t
significance of, 117
Power plant influences, environment, 6
Power plant, parameters of, 194
Power plant seismic resistance, 6
Power plant site plan
concentration field layout, 217–218, 217f–218f
conventional generating area, 219, 220f
drainage of power plant, 221
fuel transportation, 220
HV outlet orientation and passageway, 220
location and transportation, 220
power plant capacity, 216
power plant general elevation planning factors, 221–222, 222f
power plant general lay-out planning factors, 221
steam generating area, 218–219
thermal storage area, 219
water source of power plant, 220
Power system
power grid connection, 439
system communication, 439–440
system protection, 439
Tower power plants
atmospheric transmittance analysis
atmospheric aerosol attenuation, 132–135, 132t
extinction coefficient, 129, 136–137, 136f
Lambert's law, 129
molecule absorption, 130–131
molecule scattering, 131–132
monochromatic radiation, 129
cavity receiver
area of, 139–150, 139f
concentration fields, 138–139, 138f
conductive heat loss, 145–147
convective heat loss, 143–145, 143f
equivalent emittance of, 141–143, 142f
equivalent reflectance, 139–141, 140f–141f
heat loss, 139
Nusselt number, 143–145, 144f
overflowed energy, 143–145
radiation heat loss PRAD, 141–143, 142f
receiver heat loss, 147–150, 149f
reflective radiation loss, 139–141
thermal-insulating performance, 145–147
cosine loss, 119
heliostat optical code. *See* Heliostat optical code (HOC)
software, 119
specular reflectance, 127–128, 128f
Pressure increase method, 369–370
Probability density approximation parameter identification method, 168
Pyrex, 339–340

R

Radiance, 49
Radiant exposure, 49
Radiant intensity, 49
Rated thermal power of receiver, 42–44
Rayleigh scattering coefficient, 131–132
Receiver aperture area, 32

Receiver control system
control mode, 333–334
designed scope of
chemical water processing system, 332
comprehensive pump house, 333
physical control layer, 332
fire control system, 333
fire hazard inspection alarm, 333
water system, 332
logic of
central computer, 331–332
control interlock, 332, 333f
sampling data points, 332
solar irradiation and wind speeds, 331
operation mode of
initiation of, 334–335
principles, 334
technical improvement of, 335–336, 336t–337t
working mode of, 335
Receiver efficiency, 42
Receiver net collection area, 30
Receiver net thermal power, 44
Receiver peak radiant flux density, 42, 43f, 44t
Receiver system
adjustment and control principles, 321–325
molten-salt receiver, 324
saturated type water/steam receiver, 324
superheating type water/steam receiver, 323–324
cavity-type receiver, 321–322, 321f
configuration of
components, 319
drive pump, 319–320
heat-transfer medium, 320–321
receiver type selection, 320–321
thermal parameter measurement device, 320
fluid pumps/fans, 320–321
constitution, 323
control system
control mode, 333–334
designed scope of, 332–333
logic of, 331–332
discharge system and equipment of
devices, 338
mode, 337–338
principles and methods, 338–339
range, 336–337
external receiver, 321–322, 322f
Fresnel receiver, 321–322, 322f
heat-transfer loop constitution, 323
heat-transfer medium, 327
materials of
heat resistance of, 329
parabolic trough receiver, 328
pipe materials, 329
thermal-insulating materials, 328
tower-type water receiver, 328
parabolic trough receiver tube
gas permeation performance, 367–371
gas release performance, 351–356
gas release performance test, 356–367
structure, 339–342
vacuum reliability, 343–350
pipes and pumps for
feed pump, 330
pipeline and valve, 330–331
protection system, 324–325, 325f
receiver's rated power
heat-transfer process, 326
receiver power, 326
receiver quantity, 327
steam pipeline, 327
tower-type power plant, 325–326
structure, 321–325
surface coating of, 329
thermal-insulating material, of tower receiver, 329–330
tubing material, 330
Reflectance, 44
Reflection, 44
Room-temperature ionic liquid (RTIL), 393–394

S

Sand dust aerosols, 132–135
Satellite-derived solar radiation, 112
Saturated water/steam thermal storage, 400–401
Scattering radiation, 131–132
Seawater desalinization combine-cycle system, 15–16, 17f
Seismic intensity, 421–422
Sensible thermal storage system, 389–390

Sieber Kraabel model, 143–145
SimulCET, 158–159
Site selection
ecological protection, 418
land selection, 419–420
macro site selection, 418
power plant residential area, 423
principles of, 417–418
seismic intensity, 421–422
site elevation, 420–421
site location, 422–423
solar resource, 418–419, 418t
thermal storage tank, 418
Solar air receiver systems, 400
Solar altitude, 49–50
Solar collector aperture area, 31, 32f
Solar collector gross aperture area, 30
Solar collector net aperture area, 30
Solar concentrator
description of, 273
heat absorber, 274
heat-transfer medium, 274
picture of, 275, 275f
restrictive conditions, 274
schematic diagram of, 275, 275f
transparent cover, 274
Solar direct normal irradiance meter, 110–111, 111f
Solar direct normal radiation resources, in China
annual mean daily solar direct normal irradiance, 90–93
characteristics of
atmospheric circulation influences, 95
latitude and season influences, 93–94
topographic influences, 94–95, 94f
measurement of, 96–97
configuration of, 98, 99t
direct normal irradiance in Badaling, 98, 99f
geographic statistics, 97
horizontal surface, 97
irradiation meter CMP21, 97–98, 98t
Kipp & Zonen irradiation meter, 97–98, 97f
RSP-3G irradiation meter, 97–98, 98f
total/scattering irradiance, 98–99, 100f
regionalization of, 96, 96t
spatial and temporal distributions of, 93
Solar energy-generating system (SEGS), 278–279
Solar field, 30–31
Solar irradiance prediction methods
climate change influences, 88–89
estimation method for
Angstrom-Prescott model, 84
astronomic radiation, 85
coefficients a and b, 85
declination angle, 85
estimation result errors, 86–87, 87t
geographical positions of, 85, 86t
longest sunshine duration, 85
meteorological data, 84
monthly variation tendency, 87–88, 88f
primary comparative parameters, 86
Solar irradiation, 1–2, 12–13, 31
atmospheric influences
air mass number (AM), 50–51
atmospheric transparency, 51
China's topography, 52
elevation, 51
geographic latitude, 51
second-level region, 52
solar altitude, 50
sunshine duration, 51
third-level region, 52
expression, 48–49
spectrum, 49–50
Solar multiple, 46
SolarPACES, 164
Solar photovoltaic power generation, 3, 3t
Beijing Badaling CSP testing power plant, 15–16, 17f
energy storage, 15–16, 15f
Kuraymat parabolic trough/natural gas combined-cycle testing facilities, 15–16, 16f
power dispatching, 18
power grid access, 16
raw material supply, 18
seawater desalinization combine-cycle system, 15–16, 17f
Solar position
solar angle
declination angle, 53, 53f–54f, 55t
solar altitude, 56
solar azimuth angle, 57, 57f
solar constant, 56
solar flare angle, 57, 57f
solar hour angle, 54
solar time, 53–54, 56f
solar zenith angle, 56
sunshine duration, 54
tracking angle calculation

incidence angle, 58
light-receiving surface dip, 58
Solar tower power generation, 19f
concentration ratio of, 18–19
molten-salt power plant system, 19–20
Solar resource
advantages of
enormousness, 48
harmlessness, 48
long-lastingness, 48
ubiquity, 47
Beijing
Badaling, 64–68
solar elevation, 61
solar radiation, 61–63, 62t
sunrise time, 61
sunset time, 61
sunshine duration of, 63–64, 63t, 65t
sunshine percentage of, 64
disadvantages of
discontinuity, 48
instability, 48
small energy density, 48
Dunhuang
solar radiation, 74
weather conditions of, 73–74, 74t
Golmud
solar radiation, 72
sunshine duration, 73
weather conditions of, 71, 72t
Guizhou
solar radiation of, 77
weather conditions of, 76–77
Hainan
precipitation conditions, 78–79
solar radiation, 78
sunshine duration, 78
thermal characteristics of, 78, 79t
thunderstorms, 80
typhoons, 80
Harbin
basic climatic characteristics of, 80–81, 82t
solar radiation, 81–83
sunshine duration, 83
Lhasa
solar radiation and sunshine duration, 69, 71t
weather conditions, 69, 70t
Turpan
solar radiation of, 76
weather conditions, 75, 76t
Solar thermal energy generation technology, 389
advantages of, 8
annual efficiency, 10–11
annual mean performances of, 45–46, 45t
characteristics of, 10–11
environmental influences, 13
output characteristics, 14
power grid connection, 14
resource request, 12–13, 13t
concentration mode, 11–12
concentration ratio, 10–11, 11f
conversion efficiency and reliability, 2–3
design point, 28–29
molten-salt power tower generation system, 9–10, 10f
optical efficiency and cost, 1–2
optic terms, 30–41
saddle point, 10–11
system terms, 44–46
technical forms of
dish-Stirling solar power generation, 26–27
linear Fresnel reflector (LFR) solar power generation, 27
parabolic trough solar power generation, 20–26
solar power tower generation, 18–20
thermal storage costs for, 389
thermodynamic terms, 42–44
units, 4
vs. solar photovoltaic power generation, 3, 3t
Beijing Badaling CSP testing power plant, 15–16, 17f
energy storage, 15–16, 15f
Kuraymat parabolic trough/natural gas combined-cycle testing facilities, 15–16, 16f
power dispatching, 18
power grid access, 16
raw material supply, 18
seawater desalinization combine-cycle system, 15–16, 17f
water/steam power tower generation system, 9–10, 10f
Solar thermal storage system layout, 435
Solar tower power plant concentrating field, 243
design parameters, 244
air receiver, 254–255
heat-transfer fluid, 254

Solar tower power plant concentrating field (*Continued*)
receiver's aperture dimensions, 253
sun tracking, 253
temperature-monitoring picture, 254, 254f
heliostat field optimization design
concentrating field design, 246–252
heliostat concentration field design software, 245–246, 247t
operating modes of heliostat field, 244, 245t
Spectral irradiance, 49
Specular reflectance, 127–128, 128f
Spherical heliostat
concentrated solar image of, 232, 232f
focal spot's height, 232, 233f
Spinning-elevation tracking heliostat, 229, 230f–231f
Sunshine duration, 51

T

Taylor expansion method, 176
Test loop, 276, 276f
Thermal automation
automatic adjustment, 446–447
basic rules, 445
building space heating, 449
cables, 448–449
conduits, 448–449
control mode, 446
control room, 448
heating network, 450
heating station, 450
interlocking, 447
local equipment layout, 448–449
power supply, 447–448
solar tower, 449–450
steam supply, 447–448
thermal inspection, 446
thermal protection, 447
Thermal storage capacity, 44
Thermal storage reserve
principles of selecting, 214–215, 214f–215f
thermal storage capacity, 216
Thermal storage systems
alloy phase-change thermal storage material, 401, 402t
categories and constitutions of
active direct thermal storage system, 403–405
active indirect thermal storage system, 405–409
charging and discharge unit and control system, 411
composite thermal storage system, 391
evaluation indices for, 401, 403f
high-temperature synthetic oil, 410, 411f
latent thermal storage system, 390–391
low-temperature thermal storage tank, 410
passive thermal storage system, 409–410
sensible thermal storage system, 389–390
CSP plants, 401, 403f
tank fouling discharge and cleaning system, 411
thermal storage tank connecting pipeline, 411
charging and discharge equipment of, 413–414
concrete thermal storage heat exchange design
axial thermal radiation fins, 397, 397f
finite element analysis of, 397–398, 397f
optimum tubular heat exchanger, 396, 396f
pipeless straight-through heat exchanger, 396
concrete thermal storage material
embedded heat exchange steel pipe, 394, 394f
pourable ceramics, 394, 394f
thermal circulation and strength tests, 394, 395f
thermal conductivity of, 395–396, 395f
constitution of, 414
control logic of, 414–415
functionality of
development tendency of, 388–389
generating hours, 388
improved annual utilization rate, 388
variable weather, 388
inspection, facilities for, 415
molten-salt thermal storage, 393–394
phase-change material thermal storage, 398–399
room-temperature ionic liquid material, 393–394
saturated water/steam thermal storage, 400–401

selection of, 391–392, 412–413
solar air receiver systems, 400
technical requirements of, 392–393
thermal storage costs for, 389
thermal storage materials, 392
Thermal system steam sampling, 438
Time selection, of power output, 424
Toroidal heliostat
astigmatic-corrected target-aligned heliostat, 235–236, 235f
conic section (generatrix), 235–236, 236f
designed incident angle of, 235
fixed incident angle, 234
sagittal curvature radius, 234
tangential curve expression, 235–236
Toroidal surface, 234
TRNSYS model, 202, 203f
Troposphere aerosols, 132–135
True solar time, 49

W

Water/steam power tower generation system, 9–10, 10f
Water/steam receiver, 323
Water treatment
calculation of, 437–438
equipment, 437–438
receiver and evaporator makeup, 437
WinDELSOL, 119
Wind field actual measurement
gust factor, 296–298
mean wind angle, 295–296
statistical characteristics of, 296–298, 299t–300t
statistics of, 295–296, 297f
time histories of, 295–296, 296f
turbulence intensity, 296–298, 301f
wind speed and wind direction, 295–296
Wind load characteristics
concentrator
definition of, 302–303
power spectrum of fluctuating wind pressure, 308–309, 308f
pressure distribution, 305, 306f
pressure spectrum, 305
resistance and lift force, 303–305
spectrum analysis, 307–308
variation curves of, 302–303, 304f
wind load coefficients, 302–303
field measurement
actual measurement, 292–293, 293f
test instruments, 293–294, 294f
wind field actual measurement, 295–302
working conditions of actual measurement, 295
wind-induced vibration, of evacuated receiver tube
collector receiver, 314–316
definition of, 309
frame structure vortex-excited vibration, 316, 317t
natural vibration frequency, 313–314, 315f, 316t
vibration analysis, 310–313, 311f–313f
wind tunnel test, 283–292
Windproof design, 6
Wind tunnel test, 283–284
experimental results, 287–289, 288f
experiments and data processing
coordinate label of balance, 286, 286f
heliostat model picture, 284, 285f
measuring point, 287
measuring point layout of front elevation, 284, 285f
resistance coefficient, 287
wind speed and turbulence intensity, 284, 286f
mean wind resistance, 291–292
peak negative pressures, 291
peak wind pressure distribution, 289–291, 290f
rotary shaft of heliostat, 291
rules for variations, 291

FSC
www.fsc.org
MIX
Papier | Fördert
gute Waldnutzung
FSC® C083411

Kolibri 360 GmbH
Ferdinand-Jühlke-Straße 7
99095 Erfurt, Deutschland
produktsicherheit@kolibri360.de